# Animal Welfare at Slaughter

# Animal Welfare at Slaughter

## The Animal Welfare Series

Editors: Antonio Velarde, Mohan Raj

Series Editor: Xavier Manteca

5m Publishing

First published 2016



Published by
5M Publishing Ltd,
Benchmark House,
8 Smithy Wood Drive,
Sheffield, S35 1QN, UK
Tel: +44 (0) 1234 81 81 80
www.5mpublishing.com

A Catalogue record for this book is available from the British Library

ISBN 978-1-910455-69-2

Book layout by Servis Filmsetting Ltd, Stockport, Cheshire
Printed by TJ International Ltd, UK
Photos as credited in the text

# Contents

# Contributors

**Bo Algers**, Swedish University of Agricultural Sciences, Department of Animal Environment and Health, P.O. Box 234, 53223 Skara, Sweden

**Haluk Anil**, School of Veterinary Science, University of Bristol, Langford BS40 5DU, United Kingdom

**Frank Berthe**, European Food Safety Authority (EFSA), Via Carlo Magno 1/A, 43126 Parma, Italy

**Denise Candiani**, European Food Safety Authority (EFSA), Via Carlo Magno 1/A, 43126 Parma, Italy

**Antoni Dalmau**, IRTA. Animal Welfare Subprogram, Veïnat de Sies, s/n, 17121 Monells, Spain

**Luigi Faucitano**, Agriculture and Agri-Food Canada, Sherbrooke Research and Development Centre, 2000 College Street, J1M 0C8 Sherbrooke, Canada

**Carmen Gallo**, Instituto de Ciencia Animal, Facultad de Ciencias Veterinarias, Universidad Austral de Chile, Casilla 567, Valdivia, Chile

**Rebeca Garcia Pinillos**, Veterinary Surgeon, London, United Kingdom

**Marien Gerritzen**, Wageningen University and Research Centre, Livestock Research Department Animal Welfare, De Elst 1, 6708 WD Wageningen, The Netherlands

**Andrea Gervelmeyer**, European Food Safety Authority (EFSA), Via Carlo Magno 1/A, 43126 Parma, Italy

**Troy Gibson**, Department of Production and Population Health, Royal Veterinary College, Hawkshead Lane, Hertfordshire AL9 7TA, United Kingdom

**Temple Grandin**, Department of Animal Sciences, Colorado State University, Fort Collins, CO 80523-1171, USA

**Bert Lambooij**, Wageningen UR Livestock Research, P.O. Box 338, 6700 AH Wageningen, The Netherlands

**Eva Mainau**, Department of Animal and Food Science, School of Veterinary Science, Universitat

Autònoma de Barcelona, 08193 Bellaterra, Spain

**Xavier Manteca**, Department of Animal and Food Science, School of Veterinary Science, Universitat Autònoma de Barcelona, 08193 Bellaterra, Spain

**Cecilia Pedernera**, IRTA, Animal Welfare Sub-program, Veïnat de Sies, s/n, 17121 Monells, Spain

**Mohan Raj**, School of Veterinary Science, University of Bristol, Langford BS40 5DU, United Kingdom

**Déborah Temple**, Department of Animal and Food Science, School of Veterinary Science, Universitat Autònoma de Barcelona, 08193 Bellaterra, Spain

**E.M. Claudia Terlouw**, INRA UMR1213 Herbivores, Centre Auvergne-Rhône-Alpes, site de Theix, 63122 saint-Genès Champanelle, France

**Antonio Velarde**, IRTA. Animal Welfare Subprogram, Veïnat de Sies, s/n, 17121 Monells, Spain

**Hans van de Vis**, Wageningen UR Livestock Research, P.O. Box 338, 6700 AH Wageningen, The Netherlands

# Foreword

*By Temple Grandin*

*Dept. of Animal Science, Colorado State University, Fort Collins, Colorado*

The public is becoming increasingly concerned about the treatment of farm animals and knowing more about where their food comes from. Consumers become upset and highly concerned when they see videos taken by animal welfare NGOs which show animal abuse at slaughterhouses or on farms. These videos sometimes go viral and are seen by millions of people. Today it is easy to take videos because almost every mobile telephone is a video camera. Slaughterhouse managers, meat inspectors, and welfare officers must ask themselves, "How would activities in their slaughterhouse look if a video was posted online?" When animal handling is done correctly, the animals will walk quietly to the stunner. When the animal handling part of the process is conducted by well-trained employees, it is calm and low-stress.

There are parts of the stunning process that look bad even when they are done correctly. A properly stunned animal that is unconscious may have kicking reflexes. This needs to be explained to the public. I have taken many people who are outside the industry on tours of slaughterhouses. The most common comment from most people is that it was much better than they had previously imagined. The slaughter industry needs to do a better job of communicating with the public.

The purpose of this book is to provide scientifically based practical guidance for animal welfare officers, slaughterhouse managers, and meat inspectors. The species covered are cattle, pigs, sheep, and fish. Animal handling and design of races and lairages is explained in several different chapters. Each stunning method is covered in detail in four separate chapters. The controversial area of slaughter without stunning is also discussed. It is the responsibility of the readers of this book to insure that animals are handled and stunned correctly. This is important for four reasons:

1. Proper treatment improves animal welfare.
2. Correct procedures comply with legislation.
3. Maintaining high standards is the ethical thing to do to prevent pain and suffering.
4. Good treatment of animals maintains the trust of the consumer.

## IMPORTANCE OF GOOD MANAGEMENT TO MAINTAIN HIGH WELFARE STANDARDS

Slaughterhouses that have high welfare standards have managers and inspectors who care about doing things right. Places with rough handling of livestock or poor stunning practices often have managers or inspectors who do not care. My own observations in many slaughterhouses indicate that a major cause of welfare problems was either a lack of supervision of employees or poorly maintained stunning equipment. Too often, high standards that are being promoted by companies on fancy brochures and websites are not adhered to in the slaughterhouse. Continuous assessment of animal welfare with animal-based outcome measures is essential. People manage the things they measure. Scoring of animal welfare indicators with numerical scoring will help keep standards high. When numerical scores are tabulated, it is possible to determine if animal treatment at a particular slaughterhouse is becoming better or getting worse. Some of the important outcome measures are: Percentage of animals rendered unconscious with a single application of the stunner, percentage that maintain good footing and do not fall or slip during handling, percentage that remain silent in the stun box (moo, squeal, bellow), and percentage moved with no electric goad. Managers and inspectors need to put the most emphasis on activities they can directly observe. I have observed many slaughterhouse plants where the paperwork was perfect and all the records were correct but the stunner was broken. A paperwork audit does not insure good animal welfare. The best way to keep welfare standards high is for managers, inspectors, and retail meat buyers to get out of the office and observe what is actually happening in the lairage and stunning area.

chapter one

# Introduction

*Antonio Velarde[1], Déborah Temple[2], Eva Mainau[2], Xavier Manteca[2] and Mohan Raj[3]*

## Learning Objectives

- Describe animal welfare as a general concept and specifically during transport, lairage and slaughter.
- Understand the general approach to animal welfare assessment and animal-based measures.
- Be aware of the importance of education in animal welfare.

## CONTENT



[1] IRTA. Animal Welfare Subprogram, Veïnat de Sies, s/n, 17121 Monells, Spain
[2] Department of Animal and Food Science, School of Veterinary Science, Universitat Autònoma de Barcelona, 08193 Bellaterra, Spain
[3] School of Veterinary Science, University of Bristol, Langford BS40 5DU, United Kingdom

## 1.1 GENERAL CONCEPT OF ANIMAL WELFARE

Animal welfare can be defined in a number of different ways, but there is a growing consensus that whatever the definition is, it has to include three main elements (Fraser et al., 1997). The first element is the animal's adequate biological functioning (the animal's ability to cope with the environment). To cope with a challenge, the organism activates the functioning of body repair systems, immunological defences, physiological stress responses and several behavioural adaptations. Since there are various coping strategies, welfare can vary over a wide range, from very good to very bad (Broom and Johnson, 1993). The second element referring to emotional state links welfare to the absence of negative emotions such as pain or chronic fear and the presence of positive ones, and the animal's ability to express certain species-specific normal behaviours (Fraser et al., 1997). Concerns for animal welfare are generally based on the assumption that non-human animals can subjectively experience emotional states and hence can suffer or experience pleasure (Dawkins, 2000; Mendl and Paul, 2004; Boissy et al., 2007). The suffering of an individual animal is precisely what the public is concerned about and nowadays the majority of scientists accept that animal suffering, and the degree of animal consciousness that this implies, are essential aspects of animal welfare (Rushen, 2003). The third element is based on the ability to express certain species-specific normal behaviours (Hughes and Duncan, 1988). A normal behavioural pattern is usually associated with the behaviour shown by most members of a species under their natural habitat or environmental conditions. Such species-specific behaviour can be compared with the behaviour displayed by conspecifics in more confined farming conditions and, as it provides the basis for animals' requirements, it can be used for welfare assessment under farming conditions.

In general, an animal reaches the state of harmony when it is kept in an environment that allows it to satisfy its motivations. When the situation moves away from this ideal state, the animal will use a wide range of physiological mechanisms and behaviours to cope with its environment. The ability to cope successfully or not will depend on the individual itself and on how much it has to cope with. The inability to cope with an aversive situation may lead to the appearance of injuries or diseases (Broom and Johnson, 1993) and consequently to pain and suffering. Integrating these different approaches, an agreement is reached on what is needed to achieve animal welfare. It is generally accepted, then, that the welfare of a sentient animal is determined by its capacity to avoid suffering and sustain fitness (Webster, 2005). In other words, welfare comprises physical and mental health (Webster et al., 2004; Dawkins, 2004) and includes several aspects such as absence of thirst, hunger, discomfort, disease, pain, injuries, stress and the expression of normal behaviour, referred to as the five freedoms of the Farm Animal Welfare Council (1992). The five freedoms of the Farm Animal Welfare Council provide a basis for a multidimensional approach to welfare and offer a useful framework to assess the welfare of the animals. As such, it has been used as the basis for many animal protection laws worldwide.

## 1.2 ANIMAL WELFARE AT SLAUGHTER

Slaughter of animals in large numbers (billions) for human food remains a sensitive issue. At abattoirs, red meat animals are unloaded and held in pens before slaughter. During this period, animals are exposed simultaneously to a variety of stressors that may result in high levels of psychological and physical stress (Grandin, 1997), thus compromising their welfare. These potential stressors include fasting and water deprivation, mixing of unfamiliar individuals, handling by humans, exposure to a novel environment, noise, forced physical exercise and extremes of temperature and humidity. Unlike red meat animals, poultry are transported in containers, that is, crates or modules, to slaughterhouses and the containers are unloaded from the vehicles and stored for several hours depending upon the throughput rate. Under this situation, poultry are not provided with food or water and therefore prolonged lairage time is a serious welfare problem, especially under warm climatic conditions.

Humane slaughter regulations, standards or guidelines implemented worldwide, including those of the World Organization for Animal Health (OIE, 2009a,b,c), consider animals as sentient beings and therefore require that:

- animals shall be spared any avoidable pain, distress or suffering during their killing and related operations;
- no conscious animal shall be shackled or hoisted (hung upside down), excluding poultry species;
- animals must be rendered immediately unconscious or without pain (i.e. stunned) prior to slaughter and they should remain so until death occurs through blood loss (excluding religious slaughter);
- arteries supplying oxygenated blood to the brain (usually common carotid arteries) or, blood vessels from which they originate (brachiocephalic trunk), must be severed to prevent the recovery of consciousness following stunning; and
- animals must be dead before carcass dressing or any other treatment (e.g. electrical stimulation) is carried out.

The stunning methods should induce immediate loss of consciousness. If the loss of consciousness is not immediate, the induction of unconsciousness should not cause avoidable pain and suffering in animals. Therefore, stricter guidelines and procedures are needed to achieve humane stunning and slaughter (see EFSA, 2004; EFSA, 2013b). In addition, key principles to be followed in slaughterhouses include 1) do not restrain animals if not ready to stun and 2) do not stun if not ready to slaughter. There are several critical control points to be considered depending upon the species and stunning methods used; these will be addressed in the following chapters.

## 1.3 WELFARE ASSESSMENT

Welfare is multidimensional and therefore its overall assessment requires several criteria. The integrated project co-financed by the European Commission, Welfare Quality® (www.welfarequalitynetwork.net), has built on the five freedoms to establish four main welfare principles: good feeding, good housing, good health and appropriate behaviour (Blokhuis et al., 2008). Within these principles, the project highlighted 12 distinct but complementary animal welfare criteria, which provide a very useful framework for understanding the components of animal welfare:

### 1.3.1 *Principles and criteria applied to the slaughterhouse*

Good feeding includes two criteria: absence of prolonged hunger and absence of prolonged thirst. During the pre-slaughter period, animals are fasted for different durations to reduce gut content and so prevent the release and spread of bacterial contamination through faeces within the group during transport and lairage as well as through the spillage of gut contents during carcass evisceration (Faucitano and Schaefer, 2008). Fasting before slaughter, within reasonable limits, is also beneficial to the welfare of pigs as it prevents them vomiting in transit and developing hyperthermia; however, an extended fasting period causes hunger and aggressiveness (Warriss et al., 1994) and, if prolonged, animals become weak, lethargic and sensitive to cold (Gregory, 1998). Broiler chickens are usually subjected to fasting for longer durations than red meat animals. Feed withdrawal time begins when feed troughs are raised within a house before catching and crating begins and it ends when the last bird from the same flock is stunned and slaughtered. Prolonged hunger is not only a welfare problem but can also create meat quality problems, for example increasing the prevalence of dark, firm, dry meat in pigs and pale, soft, exudative-like meat in poultry.

Prolonged thirst causes stress, dehydration and loss of body condition (Gregory, 1998). At lairage, the ability to cope with dehydration varies among species and according to age. Suckling animals are particularly susceptible to dehydration because they have not learnt how to drink from a water trough and thereby fail to drink at the slaughterhouse. Poultry are not provided with water in the lairage and, as a consequence, they are subjected to prolonged thirst, especially in warm climatic conditions, and dehydration in broilers can cause weight loss and increase mortality (Warriss et al., 1992; Bianchi et al., 2005; Petracci et al., 2006).

Good housing includes comfort around resting, thermal comfort and ease of movement. Lairage at slaughter permits animals to recover from the stress and activity resulting from transport and unloading, which in some situations can be beneficial to meat quality and welfare. However, the benefit of providing animals with a resting time can be lost if the animals are subjected to poor environmental conditions. Lack of comfort around resting is likely to reduce resting time and may be a consequence of an excessive stocking density or inadequate lairage facilities, particularly inadequate flooring. Thermal comfort is defined as the range of effective temperature that provides a sensation of comfort and minimizes stress (Manteca et al., 2009). Temperatures below or above the thermoneutral zone cause cold and heat stress respectively, and even death if it is severe or prolonged. Heat stress also increases the amount of water required and can therefore increase the risk of prolonged thirst if water supply is limited (National Research Council, 1981). Maintaining thermal comfort in poultry requires forced ventilation in the lairage area, especially in hot climatic conditions (Hunter, 1998). Besides environmental conditions, stocking density determines the microclimatic conditions within crates or modules. Non-slip flooring is essential for smooth handling and movement of animals. Slipping and falling when unloading, moving to lairage and to the place of stunning, leads to poor welfare (Gregory, 1998).

Good health includes absence of injuries, disease and pain. Injuries can cause acute and/or chronic pain. This can be a consequence of rough handling (e.g. when animals are loaded and unloaded) or of the physical environment (e.g. poor flooring and inadequate design or maintenance of facilities). Fighting with other animals can also cause injury; this is more common when animals are mixed with unacquainted individuals and when animals have to compete for access to resources (e.g. water or resting space; Velarde, 2007). When moving animals at slaughter, the combination of higher speeds and poorly designed handling systems is detrimental to animal welfare because handling the animals at this rate requires considerable coercion and triggers the use of goads and sticks. Unlike red meat animals, poultry are shackled prior to stunning and slaughter. In multiple bird electrical water bath stunning, conscious birds are hung upside down on a moving metal shackle line and passed through an electrified water bath, such that the current flows through the whole body towards the shackle, the earth. Pain and suffering are associated with the stunning systems. Removing live birds manually from the old-fashioned transport crates or tipping mechanically from a modular system on to a conveyor for the purpose of shackling can be distressing and cause injuries such as dislocation of joints and broken pectoral bones. The legs of birds will be inevitably compressed during shackling and the degree of compression can be as high as 20% (Sparrey, 1994). Based on the presence of nociceptors in the skin over the legs of poultry and the close similarities between birds and mammals in nociception and pain, shackling has been reported to be a painful procedure (Gentle, 1992; Gentle and Tilston, 2000). This pain and suffering is likely to be worse in birds suffering from painful lameness due to diseases or abnormalities of the leg joint and/or bone (Butterworth, 1999; Danbury et al., 2000). This pain is also likely to be significant in birds suffering from dislocation of joints and/or fracture of bones induced by rough handling during catching, crating and uncrating. The severity of pain associated with pre-stun electric shocks needs no explanation. The complexity of commercial water bath stunning systems and physical contact between birds on the shackle line make it difficult to control the current pathway and eliminate this potential problem.

Appropriate behaviour includes the expression of social behaviour, expression of other behaviours, good human–animal relationship and a positive emotional state. During unloading, lairage or moving to the slaughter area, animals face a novel environment and handling that may cause fear. In fact, fear is an emotional state induced by the perception of a threatening or a potentially threatening situation (Boissy, 1995) and it involves physiological and behavioural changes that prepare the animal for coping with the danger (Forkman et al., 2007). General fear becomes a problem particularly when animals encounter new or unexpected stimuli (e.g. a sudden noise or movement, an unfamiliar animal), or situations (e.g. a new housing facility, transportation). This has important implications for animal management. For example, inappropriate handling, corridors/races and pen design, discontinuities in floor texture and colour, draughts and (poor) lighting may all induce fear and its undesirable consequences (Grandin, 2000).

## 1.4 ANIMAL WELFARE MEASURES AT THE SLAUGHTERHOUSE

For each one of these criteria, potential measures are identified for inclusion in welfare assessment at the slaughterhouse based on their validity, reliability repeatability, and feasibility (Velarde and Dalmau, 2012). Previous monitoring systems and legislation largely rely on examination of inputs: "what" or "how much" of different resources are given to animals (i.e. transport and lairage design, stunning equipment, space requirements, etc.). These parameters are easy to define and to measure and have a high inter- and intra-observer reliability. However, the measures have often been criticized for potentially low validity due to their indirect nature (e.g. resources may be present but inaccessible or rarely used by animals) and complex interactions with other resource and management conditions (Waiblinger et al., 2001). Thus, input measures are poor predictors of good welfare, as animals may experience the same situation or handling procedure differently depending on their genetic background and previous experiences.

Since welfare is a condition of the individual animal, wherever possible, the Welfare Quality® assessment system places its emphasis on animal-based measures (also called "outcome" measures) rather than on resources and management in an attempt to estimate the actual welfare state of the animals. Such physiological, health and behavioural measures have inherent advantages over input measures. The first advantage is clearly that, since welfare is a condition of the animal, outcome measures are likely to be the most direct reflection of their actual welfare state. They permit evaluation of welfare by directly observing the animal, regardless of how and where it is kept. Second, as they are applicable to all slaughterhouses, animal-based measures allow the welfare of animals from different slaughterhouses to be compared, and remain more transparent to stakeholders. In Welfare Quality®, resource- or management-based measures were therefore only taken into account to complement the animal-based measures or as a substitute when there were no promising animal-based measures available (Botreau et al., 2007). For example, no valid, reliable and feasible animal-based measure was found for the evaluation of prolonged thirst. In this case, it has been necessary to include resource-based measures such as the presence, number, cleanliness and functioning of drinkers in the lairage pen. Moreover, resource- and management-based measures can also be used to identify risks to animal welfare and identify causes of poor welfare so that improvement strategies can be implemented. For example, if poor stunning effectiveness has been identified as a major problem at a particular slaughterhouse, resource- and management-based indicators (such as stunning system and slaughter procedures) will be explored to identify the principal causes of the animal-based problem and the best strategies to solve it.

The measures that met the requirements for validity, reliability and feasibility, were combined and integrated into the welfare assessment protocol. Table 1.1 shows the final list of measures included in the "operational" protocols for growing pigs, beef cattle and broiler chickens, respectively, which were subsequently applied at commercial slaughterhouses. A full version of the Welfare Quality® assessment protocols can be found at http://www.welfarequalitynetwork.net/network/45848/7/0/40.

Welfare is assessed at all stages of the animal being at the slaughterhouse, from arrival from transport to stunning and slaughter. Therefore, different points in the slaughterhouse are considered, such as the unloading area, lairage, stunning area, and so on (Dalmau et al., 2009).

An important consideration to increase the repeatability and reliability of the assessment is that the measures should be simple to collect and scored, in a way that minimizes the value judgment. For this purpose, most of the animal-based measures are scored according to a 3-point scale ranging from 0 to 2. So that a score 0 is awarded where welfare is good, a score 1 is awarded where there has been some compromise on welfare, and a score 2 is awarded where welfare is poor and unacceptable. In some cases a binary (0/2 or Yes/No) is used.

### *1.4.1 Welfare assessment of pigs*

Welfare assessment starts in the unloading area, where general fear, thermoregulation behaviours, slipping and falling, sickness and dead animals are measured. The number of sick and dead animals, and number of animals shivering or panting are assessed in the lorry. Shivering is defined as slow and irregular vibration of any body part or the body as a whole. Thermal panting begins through the nose, and if prolonged, it continues as rapid open-mouthed breathing with short gasps and a lot of salivation. In addition, the presence of bedding material, length, width and height of the lorry and the total number of transported animals are also considered.

The number of animals that slip and fall during unloading is recorded as a percentage of the total number of animals in each vehicle. Slipping is defined as loss of balance without the body touching the floor, while falling is defined as loss of balance in which a part of the body other than the legs is in contact with the floor. An animal slipping while it is falling is considered only as falling. The number of animals showing general fear is assessed by means of reluctance to move and turning back. A pig shows reluctance to move when it stops walking for at least 2 seconds. Turning back occurs when the pig, that is facing the unloading area, turns its body and faces the truck. The number of lame animals is assessed when walking to the lairage area, according to a 3-point scale: 0 – normal gait; 1 – difficulties walking, but still using all legs (lameness 1); and 2 – severely lame, minimum weight bearing on affected limb (lameness 2).

At lairage, five criteria are taken into consideration. Absence of thirst is calculated based on the number of drinking points in each pen (in the case of drinking valves) or the total surface area of water

**Table 1.1** Welfare Quality® protocol to assess welfare at the slaughterhouse

| | | | Measures | | |
|---|---|---|---|---|---|
| | | Welfare criteria | Pig | Cattle | Broiler chicken |
| Good feeding | 1 | Absence of prolonged hunger | Food provision | Food supply | Feed withdrawal time |
| | 2 | Absence of prolonged thirst | Water supply | Water supply | Water withdrawal time |
| Good housing | 3 | Comfort around resting | Flooring, bedding | Flooring, bedding | As yet, no measure is developed |
| | 4 | Thermal comfort | Shivering, panting, huddling | This criterion is not applied in this situation | Panting on lorry and/or lairage |
| | 5 | Ease of movement | Slipping, falling, stocking density of lorries, stocking density of lairage pens | Slipping, falling, freezing, trying to turn, turning around, moving backwards | Stocking density in crates |
| Good health | 6 | Absence of injuries | Lameness, wounds on the body | Lameness, bruises | Wing damage, bruising |
| | 7 | Absence of disease | Sick animals, dead animals | This criterion is not applied in this situation | Dead on arrival (DOA) |
| | 8 | Absence of pain induced by management procedures | Stunning effectiveness | Stunning effectiveness | Pre-stun shock, effectiveness of stunning |
| Appropriate behaviour | 9 | Expression of social behaviours | This criterion is not applied in this situation | This criterion is not applied in this situation | This criterion is not applied in this situation |
| | 10 | Expression of other behaviours | This criterion is not applied in this situation | This criterion is not applied in this situation | This criterion is not applied in this situation |
| | 11 | Good human–animal relationship | High pitched vocalizations | Vocalizations, coercion | This criterion is not applied in this situation |
| | 12 | Emotional state | Reluctance to move, turning back | Struggling, kicking, jumping in stun box, freezing, trying to turn, turning around, moving backwards | Flapping on the line |

supplied (in the case of water troughs) per animal, their functionality and cleanliness. For the absence of hunger, the availability of feed for animals that have been held more than 12 hours in the holding pens is evaluated. The third criterion is thermal comfort. Behavioural thermoregulation measures, such as huddling, shivering or panting are also scored by using a 3-point scale: 0 – no pigs in the pen showing shivering, panting or huddling; 1 – up to 20% of pigs in the pen with the above behaviour; and 2 – more than 20% of pigs in the pen with the above behaviour. Huddling is described as a pig lying with more than 50% of its body in contact with another pig, but virtually lying on top of another pig. For the absence of disease, dead animals are recorded. The fifth criterion is comfort while resting, which is evaluated based on the space allowance. For this purpose, the length and width of the pens are measured and the number of animals counted.

During movement from the lairage pen to the stunning area, good human–animal relationship is assessed from the incidence of high pitched vocalizations (HPVs). This type of vocalization is associated with electric prod use, excessive pressure from a restraint device, stunning problems or slipping on the floor (Grandin, 1997; 2000). Afterwards, stunning effectiveness is assessed immediately after stunning and before sticking and will be covered in relevant chapters.

After slaughter, skin lesions provide valuable information regarding the management of animals on the farm of origin, during transport or at the lairage pens. The carcass is divided into five body zones and only one side is assessed (Figure 1.1): 1) ears, 2) front (from the head to the back of the shoulder), 3) middle (from the back of the shoulder to the hindquarters), 4) hindquarters and 5) legs (from the accessory digit upwards).

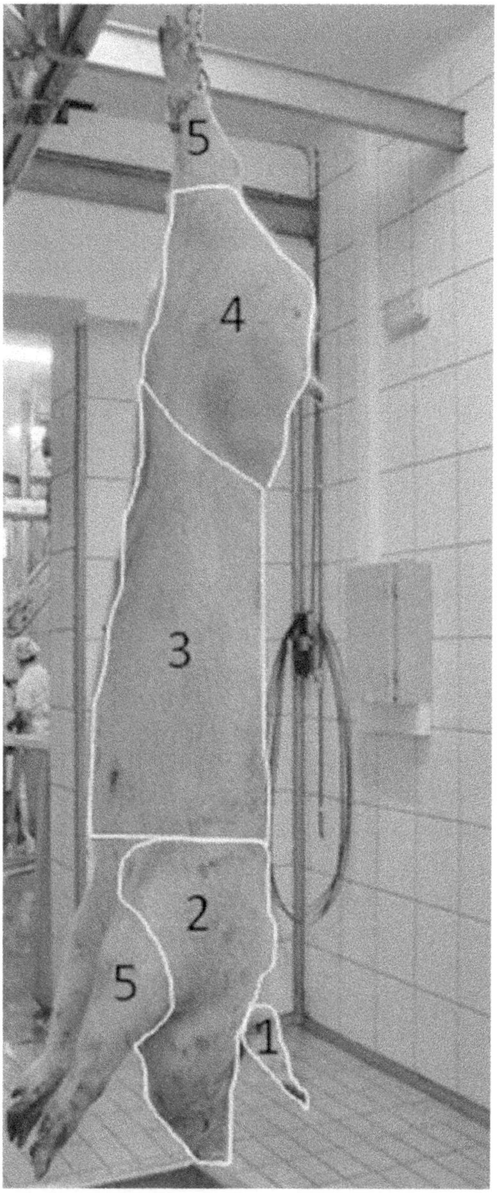


**Figure 1.1:** The five parts of the carcass for the assessment of skin lesions. 1) ears, 2) front (from the head to the back of the shoulder), 3) middle (from the back of the shoulder to the hind-quarters), 4) hindquarters and 5) legs (from the accessory digit upwards)

Each part is scored as follows: 0 – no visible skin damage, only one lesion greater than 2 cm or lesions smaller than 2 cm; 1 – between two and 10 lesions greater than 2 cm; 2 – any wound which penetrates the muscle tissue, or more than 10 lesions greater than 2 cm. The scoring of the five parts of the carcass is combined in one scoring: 0 – all body parts with a score of zero; 1 – at least one body part with a score of 1; 2 – a body part with a score of 2. Furthermore, the health status of the animals on the farm of origin is also assessed after slaughter. Internal organs are inspected for the presence of pleurisy and pneumonia in the lungs, pericarditis in the heart and white spots in the liver on a sample of 60 animals divided into three batches of 20.

### *1.4.2 Welfare assessment of cattle*

During unloading, the number of slipping, falling, freezing and turning around events or turning attempts and moves backwards per animal is assessed. Freezing is defined as when the animal refuses to move forwards or backwards within 4 seconds from being touched/coerced by the handler and the route is free in front and behind. The measure "trying to turn around" is defined as an animal that makes an unsuccessful attempt to turn (turning only the head is not considered as such). Lameness is assessed in the same way as in pigs.

At lairage, all the pens housing animals are assessed for absence of prolonged hunger, absence of prolonged thirst and comfort around resting. Absence of prolonged hunger is assessed by monitoring the feed provision in all overnight lairage pens and interviewing the staff about the type and quantity (< 2000 g per animal is considered insufficient) and checking the time animals are fed. The scoring is: 0 – no evidence of feed provision; 1 – some evidence of feed provision; 2 – clear evidence of feed provision in sufficient quantity. Absence of prolonged thirst is assessed by means of water supply in all pens. The percentage of pens with a functioning water bowl is recorded as well as the cleanliness with regard to the presence of old or fresh dirt on the inner side of the bowl. Cleanliness is scored as follows: 0 – clean: drinkers and water clean; 1 – partly dirty: drinkers dirty but fresh and clean water available; and 2 – dirty: drinkers and water dirty. Comfort around resting is assessed by space allowance and the suitability of flooring and bedding during lairage. The percentage of pens with suitable rubber, straw, wood shavings or sawdust is considered.

From lairage to stunning, two areas are considered: the corridor from the lairage to the stunning area and the stun box. While the animals are moved to the stunning area the quality of management is assessed by the number of vocalizations per animal and the prevalence of coercion observations. Vocalization is defined as an animal's vocalizing response to fear or pain-related events, such as falling, physical means of coercion, restraining and strikes by gates. Coercion is defined as the use of any item when handling live animals, including electric goad, stick, flapper, rattle or the driver's body if is contacting the animal at any time. In the stun box, fear is assessed by means of struggling, kicking and jumping. Struggling is defined as continuous struggling/panicking movements of escape such as general slipping, forward and backward movements and body trembling, lasting for more than 3 seconds, with no breaks of calm behaviour. Kicking is defined as the hind leg kicking, often as a reaction to touch/pain, and jumping as a sudden startle/flight reaction. The assessment of stunning effectiveness will be covered in relevant chapters. Bruises are assessed on the body of animals, looking at the area between the limit where the skin is taken off the carcasses and where trimming occurs. Bruises are scored according to the Australian Carcass Bruise Scoring System (Westin et al., 2009) in relation to its spread (slight–medium–heavy) and depth (i.e. if the bleeding involves any tissue other than surface muscle tissue, it is considered to be deep). Furthermore, the health status of the animals on the farm of origin is also assessed after slaughter.

### *1.4.3 Welfare assessment of broiler chicken*

After arrival and at lairage, the percentage of chickens that are panting is assessed. The average stocking density is calculated by measuring the crate size and counting the number of birds per crate. Feed and withdrawal time are taken from farm and transport records, which are kept by the slaughterhouse on arrival. During shackling, the percentage of birds showing flapping behaviour for 5 to 10 minutes is assessed. Birds may flap their wings vigorously, particularly where the line makes abrupt changes of direction. At that time, wing damage due to catching, transport and removal from transport crates can be identified by visible "dropped wings" on the slaughter transport line. In this case the wing is clearly hanging down indicating fracture or dislocation.

At the entrance to the stunning bath the percentage of birds showing pre-stun shocks for 10 minutes are assessed. A "pre-stun shock" occurs when a bird receives a premature electrical shock from the stunning bath. This can occur when the bird's wing touches splashed water or wet surfaces (electrified) at the entrance to the stunner.

The assessment of stunning effectiveness will be covered in relevant chapters. Bruising associated with catching, transport or hanging is observed in the whole carcasses. This measure assesses the bruising visible on the carcasses, distinguished from post-mortem carcass damage (which will not have caused haemorrhage into the tissues). Dead on arrival is considered: any bird which is "found dead" in the crate at unloading is considered a mortality.

## 1.5 IMPORTANCE OF EDUCATION IN ANIMAL WELFARE

Evidence shows that poor welfare is associated with quantitative and qualitative losses in the value of carcasses and meat. These losses are specifically related to animal deaths, increased bruising of carcasses and changes in the pH of the meat, resulting from poor pre-slaughter management practices. Animal handling training programmes have shown to be essential for fast and effective prevention of animal suffering during pre-slaughter management and stunning.

Effectiveness of training programmes is assessed following key impacts on participants such as changes in their reaction, skills and behaviour as well as on the animals, evaluating progress in welfare outcomes. Several studies reported the successful application of training programmes in slaughter plants. Training programmes resulted in a significant reduction in the proportion of downgraded carcasses due to bruising and carcass damage (Paranhos da Costa et al., 2014; Pilecco et al., 2013). Gallo et al. (2003) also reported a significant improvement in management conditions with a 47.7% reduction in the use of electric prods after slaughterhouse staff training. Training programmes also had a significant effect on stunning effectiveness. In a recent study realized on 15 cattle and sheep at a slaughterhouse, Gallo (2010) registered a reduced incidence of animals showing signs of consciousness after stunning (4.5 to 0.5%). Still, to ensure long-term effectiveness of the training programmes, Paranhos da Costa et al. (2012) emphasized the need to maintain a constant follow-up of the management procedures. The generally high turnover of staff is one of the difficulties to be overcome with frequent training. To motivate changes, clear and simple solutions should be provided to solve the problems faced by staff in their everyday practice.

Within Europe and since the new Regulation (EC) No. 1099/2009 came into force, many competent authorities, official veterinarians and food business operators have encountered problems with the implementation of some requirements. The main problems relate to the quality of stunning and its assessment. For this reason, recent efforts have been made to provide clear recommendations on the control measures and monitoring procedures to ensure proper stunning (EFSA, 2013a,c,d,e). Knowledge exchange becomes an efficient tool to help implement the legislation. In addition to slaughter plant staff, knowledge material and training should then address other target groups such as competent authorities, official vets and animal welfare officers.

Several key points for implementation to be effective have been highlighted by the EUWelNet project (2013, www.euwelnet.eu):

- Frequent interaction and collaboration between public and private actors and agencies is considered as a first important key point. It indeed plays a crucial role in the early identification of knowledge gaps, the dissemination of knowledge and tailor-made information and training of target groups.
- Animal welfare has to be presented in a practical and realistic manner and the performance benefits of higher welfare have to be emphasized. Training in welfare requirements works best if the industry can see economic or other advantages to what is being requested. Target groups should understand that implementation works, how it contributes to improved welfare and how it works in practice.
- Particular care should be taken to ensure that the training material and activities are available to small-scale slaughterhouses which represent a particular target group.
- Knowledge exchange strategies should actively involve the participants, promoting dialogue. Horizontal training has been particularly efficient in several training

activities such as the ones undergone by Zuin et al. (2014).

- Training has to be easily understandable and feasible, crossing social constraints and language barriers.

Knowledge exchange should definitively apply worldwide through on-line education materials or exchange programmes (Fraser, 2008). Examples of such work are given by the Humane Slaughter Association (www.hsa.org.uk) who provide training materials and opportunities in a wide range of countries, DG-SANTE who organize workshops "improving animal welfare – a practical approach" (http://ec.europa.eu/food/animals/index_en.htm) and the regional seminars for OIE focal points (http://www.oie.int/).

## 1.6 REFERENCES

Bianchi, M., Petracci, M. and Cavani, C. (2005) Effects of transport and lairage on mortality, liveweight loss and carcass quality in broiler chickens. *Italian Journal of Animal Science* 4, 516–518.

Blokhuis, H., Keeling, L., Gavinelli, A. and Serratosa, J. (2008) Animal welfare's impact on the food chain. *Trends in Food Science and Technology* 19, 75–83.

Boissy, A., Arnould, C., Chaillou, E., Désiré, L., Duvaux-Ponter, C., Greiveldinger, L., Leterrier, C., Richard, S., Roussel, S., Saint-Dizier, H., Meunier-Salaün, M.C., Valance, D. and Veissier, I. (2007) Emotions and cognition: a new approach to animal welfare. *Animal Welfare* 16, 37–43.

Boissy, A. (1995) Fear and fearfulness in animals. *Quartery Review of Biology* 70, 165–191.

Botreau, R., Bonde, M., Butterworth, A., Perny, P., Bracke, M.B.M., Capdeville, J. and Veissier, I (2007) Aggregation of measures to produce an overall assessment of animal welfare: Part 1 – A review of existing methods. *Animal* 1, 1179–1187.

Broom, D.M. and Johnson, K.G. (1993) *Stress and animal welfare.* London, UK: Chapman and Hall.

Butterworth, A. (1999) Infectious components of broiler lameness: a review. *World's Poultry Science* 55, 327–352.

Dalmau, A., Temple, D., Rodríguez, P., Llonch, P. and Velarde, A. (2009) Application of the Welfare Quality® protocol at pig slaughterhouses. *Animal Welfare* 18, 497–505.

Danbury, T. C., Weeks, C. A., Chambers, J. P., Waterman-Pearson, A. E. and Kestin, S. C. (2000) Self-selection of the analgesic drug carprofen by lame broiler chickens. *The Veterinary Record* 146, 307–311.

Dawkins, M.S. (2000) Animal minds and animal emotions. *American Zoologist* 40, 883–888.

Dawkins, M.S. (2004) Using behaviour to assess animal welfare. *Animal Welfare* 13, 3–7.

EFSA AHAW Panel (EFSA Panel on Animal Health and Welfare) (2004) Opinion of the Scientific Panel on Animal Health and Welfare (AHAW) on a request from the Commission related to welfare aspects of the main systems of stunning and killing the main commercial species of animals. *EFSA Journal* 45, 1–29.

EFSA AHAW Panel (EFSA Panel on Animal Health and Welfare) (2013a) Guidance on the assessment criteria for studies evaluating the effectiveness of stunning methods regarding animal protection at the time of killing. *EFSA Journal* 11(12), 3486, 41 pp. doi:10.2903/j.efsa.2013.3486

EFSA AHAW Panel (EFSA Panel on Animal Health and Welfare) (2013b) Scientific Opinion on monitoring procedures at slaughterhouses for bovines. *EFSA Journal* 11(12), 3460, 65 pp. doi:10.2903/j.efsa.2013.3460

EFSA AHAW Panel (EFSA Panel on Animal

Health and Welfare) (2013c) Scientific Opinion on monitoring procedures at slaughterhouses for poultry. *EFSA Journal* 11(12), 3521, 65 pp. doi:10.2903/j.efsa.2013.3521

EFSA AHAW Panel (EFSA Panel on Animal Health and Welfare) (2013d) Scientific Opinion on monitoring procedures at slaughterhouses for sheep and goats. *EFSA Journal* 11(12), 3522, 65 pp. doi:10.2903/j.efsa.2013.3522

EFSA AHAW Panel (EFSA Panel on Animal Health and Welfare) (2013e) Scientific Opinion on monitoring procedures at slaughterhouses for pigs. *EFSA Journal* 11(12), 3523, 62 pp. doi:10.2903/j.efsa.2013.3523

FAWC (1992) Farm Animal Welfare Council updates the five freedoms. *Veterinary Record* 17, 357.

Faucitano, L. and Schaefer, A. (2008) *Welfare of pigs from birth to slaughter.* Wageningen Academic Publishers, The Netherlands.

Fraser, D. (2008) Toward a global perspective on farm animal welfare. *Applied Animal Behaviour Science* 113, 330–339.

Fraser, D., Weary, D.M., Pajor, E.A. and Milligan, B.N. (1997) A scientific conception of animal welfare that reflects ethical concerns. *Animal Welfare* 6, 187–205.

Forkman, B., Boissy, A., Meunier-Salaün, M.C., Canali, E. and Jones, R.B. (2007) A critical review of fear tests used on cattle, pigs, sheep, poultry and horses. *Physiological Behaviour* 92, 340–374.

Gallo, C. (2010) Diagnóstico e implementación de estrategias de bienestar animal para incrementar la calidad de la carne de rumiantes, financiado por la Fundación para la Innovación Agraria (FIA-Chile) y la Asociación Gremial de Plantas Faenadoras y Frigoríficos de Carne (FAENACAR-Chile), 2005–2010.

Gallo, C., Teuber, C., Cartes, M., Uribe, H. and Grandin, T. (2003) Mejores en la insensibilizacion de bovinos con pistol pneumatica de propectil retenido tras cambios de equipomiento y capacitacim del personal. *Archivos de Medicina Veterinaria* 35, 159–170.

Gentle, M.J. (1992) Ankle joint (Artc. Intertarsalis) receptors in the domestic fowl. *Neuroscience* 49, 991–1000.

Gentle, M.J. and Tilston, V.L. (2000) Nociceptors in the legs of poultry: Implications for potential pain in preslaughter shackling. *Animal Welfare* 9, 227–236.

Grandin, T. (1997) Assessment of stress during handling and transport. *Journal of Animal Science* 75, 249–257. (http://www.grandin.com/references/handle.stress.html)

Grandin, T. (2000) *Livestock handling and transport* (2nd ed.). London: CAB International.

Gregory, N.G. (1998) *Animal welfare and meat science.* London: CAB International.

Hughes, B.O. and Duncan, I.J.H. (1988) The notion of ethological 'need', models of motivation and animal welfare. *Animal Behaviour* 36, 1696–1707.

Hunter, R.R., Mitchell, M.A, Carlisle, A.J., Quinn, A.D., Kettlewell, P.J, Knowles, T.G. and Warriss, P.D. (1998) Physiological responses of broilers to pre-slaughter lairage: Effects of the thermal micro-environment? *British Poultry Science* 39, 53–54.

Manteca, X., Velarde, A. and Jones, B. (2009) Animal welfare components. In F. Smulders, & B. Algers (Eds.). *Welfare of production animals: assessment and management of risks* (pp. 61–77). The Netherlands: Wageningen Academic Publishers.

Mendl, M. and Paul, E.S. (2004) Consciousness, emotion and animal welfare: insights from cognitive science. *Animal Welfare* 13, 17–25.

National Research Council (1981) Effect of environment on nutrient requirements of domestic animals (pp. 75–84). Washington: National Academy Press.

OIE (2009a) Transport of animals by land: Terrestrial Animal Health Code (18th ed.).

Paris, France: World Organization for Animal Health.

OIE (2009b) Slaughter of animals: Terrestrial Animal Health Code (18th ed.). Paris, France: World Organization for Animal Health.

OIE (2009c) Chapter 7.6 Killing animals for disease control purposes: Terrestrial Animal Health Code (18th ed.). Paris, France: World Organization for Animal Health.

Paranhos da Costa, M.J.R., Huertas, S.M., Strappini, A.C. and Gallo, C. (2014) Handling and transport of cattle and pigs in South America. In T. Grandin (Ed.), *Livestock handling and transport* (pp. 174–192). Wallingford, Oxfordshire, UK: CABI International.

Paranhos da Costa, M.J.R., Huertas, S.M., Gallo, C. and Dalla Costa, O. (2012) Strategies to promote farm animal welfare in Latin America and their effects on carcass and meat quality traits. *Meat Science* 92, 221–226.

Petracci, M., Bianchi, M., Cavani, C., Gaspari, Lavazza, P. (2006) Preslaughter mortality in broiler chickens, turkeys, and spent hens under commercial slaughtering. *Poultry Science* 85, 1660–1664.

Pilecco, M., Almeida-Paz, P.M., Tabaldi, L.A., Naas, I.A., Garcia, R.G., Caldara, F.R. and Francisco, N.S. (2013) Training of catching teams and reduction of back scratches in broilers. *Revista Brasileira de Ciência Avicola* 15 (3), 283–286.

Regulation (EC) No. 1099/2009 of 24 September 2009 on the protection of animals at the time of killing.

Rushen, J. (2003) Changing concepts of farm animal welfare: bridging the gap between applied and basic research. *Applied Animal Behaviour Science* 8, 199–214.

Sparrey, J. (1994) Aspects in the design and operation of shackle lines for the slaughter of poultry. Unpublished Mphil thesis, University of Newcastle upon Tyne, Newcastle upon Tyne, United Kingdom.

Velarde, A. (2007) Skin lesions. In A. Velarde & R. Geers (Eds.) *On farm monitoring of pig welfare* (pp.79–83). The Netherlands: Wageningen Academic Publishers.

Velarde, A. and Dalmau, A. (2012) Animal welfare assessment at slaughter in Europe: moving from inputs to outputs. *Meat Science* 92, 244–251.

Waiblinger, S., Knierim, U. And Winckler, C. (2001) The development of an epidemiologically based on-farm welfare assessment system for use with dairy cows. *Acta Agriculturae Scandinavica Section a–Animal Science suppl.* 30, 73–77.

Warriss, P.D., Brown, S.J. and Adams, M. (1994) Relationship between subjective and objective assessment of stress at slaughter and meat quality in pigs. *Meat Science* 38, 329–340.

Warriss, P.D., Bevis, E.A., Brown, S.N. and Edwards, J.E. (1992) Longer journeys to processing plants are associated with higher mortality in chickens. *Poultry Science* 33, 201–206.

Webster, A.J.F. (2005) *Animal welfare: Limping towards Eden.* Chichester, UK: Wiley-Blackwell.

Webster, A.J.F., Main, D.C.J. and Whay, H.R. (2004) Welfare assessment: Indices from clinical observation. *Animal Welfare* 13, 93–98.

Westin, R., Velarde, A., Dalmau, A. and Algers, B. (2009) Assessment of ultimate pH and bruising in cattle. In B. Forkman, L. Keeling, M. Miele and J. Roex (Eds.), *Welfare Quality Reports.* 11. Assessment of animal welfare measures for dairy cattle, beef bulls and veal calves (pp. 51–55). Cardiff, UK: University of Cardiff.

Zuin, L.F.S., Zuin, P.B., Monzon, A.G., Paranhos da Costa, M.J.R. and Oliveira, I.R. (2014) The multiple perspectives in a dialogical continued education course on animal welfare: Accounts of a team of extension agents and a manager and a cowboy from a rural Brazilian territor. *Linguistics and Education* 28, 17–27.

chapter two

# Physiology and behaviour of food animals

*E.M. Claudia Terlouw*[1] *and Troy Gibson*[2]

**Learning objectives**

- Describe the animal senses, the basic physiology of pain and stress responses, including emotional states and how they relate to animal welfare during stunning and slaughter.
- Understand the anatomy and physiology of the major farmed species and how this impacts welfare during stunning and slaughter.
- Understand the basic anatomy and physiology of the brain and how it relates to stunning and slaughter.

## CONTENT



[1] INRA UMR1213 Herbivores, Centre Auvergne-Rhône-Alpes, site de Theix, 63122 saint-Genès Champanelle, France
[2] Department of Production and Population Health, Royal Veterinary College, Hawkshead Lane, Hatfield AL9 7TA, United Kingdom

## 2.1 INTRODUCTION: ANIMALS IN THEIR ENVIRONMENT

Animals alter their behaviour to adapt to changes in their environment. The senses play an important role in informing animals of the features of their environment including potential threats. The way animals evaluate these features, including the slaughterhouse, has a considerable impact on their reactions and the ease with which they can be managed.

Most farm animal species have a wide field of vision, as the eyes are located on each side of the head. This allows the detection of potential threats in a wide area, even while grazing. Animals are capable of detecting even small changes in the environment, particularly motion (i.e. movement of equipment or personnel) and this can cause balking and refusal to enter raceways and stunning pens. In addition, farm animals have visual blind spots; often straight behind the animal where their body or ears block their vision. When objects are seen to enter or appear from a blind spot, animals can be startled and this can further complicate their management. Pigs, sheep and cattle have reduced depth perception and a limited ability to adjust, and therefore need more time than humans to distinguish clearly objects in their visual field. Sharp contrasts, changes in lighting or light reflections in water puddles or on metal are frequent causes for animals to balk in abattoirs (Grandin, 2013). Refusal to enter dark environments or to cross shadows is potentially due to their inability to distinguish clearly their environment. However, it has been suggested that when animals are given time for their eyes to adjust to the environment, they are often easier to handle and move forward (Alexander and Shillito, 1978; Philips, 1993; Lomas et al., 1998; Blackshaw, 1986).

Farm animals have a well-developed sense of smell (olfaction), but not all odours have a significant effect on their behaviour (Blackshaw, 1986). Animals may be exposed to familiar or unfamiliar odours during transport or in the abattoir, some of which evoke fear responses. For example, cattle showed fear reactions to the odours of blood, dog faeces and urine from stressed conspecifics. Meanwhile, these cattle did not react to the neutral odour of water (Boissy et al., 1998; Terlouw et al., 1998).

Hearing is well developed in farm animals and sounds may be localized by moving the head or ears (Blackshaw, 1986). The recognition of sounds may help identify the presence of danger but also gives other information. For example, cattle, pigs, sheep and chickens often vocalize to communicate socially, including in the abattoir (Grandin, 1998; Mench and Keeling, 2001; Boissy et al., 2005). The respiration and heart rate in sheep increases when exposed to loud (>100 dB) noises, suggesting that they perceive these noises as unpleasant or threatening (Ames and Arehart, 1972). To lower stress in animals the abattoir environment should exclude unfamiliar or loud noises (Grandin, 2013).

Herding or flocking instinct is a behavioural adaptation that improves survival by either minimizing the probability of predations or allowing shared vigilance. Pigs, cattle, sheep and poultry are all gregarious species. Behaviour during transport, lairage and slaughter is influenced by the space and environment that animals were exposed to during rearing, as was shown in pigs (Terlouw et al., 2009). Farm animals feel protected by the presence of conspecifics (Boissy and Le Neindre, 1997). In the abattoir, sheep flock together and prefer to move in groups. Isolation of individuals causes stress in sheep (Deiss et al., 2009) and cattle (Mounier et al., 2006; Bourguet et al., 2011).

In a stable group, animals establish a dominance hierarchy. The disruption of groups or mixing unfamiliar groups, as frequently occurs during slaughter, often leads to aggression (Terlouw et al., 2009; Mounier et al., 2006).

## 2.2 STRESS AND REACTIONS TO STRESS

The term stress is often used to describe the state of the animal when it is incapable of adapting, behaviourally and physiologically, to environmental or physical challenges (Fraser et al., 1975; Broom, 1987). Recent studies on the behaviour, physiology and anatomy of the brain have shown that many animals including some fish species are sentient, that is, they are capable of experiencing negative and positive emotions (Sneddon, 2003; Paul et al., 2005; Boissy et al., 2007). While it is important to understand the impact of physical challenges on the physiological and behavioural adaptive capacity of the animal, it is also important to take into account the emotions that animals may experience (Dantzer, 2002; Désiré et al., 2002; Dawkins, 2008).

The stress status of the animal depends on its evaluation of the environment. If the animal feels threatened, even if the threat is imaginary, it will experience negative emotions, that is, it will feel stressed (Terlouw, 2005; Veissier and Boissy, 2007). Stress may be caused by psychological factors such as fear or frustration, or physical factors, such as thermal stress, hunger or thirst. In the latter cases, homeostasis may be disturbed.

When an animal feels stressed, for example by abattoir staff or adverse weather conditions during transport, it needs to adapt and protect itself from that threat. In order to prime the body for rapid acceleration or action in response to the threat, homeostasis is altered. Changes in homeostasis in response to a stressor are governed by physiological changes, involving neurological and endocrine systems. The autonomic nervous system is a neurological system involved in vital functions and largely beyond conscious control. It has a direct influence on the activity of many organs and consists of two parts: the sympathetic and parasympathetic branches. Generally, the parasympathetic branch stimulates body functions associated with resting and digesting, while the sympathetic branch stimulates active responses. There is a balance between sympathetic and parasympathetic activity to maintain homeostasis. When an animal is faced with a stressor, there is a shift in this balance towards increased sympathetic activity and lowered parasympathetic activity to prepare the animal for increased physical activity if needed. During raised sympathetic activity, noradrenaline is released into the blood circulation from the nerve endings of the autonomic nervous system and adrenaline from the medulla of the adrenal glands.

The perception of a threatening situation will further result in the activation of the hypothalamic–pituitary–adrenal axis (HPA; Figure 2.1.). This is controlled

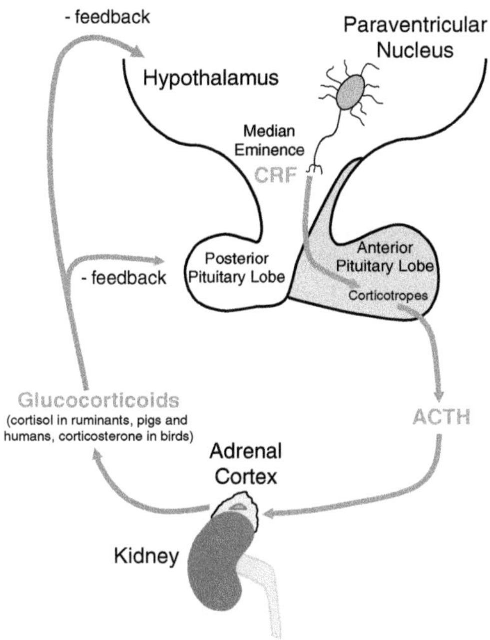


**Figure 2.1:** The basic functional anatomy of the hypothalamic–pituitary–adrenal axis (HPA). Neurons in the paraventricular nucleus release corticotropin-releasing factor (CRF) into the hypophyseal portal system, which transports CRF to the anterior lobe of the pituitary, where it stimulates the production of adrenocorticotropin (ACTH) from corticotropes. ACTH enters circulation and activates the production of glucocorticoids from the cortex of the adrenal gland. Glucocorticoids act as negative feedback on the hypothalamus and pituitary, inhibiting the synthesis and secretion of CRF and ACTH from the hypothalamus and pituitary respectively. Note: not to scale

by neurons in the hypothalamus of the brain, which secrete a hormone, called corticotropin-releasing factor (CRF). This hormone stimulates the secretion of adrenocorticotropic hormone (ACTH) into the bloodstream from the anterior lobe of the pituitary. The ACTH will subsequently stimulate the secretion of glucocorticoids (cortisol in most mammals and corticosterone in birds) from the cortex of the adrenal glands.

The increased activity of the sympathetic nervous system and of the HPA axis helps to sustain the increased physical effort and vigilance. Increased circulating glucocorticoid and catecholamine concentrations contribute to the mobilization of energy substrates, glucose and free fatty acids, in the blood. Increased circulating glucose and adrenaline concentrations allow faster metabolism needed for muscle contraction. Raised concentrations of adrenaline also increase cardiac output, cause vasoconstriction of vessels and increased blood pressure and dilation of bronchioles while functions that are less urgent at that moment, such as intestinal motility and gastric secretions, are inhibited. At slaughter, the combined effects of physical activity and increased catecholamine levels may negatively influence post-mortem muscle pH decline (Terlouw, 2005).

## 2.3 CAUSES OF STRESS

Stress may arise in animals when they expect unpleasant events based on previous experiences, associating a situation with past fear or pain. For example, pigs associated certain humans with pain and stress which resulted in a stress response when these humans were present (Terlouw et al., 2005; Hemsworth et al., 2002). Even changes in the normal situation can be perceived as stressful, especially when the changes are unexpected and are beyond the influence of the animal (Greiveldinger et al., 2007). The reason is that the animal does not know which events lie ahead and therefore they are potentially threatening. During the slaughter period, animals are subjected to various situations including herding, loading, transport, unloading, mixing and handling which may be potential causes of stress of psychological or physical origin.

### 2.3.1 Fear

Fear is a major cause of psychological stress in animals. Fear reactions may vary, depending on the stimuli, the animal and the context. They may involve defence reactions (attack, threat), avoidance (flight, escape, hiding) or movement inhibition, to remain unnoticed. Animals may vocalize and excrete pheromones[1], both helping to inform group members (Boissy, 1995; Boissy et al., 1998). The excretion of urine and faeces facilitates physical movements and consequently the expression of defence reactions while excretions may further carry the fear-related pheromones (Boissy et al., 1998). The impact of pheromones in the abattoir environment is difficult to estimate due to the range of other overwhelming stimuli. However, during transportation olfaction and pheromones play a significant role in the fear response. The behavioural fear reactions are accompanied by the classical physiological stress reactions such as increases in heart rate and in plasma catecholamine and glucocorticoid levels.

Potential causes of fear are manifold; they involve the presence of humans, aggressive conspecifics, unfamiliar situations or unexpected events. The degree of threat of a situation perceived by the animal depends partly on the characteristics of the fear-inducing stimulus such as the type of threat, but also its proximity, suddenness, duration or intensity (Boissy, 1995). It depends also on the animal's earlier experience and its genetic background. For example, fowl, cattle and pigs that have had positive experiences with humans react less fearfully to these humans but also to other humans (Jones and Faure, 1981; Boivin et al., 1994; Terlouw and Porcher, 2005) and Duroc pigs approached a human more readily than Large White pigs did (Terlouw and Rybarczyk, 2008).

When animals are in a state of excessive fear, their reactions may be pronounced and unexpected and therefore dangerous (Grandin, 1999). Moving frightened animals is also very difficult and time consuming and generally, it is more efficient to leave the frightened animals for a certain time to allow them to settle down (Grandin, 1999).

During the slaughter period, fear due to unfamiliar environments and inappropriate handling is a major cause of psychological stress. For example, cows that were more stressed than their counterparts when introduced into an unfamiliar environment during rearing were also more stressed at slaughter (Bourguet et al., 2010). Fear of humans is another important cause of stress during the slaughter period. Cattle and pigs can discriminate between individual people and approach unfamiliar humans less easily than familiar humans (Boivin et al., 1998; Terlouw and Porcher, 2005). Yet, calves and pigs can generalize positive experiences with humans: those animals that had received positive (strokes, quiet behaviour of the care taker) as opposed to negative handling (pushes, shouts) approached humans more easily even when they were unfamiliar (Lensink et al., 2000; Terlouw and Porcher, 2005).

[1] Volatile molecules that may be emitted by animals and plants and that have an effect on behaviour. Animals identify pheromones with their vomeronasal organ.

Positively handled calves were also less stressed during slaughter (Lensink et al., 2000). However, genetically selecting animals for very low fearfulness may not be beneficial because normal fear reactions are necessary for the animal to avoid danger. In addition, animals that are very unreactive may be difficult to handle. For example, bulls and pigs that were less fearful of humans were more difficult to move and these pigs received more negative interventions by abattoir personnel (Terlouw et al., 2005; Hemsworth et al., 2002; Mounier et al., 2008), which is unwanted and detrimental to animal welfare with economic consequences.

### *2.3.2 Social stress*

The presence of a stable social group is an important aspect of normal life of farm animals. Social isolation or disruption of the social group causes social stress. For example, repeated isolation of pigs resulted in increased behavioural responses indicative of stress and frustration during the isolation period (Terlouw and Porcher, 2005). Similarly, isolated cattle vocalize while heart rates increase (Boissy and Le Neindre, 1997). Isolated lambs perform high-pitched bleating and isolated chicks increase the frequency of occurrence of distress calls (Boissy et al., 2005; Marx et al., 2001).

The mixing of livestock can occur during the slaughter process at markets, transport, lairage and slaughter. This is often accompanied by agonistic interactions to establish a new dominance hierarchy. The interactions generally start within minutes after mixing of the unfamiliar animals and involve agonistic behaviour that may be more or less overt, depending on context, species and gender (Bouissou, 1980; Turner et al., 2001). They may last several hours (Bouissou et al., 1974; Brown et al., 1999). The agonistic interactions represent a negative experience for the animal, essentially due to fear and pain caused by tissue damage. In contrast to other species, when mixed, familiar sheep aggregate together, while agonistic interactions are lower than before mixing (Ruiz-de-la-Torre and Manteca, 1999).

Studies on pigs, goats, sheep, calves, cows, bulls and poultry have shown that individuals differ consistently in the way they react to different social stress-inducing situations (Terlouw et al., 2008). For example, heifers that showed more fear responses in social isolation in an open field also showed increased fear responses when confronted with an unfamiliar object or in response to a sudden event (Boissy and Bouissou, 1995). Similarly, pigs that left more easily their home pen were easier to drive through a corridor and less reactive to human approach (Lawrence et al., 1991).

Social disturbances such as separation from the pen mates or mixing of unfamiliar animals are another major cause of psychological stress during slaughter. Thus, sheep and cows that reacted more strongly to separation from the pen mates during tests, showed stronger stress reactions at slaughter (Deiss et al., 2009; Bourguet et al., 2010). The mixing of cattle and pigs of different rearing groups and even farms during the lairage period often leads to agonistic interactions. In pigs, the incidence and level of fighting often depends on the presence of a few aggressive animals in the group (Geverink et al., 1996) and are generally increased at higher stocking densities (Geverink et al., 1996), in larger compared to smaller groups (Schmolke et al., 2004) and in pigs that have been food deprived (Brown et al., 1999). Young bull calves often mount each other when stressed (Kenny and Tarrant, 1987). Maintaining the animals in their rearing group during the slaughter process reduces social stress and facilitates the slaughter procedure because of the group cohesion, as was shown for bull calves (Mounier et al., 2006). This effect was even more pronounced when the animals had been together throughout their rearing period (Mounier et al. 2006).

### *2.3.3 Environmental stressors*

Other potential causes of stress of physical origin are long and rough transport conditions, food deprivation, extreme environmental temperatures, aggression or physical trauma, which may all result in fatigue, hunger, thermal and respiratory discomfort and pain, respectively. The slaughter process is complex and often a situation may represent several potential stressors of physical and psychological origins. For example, transport is associated with a change in physical and social environment, movements of the lorry may cause physical trauma and unfavourable climatic conditions (temperature, humidity and air draughts),or water quality deterioration for fish. Recent work has reported that the presence of different stressors may exacerbate stress responses. Thus, cows and heifers that had been food-deprived and/or subjected to 5 minutes forced walking were more reactive to stressful events including handling. These results underline that at slaughter, the presence of physical stress factors may synergistically increase psychological stress (Bourguet et al., 2011).

## 2.4 PAIN

The classical picture of pain was that injury in the peripheral tissues lead to the transmission of nervous signals to certain centres of the brain. In the sixties, different types of pain were recognized, such as somatic, visceral and neuropathic pain, acute and chronic, and nociceptive and non-nociceptive pain. It was further noticed that pain could be modulated which gave rise to the gate control theory (Melzack and Wall, 1965). This theory has been further developed, and today we know that several pain modulatory systems exist (Paulmier et al., 2015). It is now well recognized that pain processing mechanisms are complex, involving sensory, motor, associative, autonomous and limbic structures (Ibinson et al., 2004; Bromm, 2001). Today, the term "pain matrix" is used, referring to the network of cortical areas through which pain is generated (Ingvar, 1999; Peyron, 2000).

Pain is thus a multidimensional phenomenon, involving a variety of different systems and processes of the body. The International Association for the Study of Pain (IASP) definition of pain states that pain is: "An unpleasant sensory and emotional experience associated with actual or potential tissue damage, or described in terms of such damage" (Merskey and Bogduk, 1994). Pain may be caused by tissue lesions or by mechanical, chemical or thermal stimulation. Such a lesion or stimulation induces a nociceptive signal which travels from its peripheral origin via the spinal cord to the brain. In the brain it reaches lower brain centres, such as the thalamus, but also parts of the cortices. The cortical regions involved are first the primary and secondary somatosensory cortices. These regions allow the interpretation of the signal in terms of the site of the body, the type and the intensity of the input. These cortices are thus involved in the sensory discrimination of pain processing (Hofbauer et al., 2001). Other regions that are activated are the anterior cingulate cortex and the insular cortex. Both of these are considered to be components of the "limbic structure" and are involved in the processing of the affective motivational dimension of pain: their activation causes the stimulation to be experienced as unpleasant (Hofbauer et al., 2001). Low-level nociceptive stimulation does not always activate the limbic structures. As indicated by the definition above, the concept of pain refers to the situation where both dimensions, the sensory (somatosensory cortices) and affective dimension (limbic cortices) are present. The pain sensation represents, therefore, a nociceptive sensation associated with a negative affective sensation (Paulmier et al., 2015).

In addition to neurological activity in the central nervous system, like any stressor, pain can cause activation of the HPA axis and a shift in the balance of the autonomic nervous system. Animals that suffer pain may adopt certain postures like abnormal lying or standing, in order to reduce the pain sensation. They may remain prostrated, or may lick or scratch the painful site. They may also vocalize (Prunier et al., 2013; Grandin, 1998).

The dual sensory and emotional components of pain make it a subjective experience, which means that the perception and experience of pain varies between individuals. Most humans are able to communicate their pain experience, but non-communicative individuals such as very young children, individuals with communicative difficulties or cognitive impairment and animals are often unable to express their pain experience verbally. Previously, it was believed that only humans were capable of experiencing pain (Flecknell, 2000). Due to the phylogenetic similarities in the central nervous system between higher animals, especially non-human mammals and humans, there can be little doubt that vertebrates are capable of experiencing pain or that they can suffer as a result of it (Barnett, 1997).

Pain during movement of animals, stunning and slaughter can arise from direct tissue trauma due to physical interventions with sticks or electrical prods during handling, or if animals hit gates or fittings of the abattoir equipment. Incorrect stunning procedures, or direct bleeding without stunning may also be painful if the animal does not lose consciousness immediately (Gibson et al., 2009a,b). These situations result in psychological stress associated with physiological and behavioural stress responses, such as avoidance reactions and increased activity of the HPA and autonomic nervous system.

## 2.5 BRAIN AND CONSCIOUSNESS

### *2.5.1 Brain anatomy*

The brain can be divided into three major parts: the cerebrum, which is the upper,

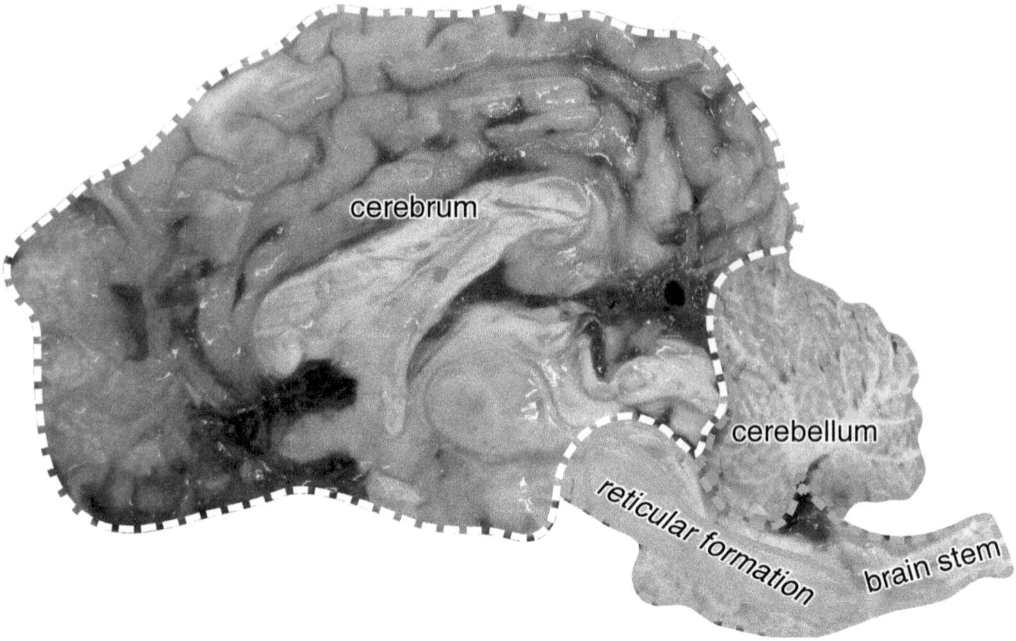


**Figure 2.2:** Right hemisphere of a cow. Some areas appear dark due to haemorrhages. Darker delineated areas represent the cerebrum, cerebellum and brainstem of the brain. The position of the reticular formation inside the brain stem is also shown.

*Photo Claudia Terlouw and Christophe Mallet*

larger part involved in complex functions, the cerebellum or hindbrain, involved in balance and coordinated movement, and the brainstem, involved in vital functions such as breathing, homeostasis, thermoregulation, heart activity and sleep/wakefulness (Figure 2.2.).

The cerebral cortex and the brainstem play a major role in consciousness in mammals. Similar structures exist also in other vertebrates, such as reptiles and birds, but they are at least partly organized differently (Dugas-Ford et al., 2012; Balanoff et al., 2013). Consciousness is a complex concept that can be distinguished into two components (Zeman et al., 1997; Laureys et al., 2004): the content (awareness of self and the environment) and the level (degree of wakefulness). The cerebral cortex is involved in the content of the conscious experience, that is, consciousness of self (the ability to recognize oneself as an individual separate from the environment and other individuals) and the environment (the ability to recognize aspects of the environment through the different senses; Laureys et al., 2004). Widespread dysfunction of the cerebrum causes loss of consciousness.

The reticular formation is a structure in the brainstem, which is associated with the level of consciousness. This structure stimulates many cortical areas via projections called the ascending reticular activating system (Munk et al., 1996; Parvizi and Damasio, 2001; Brown et al., 2012). Lesions in the reticular formation or in the ascending reticular activating system result in a comatose, unconscious state

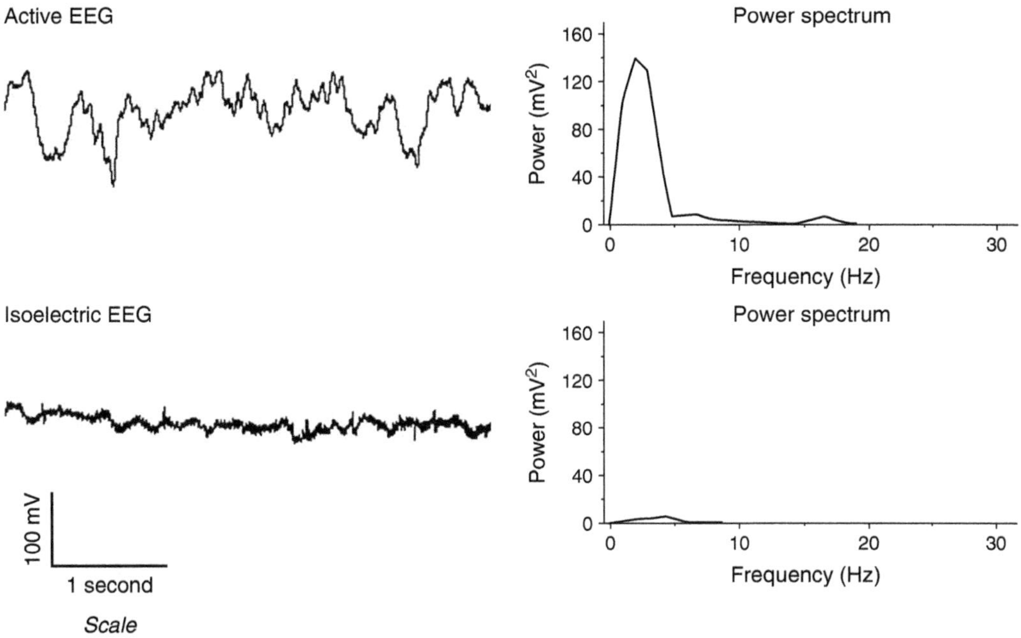


**Figure 2.3:** Example of spontaneous electroencephalographic (EEG) waveforms (left) and the corresponding EEG power spectra (right) from a calf. The top graph shows active EEG activity, while the bottom graph shows an isoelectric EEG after captive bolt stunning

characterized by the complete absence of wakefulness associated with and responsiveness to external stimuli (Laureys et al., 2004; Bateman, 2001).

## *2.5.2 Brain electrical activity and consciousness*

The normal living brain is permanently active including during sleep. All living cells maintain an electrical charge and the activity of the brain involves the permanent depolarization and repolarization of its neurons. There are a number of different electrophysiological indices that can record electrical activity of the brain. One technique used in research on stunning and killing techniques at slaughter is the recording of the electroencephalogram (EEG), where electrodes are placed on the scalp or skull to measure the spontaneous electrical activity of the brain (Figure 2.3.)[2]. Another technique is to record the cortical electrical activity, known as an electrocorticogram (ECoG), by placing the electrodes on the surface of the cerebral cortex via holes drilled into the skull. This method is more sensitive as it is not subjected to the isolating effect of the skull and skin, but also more invasive. The objective of these techniques is to determine whether the observed brain activity indicates that the animal is unconscious). The advantage of the EEG is that changes can be

[2] The EEG represents the sum of the electrical activity of populations of cortical and sub-cortical neurons located below the recording electrodes over time. The electrical signals of the EEG are extremely small (microvolts), requiring amplification, filtering and conversion from an analogue to digital signal. Once digitized, the signal can be recorded and analysed, showing the complex pattern of brainwaves.

seen immediately, while the recording is taking place. However, spatial resolution is poor, that is, EEG measurements do not differentiate between specific regions or circuits in the brain and therefore do not give detailed information of neural processes at work.

The EEG reflects changes in cortical function that can be demonstrated as an alteration in its content over time (Adrian and Matthews, 1934). There are two major classifications of brain electrical activity for EEG in humans: fast and slow waves. Fast waves (i.e. beta and gamma waves) have a high frequency (between 13 and 80 Hz) and are associated with increased vigilance/arousal. These waves reflect a fast depolarization/repolarization cycle of neurons, needed for efficient information processing in the active brain. Slow waves (delta and theta waves) have a low frequency (between 1 and 8 Hz) reflecting a slow depolarization/repolarization cycle. They are associated with somnolence, sleep and anaesthesia, which are situations where information processing is very slow or absent. Alpha waves (between 8 and 13 Hz) are associated with relaxed wakefulness in adult humans and increase with pharmacologically-induced unconsciousness. The classification of the EEG into these frequency bands was developed for the analysis of human EEG and have specific functionality associated with them. Therefore, care should be taken when making assumptions on the function of specific bands between species and different recording loci. An isoelectric EEG is a flat EEG trace, which indicates the total absence of brain activity in animals and humans. An isoelectric EEG may be reversible, but a continued isoelectric trace is generally associated with irreversible brain dysfunction leading to brain death.

Certain other forms of brain activity are also incompatible with consciousness. Sometimes, the EEG shows the presence of large synchronous low frequency (3 to 4 Hz) spikes (Blumenfeld, 2005). This occurs during epileptic seizures for example, but they can also be induced using electrical stunning techniques. During these states the brain is unable to process incoming sensory information and these spikes are associated with unconsciousness.

Another technique that is used in slaughter-based research involves evoked potentials (EPs). Examples include auditory evoked potentials (AEPs) induced by clicks or beeps near the ear, visually evoked potentials (VEPs) induced by flashes in front of the eye and somatosensory evoked potentials (SEPs) induced by electrical stimuli applied to the skin or a nerve, for example (Figure 2.4.). These can be used in the assessment of brain function during and after stunning and/or slaughter. It is important to note that evoked potentials can be present in

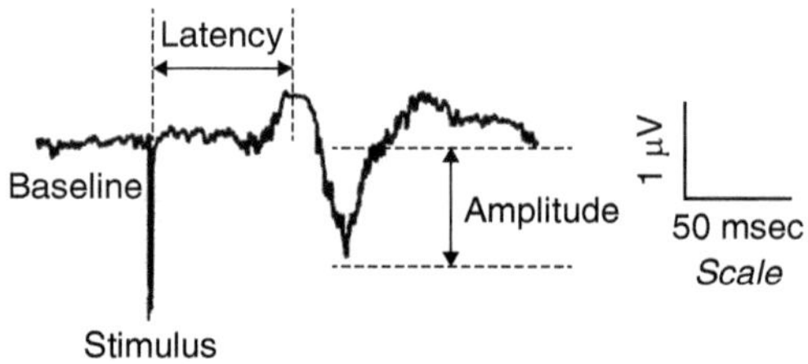


**Figure 2.4:** Example of a somatosensory evoked potential (SEP) waveform from a sheep. Evoked potentials are assessed in terms of latency from the stimulus (msec) and amplitude (µV) of peaks (positive or negative) relative to the baseline pre-stimulus recording. This evoked potential represents the averaging of 32 repetitions

unconscious animals but their absence indicates profound brain dysfunction that is incompatible with consciousness (Daly et al., 1988).

### *2.5.3 Disturbances of consciousness*

At slaughter, unconsciousness can be produced by disturbing the normal functioning of the neurones of the cerebrum, the reticular formation or of the ascending reticular activating system (Terlouw et al., 2016). The objective of stunning is to render the animal unconscious for a period sufficiently long to avoid the animal from experiencing pain or other forms of stress during and after the bleeding and subsequent procedures. Therefore, it is important to know how unconsciousness relates to the capacity to perceive pain. Studies in humans indicate that certain parts of the cerebral cortex have a central role in the conscious perception of pain stimuli (Laureys, 2005; Treede et al., 2000; Bromm, 2001; Hofbauer et al., 2001). For example, patients in the vegetative state are permanently unconscious. Although their primary cortices may function, allowing primary unscrambling of sensory, like auditory, signals, their associative cortices are unresponsive and hence there is no conscious awareness of sensory signals (Laureys et al., 2002). When vegetative patients were given a painful stimulus, cortical activation was observed in each patient. In most vegetative patients, this cortical activation is limited to the primary somatosensory cortex and does not reach the higher-order associative cortices, such as the anterior cingulate cortex and the insular cortex from which it was functionally disconnected (Schnakers et al., 2012). Hence, these patients perceive the stimulus, but do not experience the negative affective component of pain. By contrast, one study suggests that in some vegetative patients, some degree of residual pain perception may still exist (Chatelle et al., 2014). However, this is yet to be demonstrated in stunned animals. In addition, patients suffering from a condition called the "minimally conscious state", show reproducible but minimal and fluctuating signs of consciousness (Giacino et al., 2002) but are often misdiagnosed as vegetative patients. Following painful stimulation, minimally conscious patients were shown to have brain activation very similar to that measured in healthy volunteers, that is, these patients do have pain perception (Schnakers et al., 2012). These results indicate that if animals are correctly stunned, higher-order associative cortices are no longer functional and the animals are not capable of feeling pain or of experiencing other negative emotions. This is the case if dysfunction of cerebrum, the reticular formation or of the ascending reticular activating system is achieved (Terlouw et al., 2016). However, if the damage to these structures is not sufficiently widespread, unconsciousness will probably be shallow and pain perception may still be present. In addition, it has been reported that some animals after slaughter without stunning have phasic return of consciousness, where they drift in and out of consciousness until irrecoverable unconsciousness sets in (Gibson et al., 2015). Further work is needed to examine the exact relationship between these different brain states and the capacity to experience pain at slaughter.

## 2.5 REFERENCES


Adrian, E.D. and Matthews, B.H. (1934) The interpretation of potential waves in the cortex. *Journal of Physiology*, 81, 440–71.

Alexander, G. and Shillito, E.E. (1978) Maternal responses in Merino ewes to artificially coloured lambs. *Applied Animal Ethology*, 4, 141–152.

Ames, D.R. and Arehart, L.A. (1972) Physiological response of lambs to auditory stimuli. *Journal of Animal Science*, 34, 994–998.

Balanoff, A.M., Bever, G.S., Rowe, T.B. and Norell, M.A. (2013) Evolutionary origins of the avian brain. *Nature*, 501, 93–97.

Barnett, J.L. (1997). Measuring pain in animals. *Australian Veterinary Journal*, 75, 878–879.

Bateman, D.E. (2001) Neurological assessment of coma. *Journal of Neurology, Neurosurgery and Psychiatry*, 71(suppl. 1), i13–i17.

Blackshaw, J.K. (1986) Behavioural profiles of domestic animals. In: Blackshaw, J.K. (ed.) *Notes on some topics in applied animal behaviour* (3rd ed.). University of Queensland, Queensland, Australia. pp. 9–34.

Blumenfeld, H. (2005) Cellular and network mechanisms of spike-wave seizures. *Epilepsia*, 46, 21–33.

Boissy, A. (1995) Fear and fearfulness in animals. *The Quarterly Review of Biology*, 70, 165–191.

Boissy, A. and Bouissou, M.F. (1995) Assessment of individual differences in behavioural reactions of heifers exposed to various fear-eliciting situations. *Applied Animal Behaviour Science*, 46, 17–31.

Boissy, A. and Le Neindre, P. (1997) Behavioral, cardiac and cortisol responses to brief peer separation and reunion in cattle. *Physiology & Behavior*, 61(5), 693–699.

Boissy, A., Terlouw, C. and Le Neindre, P. (1998) Presence of cues from stressed conspecifics increases reactivity to aversive events in cattle: evidence for the existence of alarm substances in urine. *Physiology and Behavior*, 63, 489–495.

Boissy, A., Bouix, J., Orgeur, P., Poindron, P., Bibe, B. and Le Neindre, P. (2005) Genetic analysis of emotional reactivity in sheep: effects of the genotypes of the lambs and of their dams. *Genetics Selection Evolution*, 37, 381–401.

Boissy, A., Arnould, C., Chaillou, E., Désiré, L., Duvaux-Ponter, C., Greiveldinger, L., Leterrier, C., Richard, S., Roussel, S., Saint-Dizier, H., Meunier, S.M.C., Valance, D. and Veissier, I. (2007) Emotions and cognition: a new approach to animal welfare. *Animal Welfare*, 16, 37–43.

Boivin, X., Le Neindre, P., Garel, J.P. and Chupin, J. M. (1994) Influence of breed and rearing management on cattle reactions during human handling. *Applied Animal Behaviour Science*, 39, 115–122.

Boivin, X., Garel, J.P., Mante, A. and Le Neindre, P. (1998) Beef calves react differently to different handlers according to the test situation and their previous interactions with their caretaker. *Applied Animal Behaviour Science*, 55, 245–257.

Bouissou, M.F. (1980) Social relationships in domestic cattle under modern management techniques. *Bolletino di Zoologia*, 47, 343–353.

Bouissou, M.F., Lavenet, C. and Orgeur, P. (1974) Etablissement des relations de dominance-soumission chez les bovins domestiques. I. Nature et évolution des interactions sociales. *Annales de Biologie Animale, Biochimie, Biophysique*, 14, 383–410.

Bourguet, C., Deiss, V., Gobert, M., Durand, D., Boissy, A. and Terlouw, E.M.C. (2010) Characterising the emotional reactivity of cows to understand and predict their stress reactions to the slaughter procedure. *Applied Animal Behaviour Science*, 125(1–2), 9–21.

Bourguet, C., Deiss, V., Boissy, A.,

Andanson, S. and Terlouw, E.M.C. (2011) Effects of feed deprivation on behavioral reactivity and physiological status in Holstein cattle. *Journal of Animal Science*, 89, 3272–3285.

Bromm, B. (2001) Brain images of pain. *News in Physiological Sciences*, 16, 244–249.

Broom, D.M. (1987) Applications of neurobiological studies to farm animal welfare. *Biology of stress in farm animals: An integrated approach*. Dordrecht: Martinus Nijhoff, pp. 101–110.

Brown, S.N., Knowles, T.G., Edwards, J.E. and Warriss, P.D. (1999) Relationship between food deprivation before transport and aggression in pigs held in lairage before slaughter. *Veterinary Record*, 145, 630–634.

Brown, R.E., Basheer, R., McKenna, J.T., Strecker, R.E. and McCarley, R.W. (2012) Control of sleep and wakefulness. *Physiological Reviews*, 92, 1087–1187.

Chatelle, C., Thibaut, A., Whyte, J., De Val, M.D., Laureys, S. and Schnakers, C. (2014) Pain issues in disorders of consciousness. *Brain Injury*, 28, 1202–1208.

Daly, C.C., Kallweit, E. and Ellendorf, F. (1988) Cortical function in cattle during slaughter – Conventional captive bolt stunning followed by exsanguination compared with shechita slaughter. *Veterinary Record*, 122, 325–329.

Dantzer, R. (2002) Can farm animal welfare be understood without taking into account the issues of emotion and cognition? *Journal of Animal Science*, 80, E1–9.

Dawkins, M.S. (2008) The science of animal suffering. *Ethology*, 114(10), 937–945.

Deiss, V., Temple, D., Ligout, S., Racine, C., Bouix, J., Terlouw, C. and Boissy, A. (2009) Can emotional reactivity predict stress responses at slaughter in sheep? *Applied Animal Behaviour Science*, 119, 193–202.

Désiré, L., Boissy A. and Veissier I. (2002) Emotions in farm animals: a new approach to animal welfare in applied ethology. *Behavioral Processes*, 60, 165–180.

Dugas-Ford, J., Rowell, J.J. and Ragsdale, C.W. (2012) Cell-type homologies and the origins of the neocortex. *Proceedings of the National Academy of Sciences of the United States of America* 109, 16974–16979.

Flecknell, P.A. (2000) Animal pain – an introduction. In P, Flecknell, & A. Waterman-Pearson (Eds), *Pain management in animals* (pp. 1–8). London: W.B. Saunders.

Fraser, D., Ritchie, J.S.D. and Faser, A.F. (1975) The term "stress" in a veterinary context. *British Veterinary Journal*, 131, 653–662.

Geverink, N.A., Engel, B., Lambooij, E. and Wiegant, V.M. (1996) Observations on behaviour and skin damage of slaughter pigs and treatment during lairage. *Applied Animal Behaviour Science*, 50, 1–13.

Giacino, J.T., Ashwal, S., Childs, N., Cranford, R., Jennett, B., Katz, D.I., Kelly, J.P., Rosenberg, J.H, Whyte, J., Zafonte, R.D, Zasler, N.D. (2002) The minimally conscious state – Definition and diagnostic criteria. *Neurology*, 58, 349–353.

Gibson, T.J., Johnson, C.B., Murrell, J.C., Hulls, C.M., Mitchinson, S.L., Stafford, K.J., Johnstone, A.C. and Mellor, D.J. (2009a) Electroencephalographic responses of halothane-anaesthetised calves to slaughter by ventral-neck incision without prior stunning. *New Zealand Veterinary Journal*, 57(2), 77–83.

Gibson, T.J., Johnson, C.B., Murrell, J.C., Chambers, J.P., Stafford, K.J. and Mellor, D.J. (2009b) Components of electroencephalographic responses to slaughter in halothane-anaesthetised calves: Effects of cutting neck tissues compared with major blood vessels. *New Zealand Veterinary Journal*, 57(2), 84–89.

Gibson, T.J., Dadios, N. and Gregory, N.G. (2015) Effect of neck cut position on time to collapse in halal slaughtered cattle without stunning. *Meat Science*, 110, 310–314.

Grandin, T. (1998) The feasibility of using vocalization scoring as an indicator of poor welfare during cattle slaughter. *Applied Animal Behaviour Science*, 56, 121–128.

Grandin, T. (1999) Safe handling of large animals. *Occupational medicine*, 14, 195–212.

Grandin, T. (2013) Recommended animal handling guidelines and audit guide. AMI Foundation. http://animalhandling.org/ht/a/GetDocumentAction/i/93003. Accessed on 23/02/2015.

Greiveldinger, L., Veissier, I. and Boissy, A. (2007). Emotional experience in sheep: Predictability of a sudden event lowers subsequent emotional responses. *Physiology and Behavior*, 92, 675–683.

Hemsworth, P.H., Barnett, J.L., Hofmeyr, C., Coleman, G.J., Dowling, S. and Boyce, J. (2002) The effects of fear of humans and pre-slaughter handling on the meat quality of pigs. *Australian Journal of Agricultural Research*, 53, 493–501.

Hofbauer, R.K., Rainville, P., Duncan, G.H. and Bushnell, M.C. (2001) Cortical representation of the sensory dimension of pain. *Journal of Neurophysiology*, 86, 402–411.

Ibinson, J.W., Small, R.H., Algaze, A., Roberts, C,J., Clark, D.L. and Schmalbrock, P. (2004) Functional magnetic resonance imaging studies of pain. *Anesthesiology*, 101, 960–969.

Ingvar, M. (1999) Pain and functional imaging. *Philosophical Transactions of the Royal Society B, Biological Sciences*, 354(1387), 1347–1358. doi: 10.1098/rstb.1999.0483

Jones, R.B. and Faure, J.M. (1981) The effects of regular handling on fear responses in the domestic chick. *Behavioral Processes*, 6, 135–143.

Kenny, F.J. and Tarrant, P.V. (1987) The reaction of young bulls to short-haul road transport. *Applied Animal Behaviour Science*, 17, 209–227.

Laureys, S. (2005). Death, unconsciousness and the brain. *Nature Reviews Neuroscience*, 6(11), 899–909.

Laureys, S., Faymonville, M.E., Peigneux, P., Damas, P., Lambermont, B., Del Fiore, G., Degueldre, C., Aerts, J., Luxen, A., Franck G., Lamy, M., Moonen, G. and Maquet, P. (2002) cortical processing of noxious somatosensory stimuli in the persistent vegetative state. *Neuroimage*, 17, 732–741.

Laureys, S., Owen, A.M. and Schiff, N.D. (2004) Brain function in coma, vegetative state, and related disorders. *The Lancet Neurology*, 3, 537–546.

Lawrence, A.B., Terlouw, E.M.C. and Illius, A.W. (1991) Individual differences in behavioural responses of pigs exposed to non-social and social challenges. *Applied Animal Behaviour Science*, 30, 73–86.

Lensink, B.J., Fernandez, X., Boivin, X., Pradel, P., Le Neindre, P. and Veissier, I. (2000) The impact of gentle contacts on ease of handling, welfare, and growth of calves and on quality of veal meat. *Journal of Animal Science*, 78, 1219–1226.

Lomas, C.A., Piggins, D. and Phillips, C.J.C. (1998) Visual awareness. *Applied Animal Behavioral Science*, 57, 247–257.

Marx, G., Leppelt, J. and Ellendorff, F. (2001) Vocalisation in chicks (Gallus gallus dom.) during stepwise social isolation. *Applied Animal Behaviour Science*, 75, 61–74.

Melzack, R. and Wall, P.D. (1965) Pain mechanisms: a new theory. *Science*, 150, 971–979.

Mench, J. and Keeling, L.J. (2001) The social behaviour of domestic birds. In L.J. Keeling & H.W. Gonyou (Eds.). *Social behaviour in farm animals*. Wallingford, UK: CABI Publishing.

Merskey, H. and Bogduk, N. (1994) International Association for the Study of Pain. Task force on taxonomy. Classification of chronic pain: descriptions of chronic pain syndromes and definitions of pain terms (2nd ed.). Seattle: IASP Press. Available at: http://www.iasp-pain.org/AM/Template.cfm?Section=Pain_Definitions

Mounier, L., Dubroeucq, H., Andanson, S.

and Veissier, I. (2006) Variations in meat pH of beef bulls in relation to conditions of transfer to slaughter and previous history of the animals. *Journal of Animal Science*, 84, 1567–1576.

Mounier, L., Colson, S., Roux, M., Dubroeucq, H., Boissy, A. and Veissier, I. (2008) Positive attitudes of farmers and pen-group conservation reduce adverse reactions of bulls during transfer for slaughter. *Animal*, 2, 894–901.

Munk, M.H.J., Roelfsema, P.R., Konig, P., Engel, A.K. and Singer, W. (1996) Role of reticular activation in the modulation of intracortical synchronization. *Science*, 272, 271–274.

Parvizi, J. and Damasio, A. (2001) Consciousness and the brainstem. *Cognition*, 79(1–2), 135–159.

Paul, E.S., Harding, E.J. and Mendl, M. (2005) Measuring emotional processes in animals: the utility of a cognitive approach. *Neuroscience and Biobehavioral Reviews*, 29, 469–491

Paulmier, V., Faure, M., Durand, D., Boissy, A., Cognie, J., Eschalier, A. and Terlouw, C. (2015) Douleurs animales. 1. Les mécanismes. *INRA Productions Animales*. In press.

Peyron, R., Laurent, B. and Garcia-Larrea, L. (2000) Functional imaging of brain responses to pain. A review and meta-analysis. *Neurophysiologie Clinique*, 30, 263–288.

Philips, C.J.C. (1993) *Cattle Behaviour*. Ipswich, U.K: Farming Press Books.

Prunier, A., Mounier, L., Le Neindre, P., Leterrier, C., Mormède, P., Paulmier, V., Prunet, P., Terlouw, C. and Guatteo, R. (2013) Identifying and monitoring pain in farm animals: a review. *Animal*, 7, 998–1010.

Ruiz-de-la-Torre, J.L. and Manteca, X. (1999) Behavioural effects of social mixing at different stocking densities in prepubertal lambs. *Animal Welfare*, 8, 117–126.

Schmolke, S.A., Li, Y.Z. and Gonyou, H.W. (2004) Effects of group size on social behavior following regrouping of growing-finishing pigs. *Applied Animal Behaviour Science*, 88, 27–38.

Schnakers, C., Chatelle, C., Vanhaudenhuyse, A., Majerus, S., Ledoux, D., Boly, M., ... Laureys, S. (2010) The Nociception Coma Scale: a new tool to assess nociception in disorders of consciousness. *Pain*, 148, 215–219.

Schnakers, C., Chatelle, C., Demertzi, A., Majerus, S. and Laureys, S. (2012) What about pain in disorders of consciousness? *The AAPS Journal*, 14(3), 437–444. doi: 10.1208/s12248-012-9346-5

Sneddon, L.U. (2003) The evidence for pain in fish: the use of morphine as an analgesic. *Applied Animal Behaviour Science*, 83, 153–162.

Terlouw, C. (2005) Stress reactions at slaughter and meat quality in pigs: genetic background and prior experience: A brief review of recent findings. *Livestock Production Science*, 94, 125–135.

Terlouw, E.M.C., Porcher, J. (2005) Repeated handling of pigs during rearing. I. Refusal of contact by the handler and reactivity to familiar and unfamiliar humans. *Journal of Animal Science*, 83, 1653–1663.

Terlouw, E.M.C. and Rybarczyk, P. (2008) Explaining and predicting differences in meat quality through stress reactions at slaughter: The case of Large White and Duroc pigs. *Meat Science*, 79, 795–805.

Terlouw, E.M.C., Boissy, A. and Blinet, P. (1998) Behavioural responses of cattle to the odours of blood and urine from conspecifics and to the odour of faeces from carnivores. *Applied Animal Behaviour Science*, 57, 9–21.

Terlouw, E.M.C., Porcher, J. and Fernandez, X. (2005) Repeated handling of pigs during rearing. II. Effect of reactivity to humans on aggression during mixing and on meat quality. *Journal of Animal Science*, 83, 1664–1672.

Terlouw, E.M.C., Arnould, C., Auperin, B.,

Berri, C., Le Bihan-Duval, E., Deiss, V., Lefèvre, F., Lensink, B.J. and Mounier, L. (2008) Pre-slaughter conditions, animal stress and welfare: current status and possible future research. *Animal*, 2, 1501–1517.

Terlouw, C., Berne, A. and Astruc, T. (2009) Effect of rearing and slaughter conditions on behaviour, physiology and meat quality of Large White and Duroc-sired pigs. *Livestock Science*, 122, 199–213.

Terlouw, C., Bourguet, C. and Deiss, V. (2016) Consciousness, unconsciousness and death in the context of slaughter. Part I. Neurobiological mechanisms underlying stunning and killing. *Meat Science*, 118, 133–146. doi: 10.1016/j.meatsci.2016.03.011

Treede, R.D., Apkarian, A.V., Bromm, B., Greenspan, J.D. and Lenz, F.A. (2000) Cortical representation of pain: functional characterization of nociceptive areas near the lateral sulcus. *Pain*, 87, 113–119.

Turner, S.P., Horgan, G.W. and Edwards, S.A. (2001) Effect of social group size on aggressive behaviour between unacquainted domestic pigs. *Applied Animal Behaviour Science*, 74, 203–215.

Veissier, I. and Boissy, A. (2007) Stress and welfare: Two complementary concepts that are intrinsically related to the animal's point of view. *Physiology and Behavior*, 92, 429–433.

Zeman, A.Z.J., Grayling, A.C. and Cowey, A. (1997) Contemporary theories of consciousness. *Journal of Neurology Neurosurgery and Psychiatry*, 62, 5.

chapter three

# Reception and unloading of animals

*Luigi Faucitano*[1] *and Cecilia Pedernera*[2]

**Learning objectives**

- Describe the procedures for the management of the truck on arrival at the slaughterhouse under different ambient conditions.
- Describe the procedures for the handling of animals at unloading at the slaughterhouse.
- Describe the adequate design of dock and ramps at unloading.
- Understand the impact of these factors on animal welfare and meat quality.

## CONTENT



## 3.1 INTRODUCTION

The condition of animals (dead, fatigued or injured) on arrival at the slaughterhouse results from the cumulative and synergistic effects of a number of transport factors and for this reason it depends on how animals are prepared

[1] Agriculture and Agri-Food Canada, Sherbrooke Research and Development Centre, 2000 College Street, J1M 0C8 Sherbrooke, Canada
[2] IRTA, Animal Welfare Sub-program, Veïnat de Sies, s/n, 17121 Monells, Spain

**Table 3.1** Welfare assessment of pigs, cattle and sheep on arrival at the slaughter plant using the American Meat Institute Audit Guide (AMI, 2012)

| Criteria | Final score | | | |
|---|---|---|---|---|
| | *Excellent* | *Acceptable* | *Not acceptable* | *Serious problem* |
| Timeliness of unloading* | ≥ 95% of trailers unloaded properly | 90 to 94.9% of trailers unloaded properly | 85 to 89.9% of trailers unloaded properly | < 85% of trailers unloaded properly |
| Falls** | No falling | < 1% falling (body touches floor) | > 1% falling down | ≥ 5% falling down |
| Electric prod use*** | | | | |
| Cattle | ≤ 5% | ≤ 25% or less | 26 to 49% | ≥ 50% |
| Pigs | ≤ 10% | ≤ 25% | 26 to 79% | ≥ 80% |
| Sheep | 0% | ≤ 5% | 6 to 10% | ≥ 11% |

*Based on a 0–4 score, where 0 = ≥ 120 min (without reason), 1 = ≥ 120 min (with reason), 2 = 91–120 min, 3 = 61–90 min, 4 = 60 min of arrival. The assessment level is based on the proportion of trucks unloading in a more timely manner (score 4).
**Falls are scored in the unloading area only after all four of the animal's limbs are on the unloading ramp or dock.
***Electric prod use is scored in the unloading area only after all four of the animal's limbs are on the unloading ramp or dock.

for transport, are handled at loading at the farm and are transported. These pre-arrival factors must thus be kept under control to limit significant risk to animal welfare and economic losses. However, the benefits of maintaining good animal welfare practices at the farm and during transport can be lost due to wrong management of the truck on arrival at the slaughterhouse, poor design of unloading facilities and rough handling during unloading from transport vehicle. Based on their impact on animal welfare at this stage, factors such as the timeliness of arrival of the truck, and frequency of animals slipping and falling and electric prod use at unloading are core criteria within slaughterhouse animal welfare audit protocols (AMI, 2012; Table 3.1). This chapter presents an overview of the currently recommended practices to limit animal suffering at reception and unloading, reduce losses, ease handling and improve efficiency.

## 3.2 RECEPTION AT THE SLAUGHTERHOUSE

### *3.2.1 Pigs*

The primary recommendation is to begin unloading of animals within 30 minutes of arrival at the slaughterhouse and complete it within an hour to avoid heat and humidity rise inside the stationary truck and its negative consequences on animal welfare and meat quality of pigs (AMI, 2012). However, in commercial conditions, the waiting time to unload a truck after its arrival at the slaughterhouse is very variable ranging from 5 minutes to 4 hours (Aalhus et al., 1992; Jones,

**Figure 3.1:** The availability of more than one dock at the slaughter plant helps shorten wait time before unloading.

*(L. Faucitano, AAFC, Canada)*

1999). Previous research on pig transport showed that it can take 56 minutes for the truck to reach its maximum temperature after the stop (Haley et al., 2008). Weschenfelder et al. (2012) reported that, even after only 10 minutes wait before unloading at mild ambient temperatures (8°C), temperatures in some compartments of the transport truck can be up to 6°C warmer than the external temperature. The microclimate change inside the truck can explain the correlation between time waiting in the yard and the risk of death and/or non-ambulatory pigs at unloading reported in several studies (Haley et al., 2008; Ritter et al., 2006). Research showed that as the waiting time increases by 30 minutes, the risk of pigs dying can increase by 2.2 times (Haley et al., 2008). Driessen and Geers (2001) also reported the greatest incidence of pale and exudative pork (PSE) after 30 minutes wait when at 23°C ambient temperature or after 53 minutes wait at 11°C (70 and 45%, respectively). The presence of heat-stressed (or panting) pigs at unloading is a major risk factor for PSE pork production (van de Perre et al., 2010).

Strict coordination of truck arrivals with the predicted number of pigs in lairage, lairage capacity and speed of operation as well as a number of unloading docks allowing more than one truck to unload at the same time (Figure 3.1) may help shorten waiting times in slaughterhouses (Weyman, 1987).

If delay to unload is unavoidable, pigs must be cooled off during the wait using active ventilation and/or water sprinkling/misting in the truck. It has been shown that when ambient temperature is above 20°C, water sprinkling pigs for 5 minutes in the truck before leaving the farm and during wait time at the slaughterhouse reduced drinking behaviour in lairage, physical fatigue at slaughter (lower blood lactate levels) and water loss in meat (Fox et al., 2014; Nannoni et al., 2014). The effects of fan-assisted ventilation on the welfare of pigs in a stationary truck can be either positive (Nielsen, 1982) or none (Warriss et al., 2006). However, when active ventilation and water misting/sprinkling are combined, the transport mortality rate can be significantly decreased (Christensen and Barton-Gade, 1999) as ventilation, besides removing the excessive humidity from the interior of the sprinkled truck, increases evaporative cooling in pigs (Chevillon, 2001).

### *3.2.2 Cattle*

Under commercial conditions, vehicles often have to wait in line to unload at the slaughterhouse which creates a stressful work environment for drivers and slaughterhouse reception staff, often resulting in poor cattle welfare (Gebresenbet et al., 2004). Longer waiting times increase risk factors such as lack of ventilation, space allowance and water availability. The application of scheduling procedures for cattle delivery at the slaughterhouse is recommended to ensure that animals are unloaded quickly upon arrival (Miranda-de la Lama et al., 2014).

A number of cattle transport surveys have reported wait times before unloading at the plant of 20–30 minutes on average, with maximum waits ranging from 3 hours to overnight (FAWC, 2003; González et al., 2012; Warren et al., 2010). Longer waiting times in a stationary vehicle may expose animals to heat or cold stress depending on the variation in the internal truck environment. The main determinants of the internal thermal micro-environment in the vehicle are the external climatic conditions, ventilation type, internal air flow pattern and the total heat and moisture produced by the animals (Norton et al., 2013). In a stationary passively ventilated vehicle, such as the North American pot-belly trailer or the European trucks not used for journeys exceeding 8 hours (European Commission, 2005; Schwartzkopf-Genswein et al., 2012), the relative humidity and temperature rise quickly resulting in heat stress for animals, especially when they wait in a truck under warm conditions and at high stocking density. Severely heat-stressed cattle are recognizable for the open mouth breathing progressing to extended tongue as the body temperature steadily rises (Gaughan and Mader, 2012). Both panting and sweating are important mechanisms of evaporative heat loss, with sweating accounting for up to 80% of total evaporative heat loss (Robertshaw, 1985), and both creating a microclimate that favours dehydration (Caulfield et al., 2014; Miranda-de la Lama et al., 2012). Evaporative heat loss is diminished at high ambient humidity levels, although respiratory cooling might still be effective if the temperature of inhaled air is lower than core body temperature (Sparke et al., 2001). Climate control by monitoring of the temperature humidity index (THI: normal ≤ 74; alert ≤ 75–78; danger 79–83; emergency ≥84) is useful to assess the environmental conditions causing heat stress in cattle, as in other livestock species. However, currently there are no objective indexes of cold stress for cattle (Goldhawk et al., 2014). Bedding material, such as woodchips, sawdust, cellulose or straw, may help maintain warm temperatures in cold climates (< 10°C; Schwartzkopf-Genswein et al., 2012). Within the European Union, the welfare of farmed animals transported for commercial purposes is regulated by Council Regulation (EC) No. 1/2005 of 22 December 2004 (European Commission, 2005). This regulation includes, for journeys lasting more than 8 hours, provisions for sufficient bedding and a ventilation system that is capable of maintaining the temperature between 5 and 30°C within the vehicle, and a temperature monitoring and recording system that will alert the driver if the temperature in the livestock areas reaches these limits.

### *3.2.3 Sheep*

Due to metabolic heat production, the temperature within a vehicle transporting sheep is higher than the external temperature, especially during stationary periods, such as the wait before unloading (Cockram, 2014). Hyperthermia, as defined by a rectal temperature > 40.5°C, has been reported in conditions of high humidity at an air temperature of 33 and 40°C in sheep with fleece and shorn, respectively. In sheep, a sign of hyperthermia is rapid shallow to slower, deeper open-mouthed panting (Stockman et al., 2011). In this condition, ventilation inside the truck must be adequate. To ensure sufficient airflow while waiting in the truck, the minimum space above the top of the head should be at least either 15 or 30 cm, depending on whether sheep are exposed to forced or passive ventilation (Figure 3.2; SCAHAW, 2002). Poor airflow can be even more difficult to cope with for sheep affected by respiratory diseases, such as pneumonia and pleurisy (Green et al., 1997). When it is hot and there is no shade on arrival at the slaughterhouse, it is important that vehicles keep moving or are parked at a right angle to wind direction to ensure adequate airflow through them (Knowles, 1998). Besides hyperthermia, poor ventilation can also result in high ammonia concentrations inside the vehicle (Fisher et al., 2002), which can be aversive to sheep when levels exceed 45 ppm (Phillips et al., 2012).

**Figure 3.2:** Insufficient deck height prevents adequate airflow and produces discomfort in sheep waiting for unloading.

*(Photo courtesy of IRTA, Spain)*

### *3.2.4 Chickens and ducks*

An average of 0.2% dead-on-arrival (DOA; ranging from 0.12 to 0.46%) has been reported based on a number of chicken transport surveys conducted over the last 10 years in different regions of the world (Nijdam et al., 2004; Weeks, 2014). Hyperthermia or hypothermia are thought to be the major causes of mortality rate during transport in chickens (Petracci et al., 2006; Warriss et al., 2005), with DOAs increasing at ambient temperatures above 17°C (Warriss et al., 2005) or below 5°C (Nijdam et al., 2004). However, heat stress represents a much greater problem for chickens than cold stress (Nicole and Scott, 1990) contributing 40% of transport mortality (Bayliss and Hinton, 1990).

The risk of death at high ambient temperatures particularly increases during truck stops, including wait before unloading at the slaughterhouse (Webster et al., 1993; Weeks et al., 1997). Following arrival at the slaughterhouse, chickens can be either kept in lairage on the truck or in unloaded modules or stacks of crates. Lairage conditions can have a greater impact on the incidence of death than transport (0.9 vs. 0.5%; Knezacek et al., 2010). If chickens are kept in the truck before unloading at ambient temperatures above the critical upper temperature (29°C; Meltzer, 1983) death may occur because of hyperthermia (Webster et al., 1993). It is thus recommended not to leave a truck stationary for more than 45 minutes when the temperature of the load exceeds 30°C, unless the truck is kept in shade and air circulation is provided (CARC, 2003). Alternatively, to cool off birds during the stationary situation, fan banks with or without water sprinklers can also be used before unloading (Figure 3.3). The combination of water sprinkling or misting and ventilation prevents humidity rise in the environment and results in increased evaporative cooling allowing birds to lose body heat. Research showed that water sprinkling upon arrival followed by ventilation during 1-hour lairage in the truck reduced mortality in chickens and the production of PSE breast meat (Guarnieri et al., 2004; Vieira et al., 2010). However, in winter, cold air draught must be prevented by covering the truck sides with tarpaulin (Figure 3.4) and parking the truck in the lea of buildings or trees. Wetting bird skin must also be avoided in order to maintain the thermoregulatory function of the feathers (Weeks, 2014).

It is recommended that ducks must be unloaded as soon as possible after arrival at the slaughterhouse (Maple Leaf Farms, 2012; RSPCA, 2011). In case of delays, the loaded truck must not remain stationary for more than 15 minutes and during this wait when the ambient temperature is above 20°C it must be parked in shade and either passive (vehicle moving) or mechanical ventilation must be applied (Maple Leaf Farms, 2012; RSPCA, 2011).

**Figure 3.3:** Mechanical ventilation and water misting should be used to cool off broilers during the stationary situation before unloading.

*(Photo courtesy of I. Oliveira, NUPEA/ESALQ/USP, Brazil)*

**Figure 3.4:** Uninsulated tarps on both sides of the trailer are used to protect chickens from the cold, rain or wind.

*(Photo courtesy of T. Crowe, University of Saskatchewan, Canada)*

### *3.2.5 Rabbits*

Within the process of transporting rabbits from farm to slaughter, transport conditions appear to be more critical than journey time (María et al., 2004) and within transport conditions the procedures of loading and unloading appear to be the most critical ones (Buil et al., 2004). To avoid heat stress, especially in summer, average time before unloading should be no longer than 15 minutes (Buil et al., 2004; de la Fuente et al., 2004). Optimal ambient conditions for rabbits are in terms of 13 to 20°C ambient temperature and 55 to 65% relative humidity or temperature–humidity index (THI) lower than 27.8°C (Marai et al., 2002; Marai and Rashwan, 2004). When ambient temperatures are above 35°C or THI is higher than 28.9°C rabbits are in severe heat stress, cannot regulate their internal temperature and fall into heat prostration, while at 40°C they start panting and salivating (Marai and Rashwan, 2004). The average lethal air temperature is 42.8°C (EFSA, 2011). Furthermore, heat stress can result in greater levels of cortisol, lactate, glucose and packed-cell volume in plasma and lower initial pH in meat (de la Fuente et al., 2007; Nakyinsige et al., 2013). Although to a lesser degree than heat, cold stress may also impair the welfare of rabbits while waiting at an unprotected unloading site. De la Fuente et al. (2007) showed increased levels of CK in blood and decreased muscle glycogen concentration in rabbits exposed to temperatures ranging from −1.1 to 2.1°C. To control the ambient conditions of crated rabbits kept in the truck before unloading, the unloading site must be provided with adequate ventilation to respect their thermal comfort limits inside the crate and to remove toxic gases originating from urine and faeces. Furthermore, it must be covered and protected from the sun in summer and wind in winter (Buil et al., 2004; EFSA, 2004).

### 3.2.6 Farmed fish

It is both difficult and expensive to transport fish without imposing significant stress on them. Thus, it is usually preferable to slaughter fish at the production site. For operational reasons, however, some species, including salmon, pangasius, carp, tilapia, catfish and eel, are commonly transported to a remote point for slaughter (Lines and Spence, 2014). For example, the Atlantic salmon (*Salmo salar*) is usually transported from the sea cages to the slaughterhouse in specially designed well-boats. When the well-boat arrives at the processing plant, salmon are either transferred to the processing line or to holding cages by using a pressure-vacuum pump. If the fish are to be transferred to the processing line directly from the vessel, the vessel may stay for some hours at the quayside before the unloading process starts early next morning. In such cases, adequate water circulation should be carried out using the vessel's circulation pumps (EFSA, 2009b). Other species, like sea bream and sea bass, are normally removed from the water and transported to the processing plant in containers with water and ice using boats and/or trucks. Since sea bream and sea bass are killed by chilling in water and ice, transport is not considered to be a part of the pre-slaughter process or procedures (EFSA, 2009a). The welfare concern associated with the process of chilling live fish using ice and water slurry is dealt with adequately in another chapter dedicated to stunning and slaughter of fish.

In marked contrast to terrestrial livestock, live fish transport requires a life-support system through the marketing process. Besides the obvious requirement to provide a volume of water to contain fish, the most critical component of the life-support system is the provision of oxygen ($O_2$) to support their normal respiration (Wedemeyer, 1996a,b). Failure to do so rapidly results in hypoxia, causing unconsciousness and death. The same risk should be taken into account when large hauls of fish are kept in storage bins for hours before slaughter. In this condition, the last fish out of the storage bin are often dead before bleeding, resulting in insufficient exsanguination and, as a consequence, muscle discolouration (Margeirsson et al., 2007; Rotabakk et al., 2011). Additionally, preslaughter stress and physical activity in fish shorten the time to onset of rigor, increasing the possibility of damage to the flesh during processing (Robb, 2001).

Transport of fish out of water should

always be avoided. However, some farmed fish, including eels and pangasius, are sometimes transported in road vehicles without water. Fish transported in this way are likely to become extremely stressed not only because of the lack of water and oxygen, but also because of the pressure of other fish resting on them, rapid temperature changes, vehicle vibrations and the physical shock if they are dropped when they enter or leave the transport unit. The fact that many of these fish can survive these conditions is not an indication that it is a stress-free experience. Indeed, after a transport in these conditions, most pangasius are too exhausted to swim or maintain their equilibrium when returned to the water (Lines and Spence, 2014).

Behavioural responses and the evaluation of some physiological changes (e.g. increased levels of blood catecholamines, cortisol, glucose, lactate and haematocrit) are commonly used to indicate when fish may have suffered fear, distress and pain during the pre- and slaughter phases (Portz et al., 2006). Based on the concentrations of cortisol, glucose, lactate, sodium, chloride and osmolality in blood taken on farm and after loading, transport in the well-boat, unloading, resting, and pumping into the processing plant, pumping from the resting cages into the processing plant appear to be the most stressful pre-slaughter stages (Gatica et al., 2010)

## 3.3 UNLOADING AT THE SLAUGHTERHOUSE

### *3.3.1 Pigs*

Although unloading can be considered less stressful than loading (based on heart rate and blood lactate level; Correa et al., 2010; Edwards et al., 2010), poor handling resulting in carcass bruising and poor meat quality is inevitable at this stage, unless well-designed unloading facilities and trained personnel are used.

In practice, the most common unloading device at the slaughterhouse is the ramp or bridge, which helps the transfer of pigs from the truck ramp or lift to the dock. Keeping in mind that pigs have difficulties in descending a slope (Brown et al., 2005), ramps steeper than 20° are not recommended as they result in greater pigs' heart rate, vocalization and backing up behaviour, increased handler interventions (shouting and use of electric prods) and 3-4 times longer unloading time (Goumon et al., 2013; Ritter et al. 2008; Torrey et al., 2013a,b; Warriss et al. 1991). Higher pitched vocalization of pigs at unloading has been associated with greater risk of PSE pork production (van de Perre et al., 2010). The greater need for electric prod use observed on ramps may result in a higher incidence of fatigued pigs at unloading and lowers pork quality (Correa et al., 2013). When compared to ramps, the use of hydraulic lifts or decks reduces handling stress (lower heart rate), increases the easiness of handling and shortens off-load time (Brown et al., 2005).

To ease handling, pigs should not have closed corners to negotiate either when exiting the truck or the unloading dock. A greater heart rate and balking behaviour, and unloading time have been reported in pigs negotiating angles or bends of 45° to 90° compared with 0° and 30° (Goumon et al., 2013; Warriss et al., 1992).

Handling problems due to hesitation and refusal of pigs to go forward can also be caused by poor lighting (Figure 3.5), lack of shelter at the dock, the presence of a step (> 15 cm) or a gap between the truck deck floor and the unloading dock (AMI, 2012; SCAHAW, 2002) and the slippery ramp floor. Goumon et al. (2013) showed that a 20-cm step associated with a ramp made the work more challenging for the handler, as shown by handlers' greatest heart rate and lower easiness of handling score when pigs were moved to the ramps with step at unloading, reflecting a greater physical effort. Torrey et al. (2013b) explained the greater number of falls and slips observed at unloading in winter by the loss of balance of pigs walking on aluminium ramp floor that became slippery when temperatures dropped below freezing. Stair step dimensions of 6.6 cm rise and 26 cm tread are recommended to increase pigs' stability while descending the unloading ramp (AMI, 2012). Reducing the frequency of slips on the unloading ramp not only prevents injuries, but also improves pork quality (Rocha et al., 2016).

**Figure 3.5:** Handling problems due to hesitation and refusals of pigs to go forward can also be caused by poor lighting (view from the truck gate into the reception area).

*(L. Faucitano, AAFC, Canada)*

### *3.3.2 Cattle*

The most common lesions caused by poor handling and transport are bruising, lameness and dislocations (Minka and Ayo, 2007). These lesions are mainly associated with mishandling during loading and unloading or the use of poorly designed or maintained trailers, ramps and alleys (Miranda-de la Lama et al., 2012). To ease handling and limit the occurrence of lesions, trucks should have an integrated loading/unloading device, where the ramp angle should not be steeper than 20° for calves or 26° for adult cattle (Figure 3.6). When the slope is steeper than 10°, ramps should be fitted with foot battens or cleats which increase the animal's stability on its feet while negotiating the ramp, reducing the risk of slips and falls (EFSA, 2011). To this end, a 20-cm spacing between the cleats is recommended to ease cattle handling through the ramp (Grandin, 2014), Ideally, to eliminate external unloading ramps, the slaughterhouse dock should be built at the truck deck level. A useful indicator of ramp and floor conditions during unloading is the proportion of animals slipping. This proportion should be lower than 3% to be acceptable (Grandin and Gallo, 2007).

It has been reported that during handling at unloading, distractions, such as seeing moving people up ahead, sparkling reflections, shiny metal that jiggles or air blowing into the faces of approaching cattle, can result in cattle balking and backing up (Grandin, 1980, 1996, 1998). Like other species, cattle prefer moving from dark to light (Grandin, 1980) and

**Figure 3.6:** To ease the unloading of adult cattle the ramp angle should not be steeper than 26°.

*(Photo courtesy of B. Catanese, ISZAM, Italy)*

often balk when they enter a dark race. Installing a lamp to illuminate the unloading ramp and the corridor entrance often facilitates cattle movement, but the light must not shine into the animal's eyes (Grandin, 2008). It is recommended to move cattle in groups using flags, a plastic bag on a stick or a paddle in order to ease handling and reduce the risk of injuries (Grandin, 2008).

Special care should be taken if unloading of downer animals is the only choice by limiting suffering and placing them in separate pens where feed and water are available (Broom, 2005). The possibility of emergency killing of these animals in respect of the EU legislation (EU Regulation 1099/2009) might be considered.

### 3.3.3 Sheep

Similarly to other species, sheep are either unloaded by fixed or hydraulic ramps, which are used as ramps or elevators. Elevator ramps are common in European trucks as they are more useful for sheep handling, while folding ramps are less practical, although they are much cheaper (Miranda-de la Lama et al., 2014). As with other species, descending a ramp is more difficult for sheep than climbing it, increasing the risk of stumbling and falling, especially on steep ramps (> 26°; EFSA, 2011). Other recommended ramp features are lateral safety barriers to prevent sheep from falling or escaping during the unloading procedure and bedding material (e.g. straw and sawdust) to limit the risk of slipping and falling, and reflection of light. Furthermore, similar to other species, solid side walls should be used to prevent sheep from seeing moving objects or people along the sides which may cause balking or a fear response.

Handling at unloading is easier if it is carried out smoothly by exploiting the natural gregarious behaviour of sheep. To this end, at some slaughterhouses a model or a familiar sheep is used to lead the entire batch through the unloading process (Miranda-de la Lama et al., 2014). The recommended approach to unloading is to encourage sheep to move naturally at their own pace, letting them explore the surrounding environment before moving.

Careful handling during unloading is necessary for sheep with severe pneumonia and therefore reduced tolerance to exercise (Alley, 2002) and for sheep with lameness or other clinical conditions, such as foot rot involving necrotizing inflammation of the interdigital skin and horn of the hoof (König et al., 2011). Severe lameness can be easily recognized at this stage by the following signs: not bearing weight on one or more limbs when standing or moving, reluctance to move and difficulty or inability to stand

(Kaler and Green, 2008). As in cattle, sheep that are injured or show signs of illness or severe distress should be segregated during unloading and inspected by a veterinarian and/or trained professional and, if necessary, a procedure for emergency slaughter should be applied as soon as possible to prevent further suffering of the animal.

### 3.3.4 Chickens and ducks

In order to avoid injuries and bruises it is generally recommended to unload the modules from the truck smoothly and horizontally (Lühren, 2012). Furthermore, if a conveyor is used, its angle should be such to prevent tilting of cages or crates that causes birds to pile up (CARC, 2003).

In commercial conditions, ducks are equally unloaded from crates, container systems (mostly drawer type) or liners (Lühren, 2012) and driven to lairage pens. During unloading, ducks should be handled with care to avoid stress reactions, bone breakage or carcass contamination (EFSA, 2004). Furthermore, in case of uncrated or free ducks it is recommended to let ducks move at their own pace to avoid duck overcrowding at this stage (Maple Leaf Farms, 2012).

### 3.3.5 Rabbits

Compared with loading at the farm where rabbits can be exposed to rough handling when being placed in the transport crates (Verga et al., 2009), the unloading procedure is not as stressful as rabbits are usually unloaded in crate groups or stands using hydraulic ramps (Figure 3.7; Buil et al., 2004). However, careful handling at unloading is recommended to avoid mortality and bone fractures and bruises resulting from rough handling and falling (EFSA, 2004).

**Figure 3.7:** Rabbits are usually unloaded in crate groups or stands using hydraulic ramps. *(Photo courtesy of L. Moffat, Eyes on Animals, The Netherlands)*

### 3.3.6 Fish

At the time of transport and slaughter, fish are exposed to a range of stimuli, including crowding, handling, increased human contact, noise, transport, novel/unfamiliar environments, feed deprivation, changes in social structure and changes in environmental conditions, such as water contaminated by blood (Gregory, 2008). In their adaptation to these changes, fish experience fear, homeostasis imbalance, increased physical activity (e.g. violent escape attempts), fatigue, physical injury and psychological distress (Ashley, 2007; Conte, 2004). For this reason, fish should be slaughtered as soon as possible to improve their welfare (EFSA, 2009b).

Fish are caught and moved to slaughter by pumping or brailing. Pumping is widely used in salmon aquaculture, whereas brailing is more common for sea bass (Figure 3.8), bream and pangasius, and at small processing factories

(Lines and Spence, 2014). Pumping is a common practice to unload fish from well-boats either to a holding cage or directly in the harvest station. To avoid or, at least, limit removal of fish out of water, modern pumping systems employ either vacuum or helical screw pumps to allow live fish to be moved in water via pipe systems over large horizontal and vertical distances without damage (EFSA, 2009b). However, this method exposes fish to increasingly lower levels of water as the number of fish decreases in the well during the catching process. An alternative catching method is the use of a moving bulkhead in the well. In this procedure, which is under operator control by direct observation or by camera, the bulkhead is slowly moved up keeping the fish crowded at a constant water level (Robb, 2008).

During catching, careful handling should be applied for all fish, but special care is important for specific species, such as salmon, as they lack an abrasion-resistant keratinized integumentary system. Mishandling can, in fact, easily breach their epidermis, which is composed of a thin layer of squamous cells, exposing them to osmotic and ion regulatory disturbance and infection. To avoid disrupting the mucous coat, fish-handlers should always use wet hands or wear soft, wet gloves when handling fish (Conte, 2004). To avoid skin damage, fish-handling systems with smooth surfaces made of fibreglass and stainless steel and species-specific nets are recommended. As water has less capacity to hold oxygen at higher temperatures, seining and netting activities should be performed during the cooler portions of the day to avoid the additive effect of temperature on stress (Conte, 2004).

**Figure 3.8:** Mediterranean commercial farm of seabass crowding and brailing before slaughter. *(Photo courtesy of A. Roque, IRTA, Spain)*

## 3.4 REFERENCES


Aalhus, J.L., Murray, A.C., Jones, S.D.M. and Tong, A.K.W. (1992) *Environmental Conditions for Swine during Marketing for Slaughter – A National Review.* Technical Bulletin 1992–6E, Research Branch, Agriculture and Agri-Food Canada.

Alley, M.R. (2002) Pneumonia in sheep in New Zealand: an overview. *New Zealand Veterinary Journal* 50, 99–101.

AMI (2012) Recommended Animal Handling Guidelines and Audit Guide: A Systematic Approach to Animal Welfare (T. Grandin, ed.), American Meat Institute Foundation, Washington, D.C., 108 p.

Ashley, P.J. (2007) Fish welfare: current issues in aquaculture. *Applied Animal Behaviour Science* 104, 199–235

Bayliss, P.A. and Hinton, M.H. (1990) Transportation of broilers with special reference to mortality rates. *Applied Animal Behaviour Science* 28, 93–118.

Broom, D.M. (2005) The effects of land transport on animal welfare. *Scientific and Technical Review, World Organisation for Animal Health* 24, 683–691.

Brown, S.N., Knowles, T.G., Wilkins, L.J., Chadd, S.A. and Warriss, P.D. (2005) The response of pigs being loaded or unloaded onto commercial animal transporters using three systems. *Veterinary Journal* 170, 91–100.

Buil, T., Maria, G.A., Villarroel, M., Liste, G. and Lopez, M. (2004) Critical points in the transport of commercial rabbits to slaughter in Spain that could compromise animals welfare. *World Rabbit Science* 12, 269–279.

CARC (2003) *Recommended code of practice for the care and handling of farm animals – Chickens, Turkeys and Breeders from Hatchery to Processing Plant.* Canadian Agri-Food Research Council, Ottawa, Canada.

Caulfield, M., Cambridge, H., Foster, S.F. and McGreevy, P.D. (2014) Heat stress: A major contributor to poor animal welfare associated with long-haul live export voyages. *Veterinary Journal* 199, 223–228.

Chevillon, P. (2001) Pig welfare during pre-slaughter and stunning. In: *Proceedings of the 1st International Virtual Conference on Meat Quality.* Embrapa, Concordia, Brazil, pp. 145–158.

Christensen, L. and Barton–Gade, P. (1999) Temperature profile in double-decker transporters and some consequences for pig welfare during transport. In: *Farm Animal Welfare – Who Writes the Rules?* Occasional Publication of the British Society of Animal Science 23, 125–128.

Cockram, M.S. (2014) Sheep transport. In: Grandin, T. (ed.) *Livestock Handling and Transport.* CABI, Wallingford, UK, pp. 238–244.

Conte, F.S. (2004) Stress and the welfare of cultured fish. *Applied Animal Behaviour Science* 86, 205–223.

Correa, J.A., Torrey, S., Devillers, N., Laforest, J.P., Gonyou, H.W and Faucitano, L. (2010) Effects of different moving devices at loading on stress response and meat quality in pigs. *Journal of Animal Science* 88, 4086–4093.

Correa, J.A., Gonyou, H.W., Torrey, S., Widowski, T., Bergeron, R., Crowe, T.G., Laforest, J.P. and Faucitano, L. (2013) Welfare and carcass and meat quality of pigs being transported for 2 hours using two vehicle types during two seasons of the year. *Canadian Journal of Animal Science* 93, 43–55.

De la Fuente, J., Salazar, M.I., Ibanez, M. and Gonzalez de Chavarri, E. (2004) Effects of season and stocking density during transport on live weight and biochemical measurements of stress, dehydration and injury of rabbit at time of slaughter. *Animal Science* 78, 285–292.

De la Fuente, J., Diaz, M.T., Ibáñez, M. and González de Chavarri, E. (2007) Physiological response of rabbits to heat, cold, noise and mixing in the context of transport. *Animal Welfare* 16, 41–47.

Driessen, B. and Geers, R. (2001) Stress during transport and quality of pork. A European view. In: *Proceedings of the 1st International Virtual Conference on Pork Quality.* Embrapa, Concordia, Brazil, pp. 39–51.

Edwards, L.N., Grandin, T., Engle, T.E., Ritter, M.J., Sosnicki, A.A., Carlson, B.A. and Anderson, D.B. (2010) The effects of pre-slaughter pig management from the farm to the processing plant on pork quality. *Meat Science* 86, 938–944.

EFSA (2004) Opinion of the Scientific Panel on Animal Health and Welfare on a request from the Commission related to the welfare of animals during transport. Question number: EFSA-Q-2003-094. *EFSA Journal* 44, 1–36.

EFSA (2009a) Scientific Opinion of the Panel on Animal Health and Welfare on a request from the European Commission on species-specific welfare aspects of the main systems of stunning and killing of farmed seabass and seabream. *EFSA Journal* 1010, 1–52.

EFSA (2009b) Scientific Opinion of the Panel on Animal Health and Welfare on a request from the European Commission on species-specific welfare aspects of the main systems of stunning and killing of farmed Atlantic salmon. *EFSA Journal* 1012, 1–77.

EFSA (2011) Scientific Opinion concerning the welfare of animals during transport. EFSA Panel on Animal Health and Welfare (AHAW). *EFSA Journal* 9, 1966.

European Commission (EC) (2005) Council Regulation (EC) No. 1/2005 of 22 December 2004 on the protection of animals during transport and related operations and amending Directives 64/432/EEC and 93/119/EC and Regulation (EC) No 1255/97. *Official Journal of the European Union* 5/01/2005, 1–44.

FAWC (2003) Report on the Welfare of Farmed Animals at Slaughter or Killing – Part One: Red Meat Animals. Farm Animal Welfare Council, London, UK.

Fisher, A.D., Stewart, M., Tacon, J. and Matthews, L.R. (2002) The effects of stock crate design and stocking density on environmental conditions for lambs on road transport vehicles. *New Zealand Veterinary Journal* 50, 148–153.

Fox, J., Widowski, T., Torrey, S., Nannoni, E., Bergeron, R., Gonyou, H.W., Brown, J.A., Crowe, T., Mainau, E. and Faucitano, L. (2014) Water sprinkling market pigs in a stationary trailer. 1. Effects on pig behaviour, gastrointestinal tract temperature and trailer micro-climate. *Livestock Science* 160, 113–123.

Gatica, M.C, Monti, G.E., Knowles, T.G., Warriss, P.D. and Gallo, C.B. (2010) Effects of commercial live transportation and preslaughter handling of Atlantic salmon on blood constituents. *Archivos de Medicina Veterinaria* 42, 73–78.

Gaughan, J.B. and Mader, T.L. (2012) Body temperature and panting in cattle. *Journal of Animal Science* 90 (Suppl. 3), 253–254.

Gebresenbet, G., Ljunberg, D., Geers, R. and Van de Water, G. (2004) Effective logistics to improve animal welfare in the production chain, with special emphasis on farm-slaughterhouse system. *International Society for Animal Hygiene* 1, 37–38.

Goldhawk, C., Crowe, T., Janzen, E., González, L. A., Kastelic, J., Pajor, E. and Schwartzkopf-Genswein, K. S. (2014) Trailer microclimate during commercial transportation of feeder cattle and relationship to indicators of cattle welfare. *Journal of Animal Science* 92, 5155–5165.

González, L.A., Schwartzkopf-Genswein, K.S., Bryan, M., Silasi, R. and Brown, F. (2012) Benchmarking study of industry practices during commercial long haul transport of cattle in Alberta, Canada. *Journal of Animal Science* 90, 3606–3617.

Goumon, S., Faucitano, L., Bergeron, R., Crowe, T., Connor, M.L. and Gonyou, H.W. (2013) Effect of ramp configuration on easiness of handling, heart rate and behavior of near-market pigs at unloading. *Journal of Animal Science* 91, 3889–3898.

Grandin, T. (1980) Observations of cattle behavior applied to the design of cattle-handling facilities. *Applied Animal Ethology* 6, 19–31.

Grandin, T. (1996) Factors that impede animal movement at slaughter plants. *Journal of the American Veterinary Medical Association* 209, 757–759.

Grandin, T. (1998) Handling methods and facilities to reduce stress on cattle. *Veterinary Clinics of North America: Food Animal Practice* 14, 325–341.

Grandin, T. (2008) Engineering and design of holding yards, loading ramps and handling facilities for land and sea transport of livestock. *Veterinaria Italiana* 44, 235–245.

Grandin, T. (2014) Improving welfare and reducing stress on animals in slaughter plants. In: Grandin, T. (ed.), *Livestock Handling and Transport.* CABI, Wallingford, UK, pp. 421–450.

Grandin, T. and Gallo, C. (2007) Cattle transport. In: Grandin, T. (ed.) *Livestock*

*Handling and Transport*. CAB International, Wallingford, UK, pp. 134–154.

Green, L.E., Berriatua, E. and Morgan, K.L. (1997) The relationship between abnormalities detected in live lambs on farms and those detected at post mortem meat inspection. *Epidemiology and Infection* 118, 267–273.

Gregory, N.C. (2008) Animal welfare at markets and during transport and slaughter. *Meat Science* 80, 2–11.

Guarnieri, P.D., Soares, A.L., Olivio, R., Schneider, J.P., Macedo, R.M., Ida, E.I. and Shimokomaki, M. (2004) Preslaughter handling with water shower inhibits PSE (pale, soft, exudative) broiler breast meat in a commercial plant. Biochemical and ultrastructural observations. *Journal of Food Biochemistry* 28, 269–277.

Haley, C., Dewey, C. E., Widowski, T., Poljak, Z., and Friendship, R. (2008) Factors associated with in-transit losses of market hogs in Ontario in 2001. *Canadian Journal of Veterinary Research 72*, 377–384.

Jones, T.A. (1999) Improved Handling Systems for Pigs at Slaughter. PhD Thesis, Royal Veterinary College, University of London, UK.

Kaler, J. and Green, L.E. (2008) Recognition of lameness and decisions to catch for inspection among sheepfarmers and specialists in GB. *BMC Veterinary Research* 4, 41.

Knezacek, T. D., Olkowski, A. A., Kettlewell, P. J., Mitchell, M. A., and Classen, H. L. (2010) Temperature gradients in trailers and change in broiler rectal and core body temperature during winter transportation in Saskatchewan. *Canadian Journal of Animal Science 90*, 321–330.

Knowles, T.G. (1998) A review of the road transport of slaughter. *Veterinary Record* 143, 212–219.

König, U., Nyman, A.J. and de Verdier, K. (2011) Prevalence of footrot in Swedish slaughter lambs. *Acta Veterinaria Scandinavica* 53, 27.

Lines, J.A. and Spence, J. (2014) Humane harvesting and slaughter of farmed fish. *Scientific and Technical Review, World Organisation for Animal Health* 33, 255–264.

Lühren, U. (2012) *Overview on Current Practices of Poultry Slaughtering and Poultry Meat Inspection*. Available at: http://www.efsa.europa.eu/en/search/doc/298e.pdf (accessed on 17 October 2014).

Maple Leaf Farms (2012) *Duck Well–Being Guidelines*. Available at: http://www.mapleleaffarms.com/lib/sitefiles/File/Maple–Leaf–Farms–Duck–Well–Being–Guidelines.pdf (accessed 17 October 2014).

Marai, I.F.M. and Rashwan, A.A. (2004) Rabbits behavioural response to climatic and managerial conditions – a review. *Archives Tierzucht Dummerstorf* 47, 469–482.

Marai, I.F.M., Habeeb, A.A.M. and Gad, A.E. (2002) Rabbits' productive, reproductive and physiological performance traits as affected by heat stress: a review. *Livestock Production Science* 78, 71–90.

Margeirsson, S., Jonsson, G.R., Arason, S. and Thorkelsson, G. (2007) Influencing factors on yield, gaping, bruises and nematodes in cod (Gadus morhua) fillets. *Journal of Food Engineering* 80, 503–508.

María, G., Liste, G., Villaroel, M., Chacon, G., Garcia-Belenguer, S. (2004) The effect of transport time on the commercial rabbits in hot climates. In: *Proceedings of the AgEng 2004: Engineering the Future*. Leuven, Belgium.

Meltzer, A. (1983) The effect of body temperature on the growth rate of broilers. *British Poultry Science* 4, 489–495.

Minka, N.S. and Ayo, J.O. (2007) Effects of loading behaviour and road transport stress on traumatic injuries in cattle transported by road during the hot-dry season. *Livestock Science* 107, 91–95.

Miranda-de la Lama, G.C., Salazar-Sotelo, M.I., Perez-Linares, C., Figueroa-Saavedra, F., Villarroel, M., Sañudo, C. and Maria, G.A.

(2012) Effects of two transport systems on lamb welfare and meat quality. *Meat Science* 92, 554–561.

Miranda-de la Lama, G.C., Villarroel, M. and María, G.A. (2014) Livestock transport from the perspective of the pre-slaughter logistic chain: a review. *Meat Science* 98, 9–20.

Nakyinsige, K., Sazili, A.Q., Aghwan, Z.A., Zulkifli, I., Goh, Y.M. and Fatimah, A.B. (2013) Changes in blood constituents of rabbits subjected to transportation under hot, humid tropical conditions. *Asian Australasian Journal of Animal Science* 26, 874–878.

Nannoni, E., Widowski, T., Torrey, S., Fox, J., Rocha, L.M., Gonyou, H.W., Weschenfelder, A.V., Crowe, T. and Faucitano, L. (2014) Water sprinkling market pigs in a stationary trailer. 2. Effects on selected exsanguination blood parameters and carcass and meat quality variation. *Livestock Science* 160, 124–131.

Nicole, C.J. and Scott, G.B. (1990) Pre-slaughter handling and transport of broiler chickens. *Applied Animal Behaviour Science* 28, 57–73.

Nielsen, N.J. (1982). Recent results from investigations of transportation of pigs for slaughter. In: Moss, R. (ed.) *Transport of Animals Intended for Breeding, Production and Slaughter*. Martinus Njhoff Publisher, The Hague, The Netherlands, pp. 115–124.

Nijdam, E., Arens, P., Lambooij, E., Decuypere, E. and Stegeman, J.A. (2004) Factors influencing bruises and mortality of broilers during catching, transport, and lairage. *Poultry Science* 83, 1610–1615.

Norton, T., Kettlewell, P. and Mitchell, M. (2013) A computational analysis of a fully stocked dual-mode ventilated livestock vehicle during ferry transportation. *Computers and Electronics in Agriculture* 93, 217–228.

Petracci, M., Bianchi, M., Cavani, C., Gaspari, P. and Lavazza, A. (2006) Preslaughter mortality in broiler chickens, turkeys, and spent hens under commercial slaughtering. *Poultry Science* 85, 1660–1664.

Phillips, C.J.C., Pines, M.K. and Muller, T. (2012) The avoidance of ammonia by sheep. *Journal of Veterinary Behavior: Clinical Applications and Research* 7, 43–48.

Portz, D.E., Woodley, C.M. and Cech, J.J. (2006) Stress-associated impacts of short-term holding on fishes. *Reviews in Fish Biology and Fisheries* 16, 125–170.

Ritter, M.J., Ellis, M., Brinkmann, J., DeDecker, J.M., Keffaber, K.K., Kocher, M.E., Peterson, B.A., Schlipf, J.M. and Wolter, B.F. (2006) Effect of floor space during transport of market-weight pigs on the incidence of transport losses at the packing plant and the relationships between transport conditions and losses. *Journal of Animal Science* 84, 2856–2864.

Ritter, M.J., Ellis, M., Bowman, R., Brinkmann, J., Curtis, S.E., DeDecker, J.M., Mendoza, O., Murphy, C.M., Orellana, D.G., Peterson, B.A., Rojo, A., Schlipf, J.M. and Wolter, B.F. (2008) Effects of season and distance moved during loading on transport losses of market-weight pigs in two commercially available types of trailer. *Journal of Animal Science* 86, 3137–3145.

Robb, D.H.F. (2001) The relationship between killing methods and quality. In: Kestin, S.C. and Warriss, P.D. (eds.), *Farmed Fish Quality*. Fishing News Books, Cornwall, pp. 220–233.

Robb, D.H.F. (2008) Welfare of fish at harvest. In: Deward, J. and Branson, J. (eds.), *Fish Welfare*, Blackwell Scientific Publications, London, UK, pp. 217–242.

Robertshaw, D. (1985) Heat loss of cattle. In: Yousef, M.K. (ed.), *Stress Physiology in Livestock*, vol. 1. Basic Principles. CRC Press, Boca Raton, FL, pp. 55–66.

Rocha, L.M., Velarde, A., Dalmau, A., Saucier, L. and Faucitano, L. (2016) Can the monitoring of animal welfare parameters predict pork meat quality variation through the supply chain (from farm to slaughter)? *Journal of Animal Science* 94, 359–376.

Rotabakk, B.T., Skipnes, D., Akse, L. and Birkeland, S. (2011) Quality assessment of Atlantic cod (Gadus morhua) caught by longlining and trawling at the same time and location. *Fisheries Research* 112, 44–51.

RSPCA (2011) *RSPCA Welfare Standards for Domestic/Common Ducks*. Available at: http://www.freedomfood.co.uk/media/9309/ducks.pdf (Accessed on 17 October 2014).

SCAHAW (2002) The Welfare of Animals during Transport (Details for Horses, Pigs, Sheep and Cattle). Report of the Scientific Committee on Animal Healh and Animal Welfare, European Commission.

Schwartzkopf-Genswein, K., Faucitano, L., Dadgar, S., Shand, P., Gonzàlez, L.A. and Crowe, T. (2012) Road transport of cattle, swine and poultry in North America and its impact on animal welfare, carcass and meat quality: a review. *Meat Science* 92, 227–243.

Sparke, E.J., Young, B.A., Gaughan, J.B., Holt, M. and Goodwin, P.J. (2001) *Heat Load in Feedlot Cattle*. Meat and Livestock Australia, North Sydney, Australia, 34 pp. Available at: http://www.mla.com.au/News-and-resources/Publication-details?pubid=2883 (Accessed on 30 June 2016).

Stockman, C.A., Barnes, A.L., Maloney, S.K., Taylor, E., McCarthy, M. and Pethick, D. (2011) Effect of prolonged exposure to continuous heat and humidity similar to long haul live export voyages in Merino wethers. *Animal Production Science* 51, 135–143.

Torrey, S., Bergeron, R., Gonyou, H.W., Widowski, T., Lewis, N., Crowe, T., Correa, J.A., Brown, J. and Faucitano, L. (2013a) Transportation of market-weight pigs 1. Effect of season and truck type on behaviour with a 2 hour transport. *Journal of Animal Science* 91, 2863–2871.

Torrey, S., Bergeron, R., Faucitano, L., Widowski, T., Lewis, N., Crowe, T., Correa, J.A., Brown, J. and Gonyou, H.W. (2013b). Transportation of market-weight pigs 2. Effect of season and animal location in the truck on behavior with a 8 hour transport. *Journal of Animal Science* 91, 2872–2878.

Van de Perre, V., Permentier, L., De Bie, S., Verbeke, G. and Geers, R. (2010) Effect of unloading, lairage, pig handling, stunning and season on pH of pork. *Meat Science* 86, 931–937.

Verga, M., Luzi, F., Petracci, M and Cavani, C. (2009) Welfare aspects in rabbit rearing and transport. *Italian Journal of Animal Science* 8, 191–204.

Vieira, F.M.C., Silva, I.J.O., Barbosa Filho, J.D. and Vieira, A.M.C. (2010) Productive losses on broiler preslaughter operations: effects of distance from farms to abattoirs and of lairage times in climatized holding areas. *Revista Brasileira de Zootecnia*, 1806–9290.

Warren, L.A., Mandell, I.B. and Bateman, K.G. (2010) Road transport conditions of slaughter cattle: Effects on the prevalence of dark, firm and dry beef. *Canadian Journal of Animal Science* 90, 471–482.

Warriss, P.D., Bevis, E.A., Edwards, J.E., Brown, S.N. and Knowles, T.G. (1991) Effect of the angle of slope on the ease with which pigs negotiate loading ramps. *Veterinary Record* 128, 419–421.

Warriss, P.D., Brown, S.N., Knowles, T.G. and Edwards, J.E. (1992) Influence of width and bends on the ease of movement of pigs along races. *Veterinary Record* 130, 202–204.

Warriss, P.D., Pagazaurtundua, A. and Brown, S.N. (2005) Relationship between maximum daily temperature and mortality of broiler chickens during transport and lairage. *British Poultry Science* 46, 647–651.

Warriss, P.D., Brown, S.N., Knowles, T.G., Wilkins, L.J., Pope, S.J., Chadd, S.A., Kettlewell, P.J. and Green, N.R. (2006) Comparison of the effects of fan-assisted and natural ventilation of vehicles on the welfare of pigs being transported to slaughter. *Veterinary Record* 158, 585–588.

Webster, A.J.F., Tuddenham, A., Saville-Weeks, C.A. and Scott, G.B. (1993) Thermal stress on chickens in transit. *British Poultry Science* 34, 267–277.

Wedemeyer, G.A. (1996a) *Physiology of Fish in Intensive Culture Systems.* Chapman and Hall, New York, NY, 232 pp.

Wedemeyer, G.A. (1996b) Transportation and handling. In: Pennel, W. and Barton, B. (eds.), *Principles of Salmonid Culture. Developments in aquaculture and fisheries science* 29. Elsevier Publishing, Amsterdam, The Netherlands, pp. 727–758.

Weeks, C.A. (2014) Poultry handling and transport. In: Grandin, T. (ed.), *Livestock Transport and Handling.* CABI, Wallingford, UK.

Weeks, C.A., Webster, A.J.F. and Wyld, W.M. (1997) Vehicle design and thermal comfort of poultry in transit. *British Poultry Science* 38, 464–474.

Weschenfelder, A.V., Torrey, S., Devillers, N., Crowe, T., Bassols, A., Saco, Y., Piñeiro, M., Saucier, L. and Faucitano, L. (2012) Effects of trailer design on animal welfare parameters and carcass and meat quality of three Pietrain crosses being transported over a long distance. *Journal of Animal Science* 90, 3220–3231.

Weyman, G. (1987) *Unloading and Loading Facilities of Livestock Markets.* Council of National and Academic Awards, Environmental Studies, Hatfield Polytechnic, Hatfield, UK.

chapter four

# Lairage and handling

*Antoni Dalmau[1] and Antonio Velarde[1]*

## Learning objectives

- Describe the adequate design of lairage pens.
- Understand the impact of space allowance on animal welfare.
- Describe the needs for drinkers and feeding.
- Understand the main problems of thermoregulation and provide measures for its assessment.
- Describe the optimal lairage duration.
- Provide measures to assess welfare at lairage.

## CONTENT



## 4.1 INTRODUCTION

One of the purposes of lairage is to hold sufficient reserve of animals to avoid any interruption in the slaughter line (Velarde, 2008). Poor lairage conditions can become stress factors to the animals, as they will be allocated in a novel environment (Terlouw et al., 2008), where contact with unfamiliar persons can be aversive. To avoid the negative welfare effects of lairage, the animals should be slaughtered without undue delay after their arrival at the slaughterhouse. An exception applies for pigs, where inappropriate lairage time can

[1] IRTA. Animal Welfare Subprogram, Veïnat de Sies, s/n, 17121 Monells, Spain

have detrimental effects on meat quality (Warriss, 2003; Santos et al., 1997). A similar effect has been found in poultry. In this case a lower incidence of meat quality problems can be associated with a higher incidence of mortality rates when compared with pigs (Oba et al., 2009). Several requirements must be considered for an appropriate lairage, including space allowance, floor and wall conditions, drinkers, provision of showers and fans to improve ventilation and microclimate control (for some species in warm weather conditions), handling, provision of food when necessary, light, noise and lairage time. Animals that are transported in cages, such as broilers, layer hens, turkeys, ducks, geese, quails or rabbits remain in the cages during lairage, so the conditions are different to red meat animals.

## 4.2 LAIRAGE PENS

Holding animals in pens gives them a rest period after the physical activity associated with handling on the farm, loading and transport to slaughterhouses. If animals arrive exhausted, (e.g. due to a prolonged period of food and water deprivation, long transport distance or time, poor road conditions and adverse weather conditions) their ability to cope with the lairage conditions is impaired. The provision of some pens next to the unloading area that can be used as hospital pens will facilitate the recovery of animals arriving exhausted or with any problems other than those that require emergency killing on the lorry (e.g. downer cattle) or in the unloading bay.

The lairage pens should provide enough space to satisfy the needs of animals for comfort around resting and thermal comfort (neither too hot nor too cold) as well as providing enough space for the animal to be able to move around freely (Velarde, 2008). Cattle and pigs prefer to lie along the walls (Grandin, 1980), so long, narrow pens will stimulate the resting of the animals. Pens of different sizes will facilitate the adaptation of the group size to the space allowance without mixing of unfamiliar animals. To avoid stress problems caused by animals from one pen being disturbed by animals from another pen, the walls should be solid and high enough to avoid visual contact between unfamiliar groups of animals. The only exception is when animals are housed alone. In this case, and considering that the animal is in a new environment, isolation results in a great stress factor, so for these cases it is suggested that animals should be allowed to have visual contact with conspecifics, avoiding any physical contact. Aggressive animals should also be separated, but not visually isolated.

In pigs, the space allowance of the pens should be adequate for better access to drinkers and rest in lateral position if needed due to the environmental temperatures, and should provide a chance for subordinate animals to avoid aggression from dominant ones. Split marketing of finishing pigs in commercial farms occurs when pigs within a pen reach slaughter age with a weight variation. It involves removing the heaviest pigs from pens and marketing them one or two weeks earlier than the remaining pen mates. Studies have documented detrimental effects of this practice on welfare and carcass quality due to the mixing of unfamiliar animals during transport and at the slaughterhouse, as they fight to establish new rank orders (Fàbrega et al., 2013). In fact, mixing unacquainted pigs is usually followed by fighting and mounting behaviour, impairing animal welfare. First, aggression may result in injuries, pain and, in extreme cases, the death of the animal. Second, aggression leads to physiological stress and immunosuppression within the group. In addition, it might have detrimental effects on carcass (presence of lesions; Figure 4.1) and meat quality, with presence of pale, soft and exudative meat (PSE) or dark, firm and dry meat (DFD; Tuchscherer and Manteuffel, 2000; Turner et al., 2006; Shen et al., 2006). To avoid fighting, animals should be kept in familiar groups. Producing animals with less variation in slaughter weight would allow transporting all the animals of the farm pen at the same time without mixing unacquainted pigs. Fighting occurs mainly during the initial hours after mixing (Friend et al., 1983; Geverink et al., 1996). If mixing is unavoidable, the recommendation is to mix pigs just prior to loading or on the transport vehicle rather than later in the slaughterhouse, and to maintain in lairage the same groups from the transport. Pigs normally lie down or they are occupied with other problems not related to the hierarchies during transportation, but they are in close contact with unfamiliar animals. This seems to facilitate familiarization and recognition of new animals, which helps to reduce the intensity and duration of aggression. If the groups are mixed again during the unloading, aggression might increase at lairage with detrimental effects on welfare and meat quality. The occurrence of fighting in a pen also has detrimental effects on pigs resting in neighbouring pens as they are disturbed by the vocalization of fighting pigs (Weeks, 2008). Pigs kept in large groups (30 pigs) spend more time standing and fighting, and are involved in more agonistic interactions (bites and head knocks) than pigs kept in small groups (10 pigs; Rabaste et al., 2007). On the other hand, according to Grandin (2000), in US slaughterhouses, mixing a large group of animals, of 200 or more from three or four different farms resulted in less fighting than mixing smaller groups of 6 to 40 pigs. Transport distance and duration may affect the behaviour of animals in lairage, however, very large groups or very small groups seems to be the best option to reduce aggression. According to Faucitano (2010), space allowance has a bigger impact on pigs' social behaviour than group size. In cattle, mixing of unacquainted animals increases aggression, physical activity, bruising and DFD

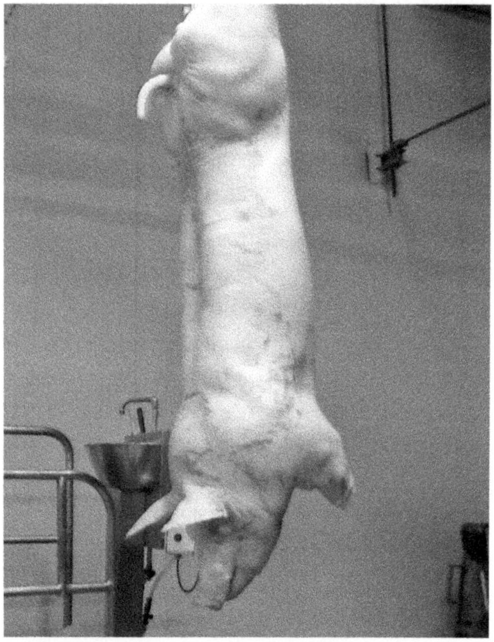

**Figure 4.1:** Presence of skin lesions in the carcass of pigs

**Figure 4.2:** Presence of bedding material in lairage during resting times for lambs

meat (Grandin, 2000). Therefore, mixed animals should be slaughtered as soon as possible to minimize welfare and meat quality problems. It is also worth noting that stress increases the susceptibility of animals to colonization as well as excretion of pathogens and, as a consequence, lairage pens could become a potential source of cross-contamination. According to Knowles (1998), when two groups of unfamiliar sheep are penned together, they stay as two groups and there are fewer aggressive interactions than when two groups of relatively familiar sheep are penned together.

Floor and wall conditions must be adequate for the species, easy to clean and with smooth and rounded surfaces to avoid injuries to the animals (Grandin, 1980). Concrete curbs between the pens prevent water in one pen from flowing into another pen. Drains should be located outside of the walking areas of the animals to avoid distraction or fear to the animals. Solid floor with an adequate slope for water/urine evacuation is preferred to slatted floor, especially if animals are not familiar with this type of floor. Floors must have a non-slip surface as slippery floors cause agitation and stress (Grandin, 1998b; Cockram and Corley, 1991).

In lairage periods longer than 12 hours, it is suggested that bedding material should be incorporated to facilitate resting of the animals. Bedding material (i.e. straw) is recommended for lambs, piglets and kid goats in cold weather conditions to facilitate thermoregulation (Figure 4.2). On the other hand, in warm weather conditions, the use of straw as bedding material increases the effective temperature up to 4°C in relation to the environmental temperature, and should be avoided.

## 4.3 STOCKING DENSITY

The lairage environment should meet the welfare needs of animals by providing enough space for the animal to be

**Figure 4.3:** High space allowance in lairage for pigs

able to move around freely. To satisfy its welfare needs, each animal should have enough space to stand up, lie down and turn around (Figure 4.3). Stocking rate refers to the area of the animals, expressed either as number of animals per unit area (head/$m^2$) or area provided per animal ($m^2$/head), also often referred to as space allowance. Stocking densities can be referred to as the live weight per unit floor area (e.g. kg/$m^2$). Lying area requirements are calculated based on body weight, but in addition, the stocking density must allow all animals access to the drinkers and, if present, to the feeders. The position of the drinkers and feeders should be accessible to the smallest animal in the group.

Pigs and poultry have great difficulty in losing heat and may therefore suffer from heat stress at ambient temperatures close to the upper limit of their thermoneutral zone and at high humidity. In pigs, lying behaviour is an important tool within behavioural adaptation for thermoregulation, and therefore related to the effective temperature of the environment. Comparing measurements of pigs' energy metabolism and animal behaviour has shown that thermal neutral or comfort behaviour for pigs occurs when they are lying touching each other. However, huddling (usually in sternal recumbency) means that it is too cold as pigs attempt to reduce heat loss. When it is too warm, pigs lie down quickly in lateral position, maintain relatively wide separation between individuals and increase their respiration rate (Santos et al., 1997). Therefore, high temperature increases the space each animal needs to dissipate heat and rest. A 100 kg live weight pig occupies 0.4 $m^2$ when lying in sternal position, and 1.0 $m^2$

in lateral recumbence. A recommendation for pigs of 90–100 kg is 0.42 $m^2$ for short lairage and 0.66 $m^2$ for long lairage time (> 3 h), for cattle of 550 kg around 1.60–1.70 $m^2$, for light lambs (< 25 kg) around 0.30 $m^2$ and for sheep in general, 0.56 $m^2$ (Weeks, 2008), although the space should be increased to 0.9–1.0 $m^2$/sheep for overnight lairage (DEFRA, 2003) and to 1.85 $m^2$ for horned cattle (Grandin, 1979). Knowles (1998) also stated that unshorn sheep had a greater space requirement than shorn sheep. In any case, the space requirements of the animals can vary depending on the temperature, humidity, lairage duration and even temperament and presence or absence of horns. High stocking densities increase aggression because the easy escape of attacked individuals is thwarted (Figure 4.4). When pigs or cattle are kept in small pens without enough space and opportunity for subordinate animals to escape, this leads to distress and physical damage, predominantly caused by dominant animals. The combination of long lairage times and high stocking density has been described as a major cause of DFD meat in pigs (Guardia et al., 2005). Therefore, for animals lairaged for long periods, a decrease in stocking density providing more space is recommended. This is especially important for mixed groups of pigs as fighting can last hours after arriving at the slaughterhouse.

## 4.4 DRINKERS

During transport, animals are usually deprived of water. Therefore, water should always be available in the lairage. The water supply system should be

**Figure 4.4:** High stocking density in lairage for pigs

designed and constructed to allow all animals easy access (i.e. in terms of comfortable drinking height and number of water troughs or drinkers) to clean water at all times, without being injured or limited in their movements, and so that the risk of the water becoming contaminated with faeces or urine is minimized. Suckling animals are particularly susceptible to dehydration because they may not have learnt how to drink from a trough (or novel/unfamiliar drinker mechanisms) and so they fail to drink the water provided at the abattoir. In such cases, an appropriate source of water should be provided. Failure to drink, even when water is available, is a particular problem in young calves (Gregory, 1998) and lambs (Jacob et al., 2006). In ruminants, the rumen holds 15 to 20% of the total body water (Dahlborn and Holtenius, 1990) and can act as a buffer against dehydration. Sheep are better adapted to drought than cattle, being able to produce concentrated urine and relatively dry faeces. There is no agreement about the number of drinkers per animal. The Welfare Quality protocols (Welfare Quality, 2009a,b) considered adequate a ratio of 12 dairy cattle, 15 beef cattle and 10 pigs per drinker (similar numbers could be considered for small ruminants). The recommendation should be that all animals have easy access to at least two drinkers or drinking places. Birds and rabbits in cages and lairaged for long periods should be provided with water through a drinking system. As the main mechanism of thermoregulation is by means of evaporation of water when panting, so a state of dehydration can be achieved easily if water is not provided after some hours in lairage. In fact, poultry and rabbits are not provided with water whilst in the lairage area because it is assumed that they will be slaughtered immediately or a short time after their arrival. In case of long lairage times it is then important to provide a system to give water to the animals. Logically, in warm conditions and long transports this point must be especially considered. Sometimes, animals do not drink immediately upon arrival in lairage pens. Knowles et al. (1993) described that sheep do not drink readily on arrival at the lairage, especially from an unfamiliar water source. Knowles (1998) describes that even at high ambient temperatures, when deprived of water for a period of at least 20 hours, transported lambs did not drink immediately when offered water.

## 4.5 FEEDING

Animals are fasted to reduce gut content during the pre-slaughter period and so prevent the release and spread of bacterial contamination through faeces within the group during transport and lairage as well as through the spillage of gut contents during carcass evisceration. Fasting before slaughter, within reasonable limits, is also beneficial to the welfare condition of the pig as it prevents pigs from vomiting in transit and developing hyperthermia. However, an extended fasting period causes hunger and aggressiveness (Warriss, 1994) and reduces live body weight and carcass yield (Warriss, 1994; Beattie et al., 2002). For instance, during the first 24 hours of fasting, pigs can lose up to 5% of body weight, around 0.2% per hour (Warris, 1993). Fasting before slaughter reduces the risk of PSE meat in pigs (López-Bote, 2001) and increases the risk of DFD meat in pigs and cattle (Gregory, 1998). The combination of prolonged fasting periods and fighting in lairage, especially during long lairage times, reduces muscle glycogen content at slaughter increasing the risk of DFD (Nanni Costa et al. 2002). In Australia, where journeys are often very long, Morris (1994) found 15% of lamb carcasses classified as DFD. According to Warriss (1994) and Warriss et al. (1998), fasting must be longer than 4 hours before transport in pigs to reduce motion sickness (Bradshaw et al., 1996) and mortality (Averós et al., 2008; Gispert et al., 2000; Guardia et al., 2004, 2005). However, the total fasting period (fasting period in the farm + transport + lairage time) should be restricted to a maximum of 18 hours. Other reports (Chevillon, 1994; Guardia et al., 2005; Kephart and Mills, 2005; Gispert et al., 1996; Bertol et al., 2005) proposed similar periods. An extended fasting period causes aggressiveness (Warriss, 1994) and has detrimental effects on meat and carcass quality (i.e. presence of scratches and bruises; Faucitano et al., 2006; Warriss et al., 1998). Information about the animals' last time of feeding will be required in the slaughterhouse to plan the time of lairage before slaughter. Wythes and Shorthose (1984) concluded that in cattle, carcass weight loss does not occur until after 24 hours of food and water deprivation. In sheep, Thompson et al. (1987) reported that feed deprivation for 24, 48, 72 and 96 hours resulted in a 3.5, 5.9, 7.3 and 7.9% loss in hot carcass weight, respectively.

The Regulation (EC) No. 1099/2009 requires that animals, which have not been slaughtered within 12 hours of their arrival, shall be fed and subsequently given moderate amounts of food at appropriate intervals. The food given must be appropriate in quantity and quality for each species, age and state and must be distributed in a way that all the animals can have good access to it, avoiding as much as possible competition among animals, as this can be another source of fighting even in unmixed groups. The food should be given by means of specific troughs and not directly on the floor. In animals in cages, such as poultry and rabbits, although it occurs rarely because animals should be slaughtered immediately after arrival, it is also important to have a system available to provide food to all the animals.

## 4.6 THERMAL COMFORT

All the animals kept in lairage should be protected from adverse weather conditions. In pigs, lairage conditions of 15–18°C and 59–65% relative humidity are recommended (Honkavaara, 1989). Cattle can cope with environmental temperatures from −10°C to 20°C and sheep, temperatures of 15 to 28°C (Wathes et al., 1983).

During hot weather, the practice of spraying pigs with cold water (10–12°C) at lairage limits the risk of hyperthermia and consequently reduces the mortality rate in lairage pens (Figure 4.5). Showering reduces aggressive behaviour and facilitates greater ease of handling into the stunning pens. The shower regime should be intermittent (i.e. once at arrival and once just before moving to stunning pen) and not for long times (i.e. 10 to 30 minutes can be enough in most cases) in order to get the greatest cooling effect and reduce activity and aggression. To optimize cooling effect, the coarseness of the spray should be

**Figure 4.5:** Use of showers in lairage for pigs

able to penetrate the hair and wet the skin. Sprinklers that create a fine mist can increase environmental relative humidity without penetrating the hair. Below 5°C, showering is not recommended as it causes thermogenesis by shivering to maintain body temperature in pigs. In other species, such as cattle and sheep, the showers are used to wash the animals. However, it must be taken into account that washing cattle and sheep has been reported as very stressful (Walker et al, 1999; Kilgour 1988) and useful only in cases where the animals are very dirty (Grandin, 2000) or under conditions with thermal stress due to very high temperatures. If animals are lairaged at a high density there will be less area from which each animal can dissipate heat because more of them are in physical contact with each other. It is under these circumstances that hyperthermia is likely to develop as well, and in sheep, for instance, the problem will be exacerbated if animals are fully fleeced (Knowles, 1998).

Pigs and poultry have great difficulty in losing heat and suffer from heat stress at ambient temperatures close to the upper limit of their thermoneutral zone and at high humidity (Hunter et al., 1998). Heat stress increases the risk of mortality and deteriorates meat quality. The causes of mortality in broilers seem to be related to acute (14%) and congestive (37%) heart failure and trauma (32%; Petracci et al., 2006). Birds are able to regulate their body temperature by means of panting and losing heat through their skin and feather cover (Figure 4.6). In cages, the ability to lose temperature is compromised due to the proximity between birds. The increase in panting (less effective in reducing body temperature in environments with high humidity) produces an alteration in the acid-base balance and dehydration. If the body temperature increases above 4°C, the birds will die. The environmental temperature inside the cages increases within the first hour of lairage, after which time an elevated equilibrium temperature is reached. The temperature is higher in the cages on the top than at the bottom due to rising heat. Dissipation of water vapour and heat from the cages is the key point to reduce thermal stress. For instance, less than 0.1 m/s airflow through each cage is not enough for heat to be dissipated (MAFF, 1998). The increased environmental temperature in a full module of approximately 9°C compared to lairage temperature indicates that, even with low humidity, there may be a potential heat stress problem for lairage temperatures of 17°C or more (MAFF, 1998). Extraction of water vapour and heat from the modules is more effective at reducing heat stress than attempting to ventilate the entire lairage (MAFF, 1998). Although Oba et al. (2009) and Guarnieri et al. (2004) reported the advantages of showers at lairage for broilers, according to MAFF (1998), the use of water sprays or misting systems, while reducing air temperature by 2 or 3°C, raises humidity, saturating the air and inhibiting the birds' heat loss through panting. Therefore, showers are only recommended if the environment around the animals can be maintained with low humidity.

**Figure 4.6:** Chick panting due to thermal stress on arrival in the slaughterhouse

## 4.7 LIGHT, NOISE AND AIR CONDITIONS

Lighting and artificial illumination in the lairage area should be enough to inspect the animals and allow them to move without fear and distress (Figure 4.7). Animals are very sensitive to high-frequency sounds, and noise is a potential stressor. Vocalizations of stressed animals, combined with human shouting, which is particularly abhorrent to animals (Weeks, 2008), provoke a stressful auditory environment. In a UK survey of cattle, sheep and pig lairage conditions, the pig's noise at lairage was reported as the loudest (Weeks et al., 2009). Sound levels measured in lairage can range from 76 to 108 dB on average, with the highest levels (120 dB) recorded in the prestun area (Talling et al., 1996; Rabaste et al., 2007). Sudden and high-pitched sounds, such as gates clanging and pig vocalization (80–103 dB; Weeks 2008; Weeks et al., 2009) can be a source of stress, as shown by increased blood lactate, creatine phosphokinase (CPK) cortisol levels and increased heart rate (Talling at al., 1996; Kanitz and Tuchscherer, 2005). More specifically, Spensley et al. (1995) found that novel noises ranging from 80 to 89 dB increased heart rate in pigs. In addition, Talling et al. (1998) described that intermittent sounds are more disturbing to pigs than continuous sound. High-pitched sounds from hydraulic systems are very disturbing to cattle (Grandin, 2000). Eldridge et al. (1989) studied cattle behaviour in overnight lairage, and reported that cattle housed close to a noisy environment (next to unloading facilities) exhibited more movement and had higher carcass bruise scores than cattle held in quieter pens. Slaughterhouses with high concrete ceilings and precast concrete

**Figure 4.7:** Light conditions in sheep during lairage

walls have more echoes and noise than slaughterhouses built from foam-core insulation board (Grandin, 2000).

The quality of the air during long lairage periods is another important aspect to take into account. High ammonia levels in an enclosed building is detrimental to poultry, pig and cattle welfare (Jones et al., 2005; Kristensen et al., 2000; Kristensen and Wathes, 2000; Ritter et al., 2006). An experimental choice test, in which pigs were exposed to different concentrations of ammonia (10, 20 and 40 ppm), showed that ammonia levels higher than 10 ppm are aversive to pigs (Wathes et al., 2002). Hence, exposure to moderate levels of ammonia (10 ppm) in lairage for a short period (1 hour) may be still acceptable on animal welfare grounds (Weeks, 2008). For animals transported in modules or crates, the space allocated between columns of modules or crates determines the efficacy of ventilation and air movement: the wider the space, the better the ventilation.

## 4.8 LAIRAGE TIME

Lairage time should be optimum in pigs to reduce the incidence of PSE meat. The negative welfare consequences of lairage can be reduced if animals are unloaded as quickly as possible after arrival and subsequently slaughtered without undue delay. Several studies (Warriss et al., 1992; Brown et al., 1999; Pérez et al., 2002; Honkavaara, 1989) concluded, on the basis of cortisol levels in slaughter blood samples collected from pigs, that after 2 to 3 hours in lairage, basal levels are reached again. Therefore, this period of resting is the most recommended to recover pigs from stress (Grandin, 1994; Warris et al., 1998; Young et al., 2009, Pérez et al., 2002). However, Van de Perre et al. (2010) recommend a lairage time of 2–4 hours in summer and less than 2 hours in winter. Nevertheless, this might depend on the country, season and genotype of pigs. For instance, Fraqueza et al. (1998), in Brazil, described that pigs subjected to heat stress should not be allowed to a lairage longer than 30 minutes to avoid problems of PSE meat, and Santos et al. (1997), also in Brazil, do not recommend lairage at high temperatures. On the other hand, other studies show little effect of lairage duration on meat quality when the pig population is free of the halothane gene and the handling of animals before (i.e. from farm to abattoir) and after the lairage is not detrimental to the welfare of animals (good weather conditions, handling and environmental temperatures). Aaslying et al. (2001) and Geverink et al. (1996) do not recommend lairage in those cases. In broilers, although long lairage time reduces the incidence of PSE meat, this practice is not recommended because it increases mortality rates (Oba et al., 2009). In rabbits, although the level of stress after 8 hours of lairage is lower than with 2 hours, no differences were found in relation to meat quality (Liste et al., 2009). However, the authors found more bruises in rabbits with 2-hour lairage than with 8-hour. Their hypothesis is that during the capture process prior to stunning, the rabbits lairaged for short period were more stressed and probably more reactive and more likely to develop bruises. However, further studies are needed for this species. In ruminants, the presence of meat quality problems due to short lairage time is rarely described.

Long lairage periods might demand a great mobilization of metabolic energy in all species due to the fact that animals are in a novel situation and usually without food. Under this situation, prolonged periods of pre-slaughter stress can increase the incidence of DFD meat in pigs (Fernandez and Tornberg, 1991; Pérez et al., 2002, Warriss et al., 1998) and the incidence of dark cutting in beef cattle (Warner et al., 1998), with few effects on eating quality or tenderness. In addition, the longer the lairage period, the higher the number of bruises in pigs and cattle carcasses (Brown et al., 1999; Faucitano, 2001, Warriss et al., 1998), dead chickens before slaughter (Nijdam et al., 2004), and physiological stress indicators in rabbits (Liste et al., 2009). A long lairage period increases losses in carcass weight (Gispert et al., 2000) and impairs the production of specific products, such as cured ham (Nanni Costa et al., 2002). As mentioned previously, if animals are not fed, weight loss increases due to the prolonged fasting periods.

## 4.9 MONITORING WELFARE

In lairage, several measures can be used to assess the welfare of the animals. The first animals to be assessed are those housed in the hospital or isolation pens, as they need special care and even emergency killing when no suitable alternatives are available to provide optimum welfare for them. Both red meat animals and poultry can be assessed for cleanliness at arrival to lairage (Figure 4.8). Dirty animals are due to either poor litter conditions in the poultry barn or muddy feedlots, heat stress in pig farms or other factors affecting all the species, such as stocking density and general cleanliness of the farm and even lairage pens (Figure 4.9). For instance, a 4-point scoring system can be used according to Grandin (2010): 1 – clean animal or bird. Birds must have completely clean feet and cattle may have soil below the knee; 2 – legs are soiled; 3 – legs and belly/breast is soiled; 4 – legs, belly/breast and sides of body are soiled. The percentage of thin and emaciated animals (presence/absence) can also be scored in the slaughterline (Ritter et al., 1999). Fear induces physiological changes in the animals such as tachycardia, increased respiration rate and elevated body temperature, but these indicators are not feasible to be assessed at the slaughterhouse. Behavioural changes are more evident and include heightened alertness, immobilization, aggression and escape/avoidance, so these could be easily included in a monitoring schema. Vocalization is a feasible indicator for the affective state in cattle (Manteuffel et al., 2004; Grandin, 1998a). An increased frequency of vocalization is often correlated with an increase in the physiological stress response (Hemsworth et al., 2011). Watts and Stookey (1998) described that cattle vocalization increased when isolated and suggested it as a good measure to assess severe stress in this species. In contrast, the presence of vocalizations is not a good indicator for sheep as they tend not to show any outward vocal or behavioural display indicative of pain or distress and tend to "suffer in silence".

In the Welfare Quality protocols (Welfare Quality, 2009a,b) aspects such as the absence of thirst and hunger are assessed at lairage. For this purpose, the number of drinking points in each pen (in

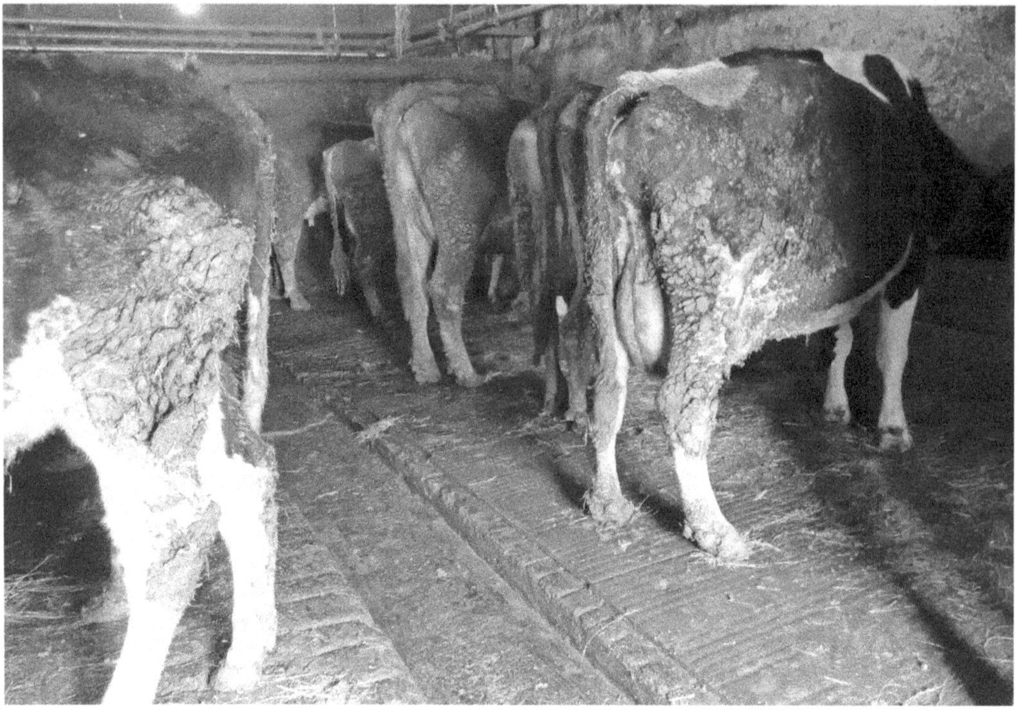

**Figure 4.8:** Dirty pens with excess of water in lairage for cattle

**Figure 4.9:** Resting behaviour at lairage for cattle

the case of drinking valves) or the total surface area of water supplied (in the case of water troughs) per animal, their functionality and cleanliness, are calculated. Cleanliness is scored as follows: 0 – clean: drinkers and water clean at the moment of inspection; 1 – partly dirty: drinkers dirty but water fresh and clean at moment of inspection; and 2 – dirty: drinkers and water dirty at moment of inspection. The availability of feed when animals are held more than 12 hours is evaluated. In cattle, feed provision is assessed by asking staff about the type and amount (< 2000 g per animal is considered insufficient) of feed and checking the time animals are fed. The scoring is: 0 – no evidence of feed provision; 1 – some evidence of feed provision; 2 – clear evidence of feed provision in sufficient quantity. Another important criterion is thermal comfort, especially in pigs (Velarde and Dalmau, 2012). Behavioural thermoregulation measures, such as huddling, shivering or panting are scored by using a 3-point scale: 0 – no pigs in the pen showing shivering, panting or huddling; 1 – up to 20% of pigs in the pen with the above behaviour; and 2 – more than 20% of pigs in the pen with the above behaviour. Huddling is described as a pig lying with more than 50% of its body in contact with another pig, but virtually lying on top of another pig. Panting is also a suitable indicator of thermal stress in poultry and rabbits. In broilers in cages, rectal temperatures at intervals throughout lairage can assess heat stress indicated by a body temperature above 42°C. Finally, other measures to be considered are mortality rates in lairage and space allowance (for this purpose, the length and width of pens are measured and the number of animals counted), and the presence and amount of bedding material, if needed. In this case, the percentage of pens with suitable rubber, straw, wood shavings or sawdust is considered (Velarde and Dalmau, 2012).

## 4.10 REFERENCES


Aaslying, M.D., Barton Gade, P. 2001. Low stress pre-slaughter handling effect of lairage time on the meat quality of pork. *Meat science* 57: 87–92.

Averós, X., Knowles, T.G., Brown, S.N., Warris, P.D., Gosálvez, L.F. 2008. Factors affecting the mortality of pigs being transported to slaughter. *Veterinary Record* 163: 386–390.

Beattie V.E., Burrows, M.S., Moss, B.W., Weatherup, R.N. 2002. The effect of food deprivation prior to slaughter on performance, behaviour and meat quality. *Meat science* 62: 413–418.

Bertol, T.M., Ellis, M., Ritter, M.J., McKeith, F.K. 2005. Effect of feed withdrawal and handling intensity on longissimus muscle glycolytic potential and blood measurements in slaughter weight pigs. *Journal of Animal Science* 83: 1536–1542.

Bradshaw, R.H., Parrott, R.F., Goode, J.A., Lloyd, D.M., Rodway, R.G., Broom, D.M. 1996. Behavioural and hormonal responses of pigs during transport: effect of mixing and duration of journey. *Animal science* 62: 547–554.

Brown, S.N., Knowles, T.G., Edwards, J.E., Warriss, P.D. 1999. Relationship between food deprivation before transport and aggression in pigs held in lairage before slaughter. *Veterinary Record* 145: 630–634.

Chevillon, P. 1994. Le controle des estomacs de porcs à l'abbatoir: miroir de la mise à jeun en élevage. *Techni-porc* 17: 23–30.

Cockram, M.S., Corley, K.T.T. 1991. Effect of pre-slaughter handling on the behaviour and blood composition of beef cattle. *British Veterinary Journal* 147: 444–454.

Dahlborn, K., Holtenius, K. 1990. Fluid absorption from the rumen during rehydration in sheep. *Experimental Physiology* 75: 45–55.

DEFRA 2003. *Code of recommendations for the welfare of livestock. Sheep.* London, UK: Defra Publications.

Eldridge, G.A., Warner, R.D., Winfield, C.G., Vowles, W.J. 1989. Pre-slaughter management and marketing systems for cattle in relation to improving meat yield, meat quality and animal welfare. Melbourne, Australia: Department of Agriculture and Rural Affairs.

Fàbrega, E., Puigvert, X., Soler, J., Tibau, J., Dalmau, A. 2013. Effect of on farm mixing and slaughter strategy on behaviour, welfare and productivity in Duroc finished entire male pigs. *Applied Animal Behaviour Science* 143: 31–39.

Faucitano, L. 2001. Causes of skin damage to pig carcasses. *Canadian Journal of Animal Science* 81: 39–45.

Faucitano, L. 2010. Invited review: Effects of lairage and slaughter conditions on animal welfare and pork quality. *Canadian Journal of Animal Science* 90: 461–469.

Faucitano, L., Saucier, L., Correa, J.A., Méthot, S., Giguère, A., Foury, A., Mormède, P., Bergeron, R. 2006. Effects of feed texture, meal frequency and pre-slaughter fasting on carcass and meat quality, and urinary cortisol in pigs. *Meat Science* 74: 697–703.

Fernandez, X., Tornberg, E. 1991. A review of the causes of variation in muscle glycogen content and ultimate pH in pigs. *Journal of Muscle Food* 2: 209–235.

Fraqueza, M.J., Roseiro, L.C., Almeida, J., Matias, E., Santos, C. and Randall, J.M. 1998. Effects of lairage temperature and holding time on pig behavior and on carcass and meat quality. *Applied Animal Behavioural Science* 60: 317–330.

Friend, T.H., Knabe, D.A., Tanksley, T.D., Jr., 1983. Behavior and performance of pigs grouped by three different methods at weaning. *Journal of Animal Science* 57: 1406–1411.

Geverink, N.A., Engel, B., Lambooij, E., Wiegant, V.M. 1996. Observations on behaviour and skin damage of slaughter pigs and treatment during lairage. *Applied Animal Behavioural Science* 50: 1–13.

Gispert, M., Guàrdia, M.D., Diestre, A. 1996. La mortalidad durante el transporte y la espera en porcinos destinados al sacrificio. *Eurocarne* 45: 73–79.

Gispert, M., Faucitano, L., Guárdia, M.D., Oliver, M.A., Coll, C., Siggens, K., Harvey, K., Diestre, A. 2000. A survey on pre-slaughter conditions, halothane gene frequency, and carcass and meat quality in five Spanish pig commercial abattoirs. *Meat Science* 55: 97-106.

Grandin, T. 1979. Designing meat packing plant handling facilities for cattle and hogs. *Transactions of the American Society of Agricultural Engineers* 22: 912–917.

Grandin, T. 1980. Designs and specifications for livestock handling equipment in slaughter plants. *International Journal of the Study of Animal Problems* 1: 178–299.

Grandin, T. 1994. Farm animal welfare during handling, transport and slaughter. *Journal of the American Medical Association* 204: 372–377.

Grandin, T. 1998a. The feasibility of using vocalization scoring as an indicator of poor welfare during cattle slaughter. *Applied Animal Behavioural Science* 56: 121–128.

Grandin, T. 1998b. Objective scoring of animal handling and stunning practices in slaughter plants. *Journal of the American Veterinary Medical Association* 212: 36–93.

Grandin, T. 2000. Chapter 20. Handling and welfare of livestock in slaughter plants. In: T. Grandin (Ed.). *Livestock handling and transport* (pp. 409–439). Fort Collins, CO: CAB International.

Grandin, T. 2010. Auditing animal welfare at slaughter plants. *Meat Science* 86: 56–65.

Gregory, N.G. 1998. *Animal welfare and*

*meat science*. Wallingford, UK: CAB International.

Guardia, M.D., Estany, J., Balasch, S., Oliver, M.A., Gispert, M., Diestre, A. 2004. Risk assessment of PSE conditions due to pre-slaughter conditions and RYR1 gene in pigs. *Meat Science* 67: 471–478.

Guardia, M.D., Estany, J., Balasch, S., Oliver, M.A., Gispert, M., Diestre, A. 2005. Risk assessment of DFD meat due to pre-slaughter conditions in pigs. *Meat Science* 70: 709–716.

Guarnieri, P.D., Soares, A.L., Olivo, R., Schneider, J.P., Macedo, R.M., Ida, E.I., Shimokomaki, M. 2004. Preslaughter handling with water shower spray inhibits PSE (Pale, Soft, Exudative) broiler breast meat in a commercial plant. Biochemical and ultrastructural observations. *Journal of Food Biochemistry* 28: 269–277.

Hemsworth, P.H., Rice, M., Karlen, M.G., Calleja, L., Barnett, J.L., Nash, J., Coleman, G.J. 2011. Human–animal interactions at abattoirs: relationships between handling and animal stress in sheep and cattle. *Applied Animal Behavioural Science* 135: 24–33.

Honkavaara, M. 1989. Influence of lairage on blood composition of pig and on the development of PSE pork. *Journal of Agricultural Science of Finland* 61: 433–440.

Hunter, R.R., Mitchell, M.A, Carlisle, A.J., Quinn, A.D., Kettlewell, P.J, Knowles, T.G., Wariss, P.D. 1998. Physiological responses of broilers to pre-slaughter lairage: Effects of the thermal micro-environment? *British Poultry Science*, Vol. 39. pp. 53–54.

Jacob, R.H., Pethick, D.W., Ponnampalam, E., Speijers, J., Hopkins, D. L. 2006. The hydration status of lambs after lairage at two Australian abattoirs. *Australian Journal of Experimental Agriculture* 46: 909–912.

Jones, E.K.M., Wathes, C.M., Webster, A.J.F. 2005. Avoidance of atmospheric ammonia by domestic fowl and the effect of early experience. *Applied Animal Science Behaviour* 90: 293–308.

Kanitz, E., Tuchscherer, W.O.M. 2005. Central and peripheral effects of repeated noise stress on hypothalamic–pituitary–adrenocorticalaxis in pigs. *Livestock Production Science* 94: 213–224.

Kephart, K.B., Mills, E.W. 2005. Effect of withholding feed from swine before slaughter on carcass and viscera weights and meat quality. *Journal of Animal Science* 83: 715–721.

Kilgour, R. 1988. Behaviour in the pre-slaughter and slaughter environments. In: Proceedings of the International Congress of Meat Science and Technology, Part A. Brisbane, Australia, Pp. 130–138.

Knowles, T.G. 1998. A review of the road transport of slaughter sheep. *Veterinary Record* 143: 212–219

Knowles, T.G., Warriss, P.D., Brown, S.N., Kestin, S.C., Rhind, S.M., Edwards, J.E., Anil, M.H., Dolan, S.K. 1993. *Veterinary Record* 133: 287

Kristensen, H.H., Wathes, C.M. 2000. Ammonia and poultry: A review. *World Poultry Science Journal* 56: 235–243.

Kristensen, H.H., Burgess, L.R., Demmers, T.G.H., Wathes, C.M. 2000. The preferences of laying hens for different concentrations of ammonia. *Applied Animal Behaviour Science* 68: 307–318.

Liste, G., Villaroel, M., Chacon, G., Sanudo, C., Olleta, J.L., Garcia-Belenguer, S., Alierta, S., Maria, G.A 2009. Effect of lairage duration on rabbit welfare and meat quality. *Meat Science* 82: 71–76.

López-Bote, C.J. 2001. Efecto de la alimentación sobre la composición y atributos de la calidad de la carne (Cap 6). In: S. Martin Bejaramo (Ed.). *Enciclopedia de la carne y de los productos cárnicos*. Vol I: 189–201. Plasencia, Spain: Martin and Macias.

MAFF. 1998. *Guide to the alleviation of thermal stress in poultry in lairage*. London, UK: MAFF Publications.

Manteuffel, G., Puppe, B., Schon, P.C. 2004. Vocalization of farm animals as a measure of welfare. *Applied Animal Behaviour Science* 88: 163–182.

Morris, D. G. 1994. Literature review of welfare aspects and carcass quality effects in the transport of cattle, sheep and goats (LMAQ.011). Report prepared by the Queensland Livestock and Meat Authority Spring Hill, Queensland, Australia.

Nanni Costa, L., Lo Fiego, D.P., Dall'Ollio, S., Davoli, R., Russo, V. 2002. Combined effects of pre-slaughter treatments and lairage time on carcass and meat quality in pigs of different halothane genotype. *Meat Science* 61: 41–47.

Nijdam, E., Arens, P., Lambooij, E., Decuypere, E., Stegeman, J.A. 2004. Factors influencing bruises and mortality of broilers during catching, transport and lairage. *Poultry Science* 83: 1610–1615.

Oba, A., Almeida, M., Pinheiro J.W., Ida, E.I., Machi, D.F., Soares, A.L., Shimokomaki, M. 2009. The effect of management of transport and lairage conditions on broiler chicken breast meat quality and DOA (death of animal). *Brazilian Archives of Biology and Technology* 52: 205–211.

Pérez, M.P., Palacio, J., Santolaria, M.P., Aceña, M.C., Chacón, G., Verde, M.T., Calvo, J.H., Zaragoza, P., Gascón, M., Garcia-Belengué, S. 2002. Influence of lairage time on some welfare and meat quality parameters in pigs. *Veterinary Research* 33: 239–250.

Petracci, M., Bianchi, M., Cavani, C., Gaspari, Lavazza, P. (2006) Preslaughter mortality in broiler chickens, turkeys, and spent hens under commercial slaughtering. *Poultry Science* 85: 1660–1664.

Rabaste, C., Faucitano, L., Saucier, L., Foury, D., Mormède, P., Correa, J. A., Giguère, A., Bergeron, R. 2007. The effects of handling and group size on welfare of pigs in lairage and its influence on stomach weight, carcass microbial contamination and meat quality variation. *Canadian Journal of Animal Science* 87: 3–12.

Ritter, L.A., Xuc, J., Dial, G.D., Morrison, R.B., Marsh, W.E. 1999. Prevalence of lesions and body condition scores among female swine at slaughter. *Journal of the American Veterinary Medical Association* 214: 525–528.

Ritter, M.J., Ellis, M., Brinkman, J., DeDecker, J.M., Keffaber, K.K., Kocher, M.E., Peterson, B.A., Schlipf, J.M., Wolter, B.F. 2006. Effect of floor space during transport of market-weight pigs on the incidence of transport losses at the parking plant and the relationships between transport conditions and losses. *Journal of Animal Science* 84: 2856–2864.

Santos, C., Almeida, J.M., Matias, E.C., Fraqueza, M.T., Rosaro, C., Sardina, L. 1997. Influence of lairage environmental conditions and resting time on meat quality in pigs. *Meat Science* 45 (2): 253–262.

Shen, Q.W., Means, W.J., Thompson, S.A., Underwood, K.R., Zhu, M.J., McCormick, R.J., Ford, S.P., Du, M. 2006. Pre-slaughter transport, AMP-activated protein kinase, glycolysis, and quality of pork loin. *Meat Science* 74: 388–395.

Spensley, J.C., Wathes, C.M., Waran, N.K., Lines J.A. 1995. Behavioural and physiological responses of piglets to naturally ocurring sounds. *Applied Animal Behaviour Science* 44: 277.

Talling, J.C., Waran, N.K., Wathes, C.M., Lines, J.A. 1996. Behavioural and physiological responses of pigs to sound. *Applied Animal Behaviour Science* 48: 187–202.

Talling, J.C., Waran, N.K., Wathes, C.M., Lines, J.A. 1998. Sound avoidance by domestic pigs depends on characteristics of the signal. *Applied Animal Behaviour Science* 58: 255–266.

Terlouw, E.M.C., Arnould, C., Auperin, B., Berri, C., Le Bihan-Duval, E., Deiss, V., Lefevre, F., Lensink, B.J., Mounier, L., 2008. Pre-slaughter conditions, animal stress and welfare: current status and possible future research. *Animal* 50: 1501–1517.

Thompson, J., O'Halloran, W., McNeil, L.D.,

Jackson-Hope, N., May, T. 1987. The effect of fasting on live weight and carcass characteristics in lambs. Meat *Science* 20: 293–309.

Tuchscherer, M., Manteuffel, G., 2000. The effect of psycho stress on the immune system. Another reason for pursuing animal welfare (Review). *Archiv für Tierzucht* (*Archives of Animal Breeding*) 43: 547–560.

Turner, S.P., Farnworth, M.J., White, I.M.S., Brotherstone, S., Mendl, M., Knap, P., Penny, P., Lawrence, A.B., 2006. The accumulation of skin lesions and their use as a predictor of individual aggressiveness in pigs. *Applied Animal Behaviour Science* 96: 245–259.

Van de Perre, V., Ceustermans, A., Leyten, J., Geers, R. 2010. The prevalence of PSE characteristics in pork and cooked ham – Effects of season and lairage time. *Meat Science* 86: 931–937.

Velarde, A. 2008. How can lairage conditions be improved? Animal welfare at slaughter and killing for disease control – Emerging issues and good examples (pp.31–36). Hindåsgården (Sweden), 1–3 October 2008.

Velarde, A., Dalmau, A. 2012. Animal welfare assessment at slaughter in Europe: Moving from inputs to outputs. *Meat Science* 92: 244–251.

Walker, P., Warner, R., Weston, P., Kerr, M. 1999. Effect of washing cattle pre-slaughter on glycogen levels of m. semitendinosus and m. semimembranosus. In: International Congress of Meat Science and Technology (Paper n° 4-P27, pp. 289–299). Yokohama, Japan.

Warner, R.D., Truscot, T.G., Eldridge, G.A., Franz, P.R. 1998. A survey of the incidence of high pH beef meat in Victorian abattoirs. In: 34th International Congress of Meat Science and Technology (pp. 150–151). Brisbane, Australia.

Warriss, P.D. 1993. Ante mortem factors which influence carcass shrinkage and meat quality. Proceedings of 39th International Congress of Meat Science and Technology (pp. 51–65), Calgary, Canada. Agriculture Canada: Ottawa, Canada.

Warriss, P.D. 1994. Ante mortem handling of pigs. In: Cole, D.J.A., Wiseman, J. Varley, M.A. (Eds.). *Principles of pig science* (pp. 425–432). Nottingham, UK: Nottingam University Press.

Warriss, P. D. 2003. Optimal airage times and conditions for slaughter pigs: a review. Vet. Rec. 153: 170–176.

Warriss, P. D., Brown, S. N., Edwards, J. E., Anil, M. H. and Fordham, D. P. 1992. Time in lairage needed by pigs to recover from the stress of transport. *Veterinary Record* 131: 194–196.

Warriss, P.D., Brown, S.N., Edwards, J.E., Knowles, T.G. 1998. Effect of lairage time on levels of stress and meat quality in pigs. *Animal Science* 66: 255–261.

Wathes, C.M., Jones, C.D.M., Webster, A.J.F. 1983. Ventilation, air hygiene and animal health. *Veterinary Record* 113: 554–559.

Wathes, C.M., Jones, J.B., Kristensen, H.H., Jones, E.K.M., Webster, A.J.F. 2002. Aversion of pigs and domestic fowl to atmospheric ammonia. *American Society of Agricultural and Biological Engineers* 45: 1605–1610.

Watts, J.M., Stookey, J.M. 1998. Effects of restraint and branding on rates and acoustic parameters of vocalization in beef cattle. *Applied Animal Behaviour Science* 62: 125–135.

Weeks, C.A. 2008. A review of welfare in cattle, sheep, and pig lairages, with emphasis on stocking rates, ventilation and noise. *Animal Welfare* 17: 275–284.

Weeks, C.A., Brown, S.N., Lane, S., Heasman, L., Benson, T., Warriss, P.D. 2009. Noise levels in lairages for cattle, sheep, and pigs in abattoirs in England and Wales. *Veterinary Record* 165: 308–314.

Welfare Quality. 2009a. Welfare Quality assessment protocol for pigs (sows and piglets, growing and finishing pigs).

Lelystad, Netherlands: Welfare Quality Consortium.

Welfare Quality. 2009b. Welfare Quality assessment protocol for cattle. Lelystad, Netherlands: Welfare Quality Consortium.

Wythes, J. R., Shorthose, W.R. 1984. Marketing cattle: Its effects on liveweight, carcasses and meat quality. Australian Meat Research Corporation Review No. 46., Sydney, Australia: Australian Meat Research Committee.

Young, J.F., Bertram, H.C., Oksbjerg, N. 2009. Rest before slaughter ameliorates preslaughter stress-induced increased drip loss but not stress-induced increase in the toughness of pork. *Meat Science* 83: 634–641.

chapter five

# Practical methods to improve animal handling and restraint

*Temple Grandin*[1]

**Learning objectives**

- Provide an understanding of basic behavior principles of moving cattle, pigs, and sheep such as flight zone and point of balance.
- Learn how to make simple modifications of races, stun boxes, and alleys to improve animal movement.
- Correct mechanical problems with restraint devices that cause stress, injury, or bruises.
- Be able to identify on-farm factors that are associated with handling problems at the abattoir.

## CONTENT



[1] Department of Animal Sciences, Colorado State University, Fort Collins, CO 80523-1171, USA

## 5.1 INTRODUCTION

Cattle, pigs, and sheep will move more easily through the lairage and races in an abattoir when people understand and use certain species-specific behavioral principles. This will improve both animal welfare and meat quality. If animals constantly balk, refuse to move, turn around or back up in the race, slaughterhouse employees are more likely to use force and harsh methods such as multiple shocks with electric goads to move them. Multiple shocks from an electric goad during the last few minutes before stunning or cattle becoming agitated shortly before slaughter resulted in tougher beef (Warner et al., 2007; Gruber et al., 2010; Ferguson and Warner, 2008). In pigs, rough handling, electric goads, or jamming in the single file race resulted in higher blood lactate levels and poorer meat quality (Dokmanovic et al., 2014; Edwards et al., 2010a,b). Chapter 14 in this book presents a complete review of the effects of pre-slaughter stress, on meat quality.

## 5.2 BASIC BEHAVIORAL PRINCIPLES OF ANIMAL HANDLING

### *5.2.1 Flight zone effects on behavior*

People working with animals need to understand the flight zone (Grandin, 1980; Grandin and Deesing, 2008). An animal that is trained to lead has no flight zone. Wild or extensively raised animals may have a large flight zone. If people stand inside the flight zone, the animals may become agitated. They may react by moving to the back of a lairage pen, rearing up in a race, or jumping and constantly moving while standing in the single file race. Handlers should back up and remove themselves from the flight zone if animals become agitated. The animal will usually become calmer and less active when the person backs away. When a person stands close to a high flight zone animal, the installation of a solid barrier will reduce reactivity (Muller et al., 2008).

### *5.2.2 Point of balance principles for moving cattle, pigs, and sheep*

When cattle, pigs, or sheep are being moved through a single file race, the point of balance is at the shoulder (Figure 5.1; Grandin, 1980, 1993). The animal will move forward when a person

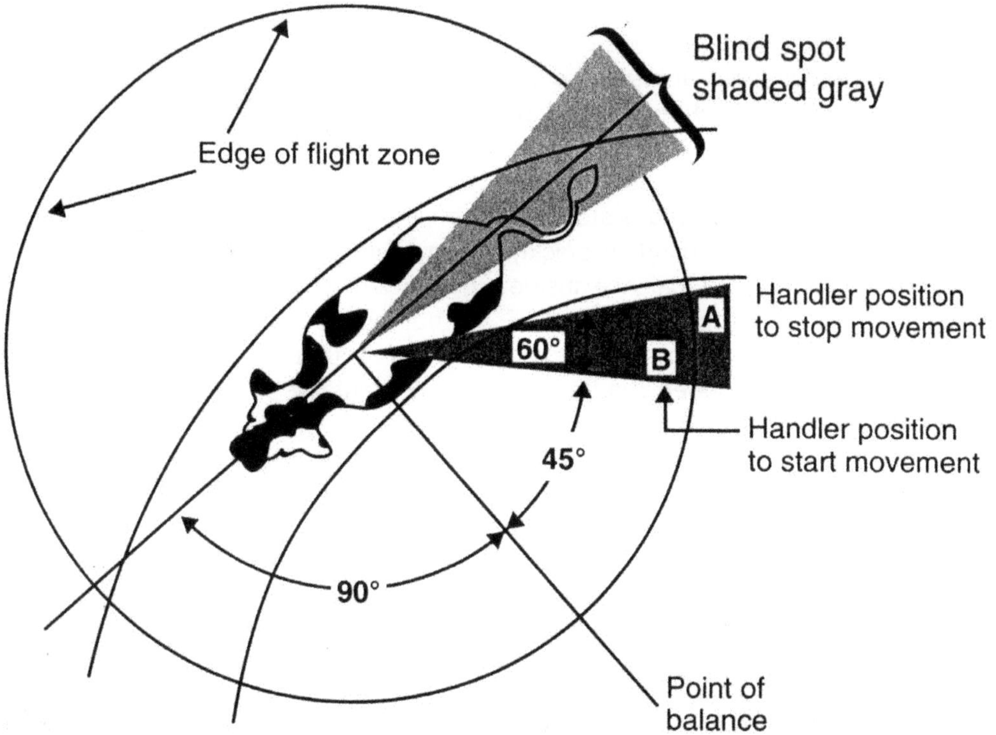


**Figure 5.1:** Flight zone diagram. Handlers need to understand that the size of the flight zone varies depending on the wildness or tameness of the animal. If an animal rears up in a race or becomes agitated it is usually due to a person standing too close to it. The person should back up and get out of the flight zone. To move an animal forward, a handler must be behind the point of balance

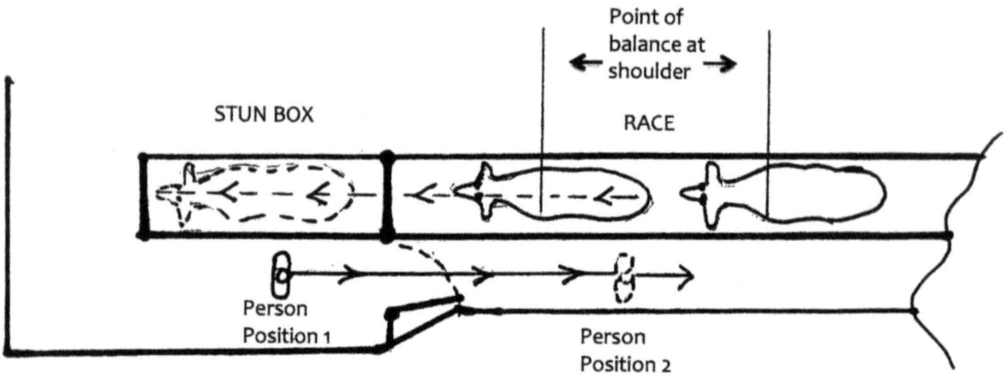


**Figure 5.2:** Animals will move forward when the handler walks back by them in the opposite direction of desired movement. The handler moves from position 1 to position 2 to move one animal into the stun box. The handler passes the point of balance at the shoulder on the first animal, but stops in front of the point of balance of the second animal to stop it from moving forward

is behind the point of balance and move backwards if a person stands in front of the point of balance (Kilgour and Dalton, 1984). A common mistake made by handlers when moving an animal through a single file race is to stand in front of the point of balance and poke it on the rear. Figure 5.2 shows an easy method to move a single animal into a stunning box or restrainer. The handler quickly walks past the point of balance inside the flight zone in the opposite direction of desired movement. The animal will usually move forward when the handler moves past the point of balance at the shoulder.

When animals are being moved out of a lairage pen, the point of balance may move forward and be slightly past the eye. When animals are being removed from a lairage pen, the same principle that is used for moving animals through the single file race can be used. The handler moves inside the flight zone in the opposite direction of desired movement. When the handler moves along one side of a lairage pen from the entrance gate towards the back of the pen, the animals will usually move towards the gate when the handler walks past them.

### *5.2.3 Animals go back to where they came from*

Cattle, pigs, and sheep all have a natural behavior to go back to where they came from. This is why animals will sometimes get back on a truck if the lairage pen gate is not opened or illuminated. Handling facilities with a full half circle crowd pen take advantage of this natural behavior. Figures 5.3, 5.4, and 5.5 show a round crowd pen and race systems for cattle, sheep, and pigs. Figure 5.3 is currently being used in a cattle slaughterhouse that processes approximately 40 cattle per hour. A sheep handling system similar to that in Figure 5.4 is being used in a British abattoir (Bates et al., 2014). The author visited both of these slaughterhouses and the animals move quietly and easily through these systems (Grandin 1997).

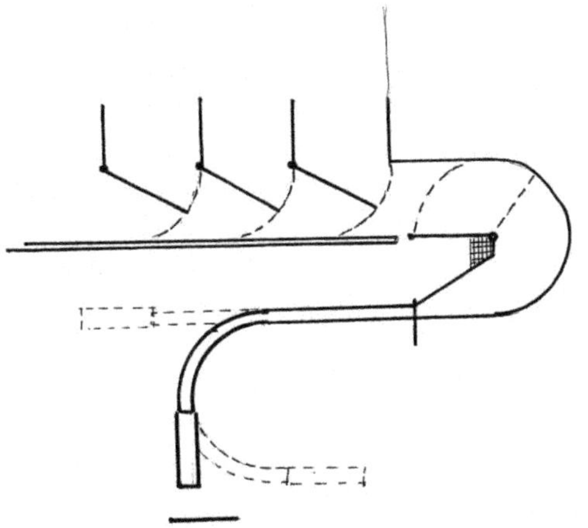

**Figure 5.3:** System for moving cattle to the stun box. This features a full half circle round crowd pen that takes advantage of the natural behavior of cattle to go back to where they came from. When cattle enter the crowd pen, the gate is moved from position 1 to position 2. The handler moves to the small catwalk located at the gate pivot point. The cattle will circle around the handler to enter the single file race. This system will easily handle 30 to 60 cattle per hour

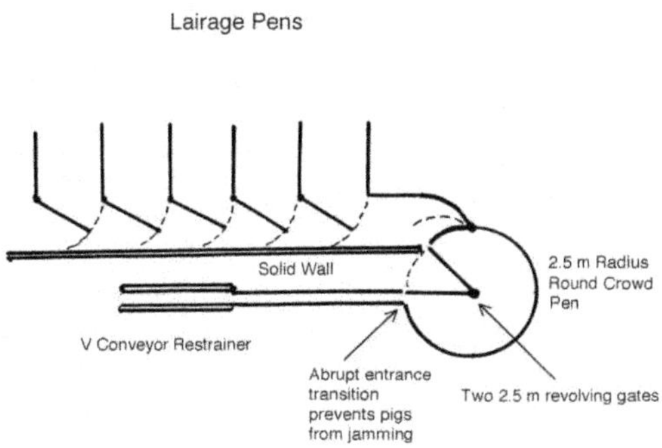


**Figure 5.4:** System for moving pigs through a single file race to an electric stunner. Crowd pens for pigs must have an abrupt entrance because pigs will jam in a funnel. This system also takes advantage of the natural instinct of pigs to go back to where they came from and it has two gates that continuously revolve

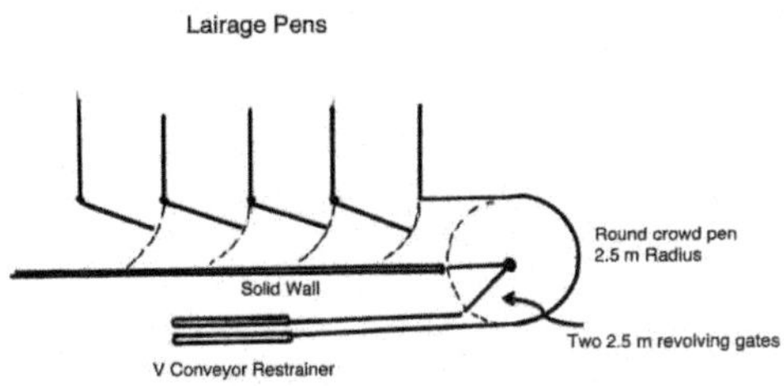


**Figure 5.5:** System for moving sheep with two gates that continuously revolve. A funnel-shaped entrance will work well for sheep and the half circle crowd pen takes advantage of the natural tendency of sheep to go back to where they came from

### *5.2.4 Lone animal becomes stressed*

All species may become highly stressed when they are alone and separated from their herd or flock mates (Bates et al., 2014; Grandin, 2014). Animals that are last in line in a race may be more stressed (Stockman et al., 2012). This is especially a problem in very small slaughterhouses because the last animal in line may spend five to ten minutes alone. A lone animal is also more likely to injure itself or become agitated and run over people in an attempt to return to its herd mates. Single lone cattle are a major cause of human injuries (Grandin, 2014). People should not get in a small confined area with a single agitated bovine.

### *5.2.5 Behavioral signs of fear and stress*

Animals definitely can experience fear (LeDoux, 1996; Rogan and LeDoux, 1996; Panksepp, 2011). One indicator of fear is when cattle show visible eye white (Core et al., 2009; Janczak and Braastad, 2004). Other signs of behavioral agitation are defecation (Romeyer and Bouissou, 1992), and tail twitching in cattle. In cattle and pigs, vocalizations (moos, bellows, squeals) during handling and restraint at a slaughterhouse are indicators of stress (Dunn, 1990; Hemsworth et al., 2011; Warriss et al., 1994). Cattle vocalizations (moo or bellow) were associated with either equipment problems, rough handling, excessive pressure from a restraint device, electric goads, gates slammed on cattle, or other obvious aversive events (Grandin, 1998, 2011; Bourquet et al., 2010). The number of animals vocalizing was counted in the stun box and during entry into the stun box. Vocalizations of cattle waiting in the lairage were not tabulated.

### *5.2.6 Effects of previous experiences and genetics on handling*

Animals with a more excitable temperament will often become more agitated when suddenly exposed to a novel situation compared to animals with a calmer temperament (Grandin and Deesing, 2013). Animals that become more agitated and temperamental on the farm may become more agitated and have stronger physiological responses when they arrive at the slaughterhouse (Bourquet et al., 2011). Bourquet et al. (2010) and Lewis et al. (2008) stated that the animals may be reacting to the novelty of an environment of either a slaughterhouse or farm handling system. Working with young, extensively raised cattle to get them accustomed to handling will help produce animals that will be easier to handle in the future (Fordyce 1987; Fordyce et al., 1988; Becker and Lobato, 1997; Binstead, 1977; Probst et al., 2013).

## 5.3 METHODS FOR IMPROVING HANDLING AND ANIMAL MOVEMENT

The first step in improving animal movement is to correct mistakes that people make while handling and/or moving animals. Training employees and stopping obvious handling mistakes will make it possible to determine if the problems with animal movement in a particular slaughterhouse are due to people making mistakes or a true fault with the design, layout and/or construction of stun box, race, lairage, or unloading ramp.

### *5.3.1 Step 1: correct handling mistakes and train employees*

- *Use flight zone and point of balance principles.* These principles were described in sections 5.2.3 and 5.2.4.
- *Stop all yelling and whistling; yelling is highly stressful to livestock* (Hemsworth et al., 2011; Waynert et al., 1999; Pajor et al., 2003).
- *Stop constant use of electric goads.* They should never be a person's primary driving tool. A flag, paddle, pig board, or other materials should be a person's primary driving aid. An electric goad should only be picked up when animals refuse to move and then put away. In most abattoirs, the only place an electric goad might be needed is at the entrance of the stun box or conveyor restrainer. Driving aids must never be applied to sensitive parts of the animals such as genitals, rectum, eyes, or nose.

- *Fill the crowd pen half full with cattle or pigs.* Animals need room to turn if they get turned around in the wrong direction. This will require moving smaller bunches of cattle or pigs. Good handling will require more walking to go back and forth between the crowd pen and the lairage.

### 5.3.2 Step 2: Move cattle and pigs in small groups and use following behavior

Take advantage of the natural behavior of animals to follow the leader. For cattle and pigs, handlers should be trained to wait until the single file race is almost empty before bringing the next small group of pigs or cattle into the crowd pen that leads to the race. The animals will follow the leader into the race and pass through the crowd pen without stopping. If the crowd pen is filled with pigs or cattle when the single file race is full, they may turn around and attempt to return to the lairage because they have no place to go. The crowd pen should be used as a "passing through" pen. People handling animals are often reluctant to wait until the single file race is partially empty to take advantage of following behavior because they worry that the stun box may run out of animals. Several weeks of practice and training may be required to train handlers to determine the exact timing of bringing up the next bunch of animals from the lairage.

### 5.3.3 Step 3: Move sheep in continuous flow

- Sheep have very strong following behavior. They can be handled in larger groups in continuous flow. They will follow when they see other sheep ahead. Following behavior is so strong in sheep that a person may have difficulty holding a sheep to prevent it from following other sheep.
- *Leader sheep.* The use of trained leader sheep can facilitate unloading of sheep off trucks and movement up the races. The author has observed that leader sheep are most effective if each leader is trained for a specific job. There is the Judas lamb that leads the lambs up the stunning race and then another specialist that goes on the trucks and helps unload the sheep. For biosecurity reasons, the leaders should live in the lairage. New leaders are trained by having a ewe train her lamb.

## 5.4 SIMPLE MODIFICATION OF FACILITIES TO IMPROVE ANIMAL MOVEMENT

When animals refuse to move easily through a handling facility, a common response from many slaughterhouse managers is to suggest tearing out the entire facility and making major modifications. In many slaughterhouses, simple changes will often improve animal movement. Making many small changes to both the facility and animal handling methods can often provide big improvements. The author has worked with many older slaughterhouses, and in a high percentage of them, greatly improving animal movement did not require rebuilding the entire facility.

**Table 5.1** Common distractions that may cause animals to balk and refuse to move through races, alleys, or into stun box or restrainer

| Distractions | Remedy |
|---|---|
| Reflections on shiny metal. | Move overhead lamps or dull the surface of the metal to eliminate reflections. Another option is to experiment with turning off some lights. |
| Air blowing into the faces of approaching animals. | Change air flow direction. |
| Seeing people walking by or moving equipment. | Install a solid shield, experiment with cardboard. |
| Chains hanging down in a race, hose on floor, coat on fence. | Remove them. |
| Gates that jiggle. Especially a problem at stunbox/restrainer entrance or single file race entrance. | Stop gate movement. |
| High contrast walls or changes in flooring type. | Make the floor or walls all the same material and color. |
| Stun box, restrainer entrance, or race entrance is too dark. | Add indirect lighting. Animals will move towards the light. Experiment with portable lights. |
| Flapping paper towels. | Remove them. |
| Vertical sliding stun box entrance door is too low and animals bump their shoulders. | Raising the bottom edge of the door will facilitate entry. Cattle may enter more easily if the door provides 30 cm or more clearance above the shoulder. |

### *5.4.1 Remove distractions that make animals balk*

A calm animal will stop and point its eyes and ears towards distractions that may cause it to stop moving. To locate a distraction that is causing balking, a person should look up the alleys, races, and stunbox/restrainer entrances to locate the distractions (Grandin and Deesing, 2008). The person should get their eyes at animal eye level. Table 5.1 includes a list of common distractions.

### *5.4.2 Remove or tie open backstop (anti back-up gates)*

When animals are constantly backing up in a single file race or refusing to move forward, a common mistake is to add additional one-way backstop gates to prevent animals from backing up.

**Table 5.2** Case studies of trouble-shooting handling problems in two slaughterhouses: list of modifications to improve animal movement

| Slaughterhouse 1 | Slaughterhouse 2 |
|---|---|
| • Moved smaller bunches of animals<br>• Installed light on stun box entrance<br>• Removed paper towel dispenser with waving towel<br>• Raised height of a vertical slide door at the stunning race entrance so cattle would not bump their backs | • Moved smaller bunches of animals<br>• Installed light on stun box entrance<br>• Blocked sunbeams in alley with plastic sheet<br>• Used cardboard shields to prevent approaching animals from seeing people<br>• Changed sequence of opening and closing alley gates between the lairage and the single file race |

In many facilities, the author has eliminated distractions and has been able to tie open most of the backstops. If animals are backing up, there is a problem that needs to be corrected. Adding more backstops is treating the symptom of a handling problem and not the cause. One-way backstop gates at race entrances are a major cause of balking and refusing to enter. Equip a backstop at the entrance of a race with a remote control to hold it open for entering animals. This may be either a simple rope or a powered device. This will facilitate the entry of cattle or pigs.

### *5.4.3 Experiment with cardboard and portable lights*

Try adding, moving, and turning off lamps. Many times the author has gone into an abattoir and improved handling by experimenting with portable lamps and large pieces of cardboard. A lamp installed at either a race or stun box entrance improved animal movement (Grandin 1982, 2001, 2010; van Putten and Elshof, 1978). Cardboard can be used to prevent approaching animals from seeing people, moving conveyors, and other distractions. Experiment with cheap materials to avoid the cost of expensive modifications that are not necessary. Table 5.2 shows two case studies at two different slaughterhouses. Both slaughterhouses were older facilities. To have easy, calm animal movement may require locating and fixing five or six problems.

### *5.4.4 Animals refuse to enter the stun box, or conveyor restrainer*

*Air currents.* The most common cause of balking is air currents blowing towards approaching animals (Grandin, 1996). Air curtains and fans blowing on livestock to keep flies out may cause animals to back up and refuse to move forward.

*Seeing people through the front of the box or restrainer.* Animals will not enter the stun box or restrainer, if people or moving equipment is visible through the front of a stun box head holder (restrainer) or through the end of a conveyor restrainer (Grandin, 1996). A wall or curtain should be installed to block vision of people or equipment. This should have a blank white surface and be placed about 1.2 meters in front of the head holder.

*Objects that move.* Cattle, pigs, and sheep are very sensitive to small things that move. Some of the common ones that make animals balk are steel floors that jiggle, plastic strip curtains, and paper towels. Sometimes a very slight movement will stop the animals. Gates that jiggle may also stop movement.

*Reflections cause balking.* The increased use of shiny stainless steel equipment has increased problems with reflections that cause animals to balk and refuse to move. When new stainless steel equipment is being purchased, it should be specified to have a dull non-reflective finish. Reflections can sometimes be made to disappear by moving overhead lamps 1 meter off the centerline of either the box, race, or restrainer. Sometimes turning off lights will help. People need to experiment with adding lamps, moving existing lamps, and turning off different existing lamps.

*Animals dislike dark places.* Experiment with portable lights to light up a race, stun box, or restrainer entrance that is too dark. Try positioning the lamp in many different locations. The principle is to have indirect lighting. It must never shine directly into the eyes of approaching animals.

*Frightened by sudden noise.* Intermittent noise such as air hissing and metal clanging and banging frightens and startles animals (Grandin, 1996). Animals are very sensitive to intermittent sound stimuli or high-pitched noises (Talling et al., 1998; Lanier et al., 2000). Air hissing from valves on pneumatic equipment must be silenced. Commercially available muffling devices should be installed on air exhausts to stop sudden hissing. These devices must be periodically checked and replaced because they become worn and lose the ability to stop noise. To eliminate metal-on-metal clanging of stun box doors and sliding panels on head holders, they can be constructed from either a metal plate sliding in a plastic track or the entire door can be constructed from a solid piece of very heavy plastic board. To prevent a plastic cattle stun box door from warping or bending, extremely heavy, thick panels over 3 cm or greater thickness must be used. Plastic doors for sheep or pigs can be made from thinner plastic.

## 5.5 COMMON MECHANICAL PROBLEMS WITH STUN BOXES, HEAD HOLDERS, POWERED GATES, AND CONVEYOR RESTRAINERS THAT ARE DETRIMENTAL TO ANIMAL WELFARE

Often, very simple modifications will correct serious welfare problems that occur in stunning boxes, head holders, and other devices used to restrain cattle and other large animals. One of the most common problems is poorly designed and constructed, or worn floors that cause slipping or falling. Other common problems are bruises or injuries caused by equipment and excessive pressure applied by restraint devices (Grandin, 1981).

### 5.5.1 *Powered gates must never drag animals or knock them over*

Powered gates are sometimes used to move pigs and cattle through alleys or into equipment such as a $CO_2$ stunner. From a welfare standpoint, it is never acceptable for a powered gate to knock an animal down, cause it to fall, or drag it along the floor. These problems must be corrected.

The author has observed welfare improvements in pig and cattle slaughterhouses where powered gates were converted from fully automatic operation to semi-automatic. In these abattoirs, a person controlled forward movement of the gate against the animals by pressing a switch. After the animals had been moved, the gate automatically returned to the start position for the next group of pigs. One innovative pig slaughterhouse used powered gates in their $CO_2$ system that had a proximity switch. To move the powered gate forward, the handler grabbed the steel frame and the proximity switch turned on the gate to move it forward. The handler had to walk with the gate because it would stop when the handler let go of the frame. This system eliminated knocking over or dragging pigs, because the handler had to walk with the powered gate and hold the frame to keep the gate moving.

### 5.5.2 *Animals refuse to stand still in stun box or restrainer*

After the animal enters the box, it may fail to settle down and stand still. The most common cause of this problem is slippery floor. The animal's feet should be observed from the shackler side of the box. Animals that are restless are often slipping. Either a front foot or a back foot will be moving with repeated small rapid slips. A non-slip floor should be installed. When the box was new, the floor may have been non-slip. When the floor wore out, the animal started slipping. Steel floors with a diamond or checker pattern wear out quickly. Checker or diamond plate will not provide a sufficient non-slip floor. Below is a list of non-slip flooring alternatives for stun boxes or restrainers:

- Deeply grooved concrete.
- Steel floor with steel rods welded to it. Rods must be welded flush with the floor. Never overlap the rods.
- Rubber mats with a ribbed or deeply textured surface.
- The floor of the box should be flat. Floors that are stepped or sloped on a steep angle should not be used.

### 5.5.3 *Reducing vocalization*

When cattle or pigs vocalize in a stun box, head holder, or conveyor restrainer, this is associated with physiological measures of stress (Hemsworth et al., 2011; Dunn, 1990; Warriss et al., 1994). Vocalization such as squeals, bellows, or moos can be easily measured. Animals can be scored as either vocalizing when they are being restrained or silent. In cattle, vocalization was associated with obvious aversive events such as excessive pressure from a restraint device, sharp edges, electric goads, or slipping on the floor (Grandin, 1998, 2001; Bourquet et al., 2010; Hemsworth et al., 2011). Reducing pressure applied by a head restraint reduced vocalization in cattle from 23% of the animals to 0%.

Munoz et al. (2012) also reported that excessive pressure applied to either the body or the head increased vocalization. When head holders and other restraining devices are used correctly, the percentage of cattle or pigs that vocalize during restraint can be reduced to 5% or less (Grandin and Johnson, 2005; 2012; Barbalho, 2007). There is a problem that must be corrected if cattle or pigs vocalize in direct response to the application of a head holder or body restraint.

Vocalization scoring cannot be used with sheep to locate problems with restraint devices. Sheep vocalize when separated from the flock or when they are isolated (Deiss et al., 2009; Ligout et al., 2011). They do not vocalize in an immediate response to aversive or painful handling procedures. In the wild, a sheep's defense from predators is bunching together and an injured sheep remains silent to avoid attracting predators (Dwyer, 2004).

### *5.5.4 Pressure-limiting devices on restraint devices*

Head holders, body restraint, and other devices designed to press against an animal should be equipped with pressure-limiting devices and move steadily. Sudden jerky motion may cause animals to become agitated (Grandin, 1992). A head holder or body restraint should automatically stop applying pressure before it causes injury, bruises, or vocalization. On a typical restraint box, less air or hydraulic pressure is required to operate a head holder compared to a heavy door. One easy method to limit pressure applied by a head holder is to use smaller diameter pneumatic cylinders. Hydraulic systems will require a separate pressure relief valve set at a lighter setting for the head holder. Vocalization scoring is an easy method to detect problems with excessive pressure. If a bovine vocalizes (moo or bellow) immediately after its head or body is restrained, it is most likely due to excessive pressure, or a sharp edge or pinching.

### *5.5.5 New head holder design concept to reduce electric goad use*

When head holders are poorly designed, increased stress from electric goad use may result (Ewbank et al., 1992). Many systems have a rear pusher gate to push cattle forward into a head holder. A new design eliminates the rear pusher gate as the entire head holder is mounted on a track so it can be moved backwards towards the animal to get its head in the correct position.

### *5.5.6 Upright restraint devices are recommended*

Restraint devices that hold animals in a comfortable upright position are less stressful than inverted restraint. Sheep were more willing to enter an upright restrainer compared to one that inverted them (Rushen, 1986). The percentage of cattle vocalizing was greater when the cattle were inverted (rotated) on their backs (Velarde et al., 2014; Dunn, 1990). Restraining calves by suspending them by one back leg was more stressful than upright restraint (Westervelt et al., 1976). There are some well-designed boxes that invert cattle that can achieve a vocalization score of 5% or less. The best designs prevent the animal's body from slipping during rotation. Another advantage of upright boxes is that the cost to build them is much lower.

### *5.5.7 Correcting conveyor restrainer problems*

When a V conveyor restrainer is used for either pigs or sheep, both sides must run at the same speed. When the two conveyors run at different speeds, the animals may struggle. The angle of the two conveyors may also cause struggling if it is too steep.

A common problem with conveyor restrainers is that pigs or cattle refuse to enter. This is often caused by the visual cliff effect at the entrance. Farm animals are able to perceive the visual cliff (Lemmon and Patterson, 1964). Since the conveyor restrainer is above the floor, the entrance must be designed so the animals do not see the steep drop off. Avoid having light come up through the bottom of the conveyor. Another approach is to install a false floor that is just below where the animal's feet would touch. This provides the illusion of a solid floor to walk on (Grandin, 2003).

### *5.5.8 Prevent bruises caused by stun box doors*

Poorly designed or carelessly operated stun boxes or restraint devices can cause bruises and injuries. Bruises can also occur in unconscious animals after stunning and prior to bleeding (Meischke and Horder, 1976). In one slaughterhouse, a high percentage of bruises were caused by the vertical slide tailgate on a cattle stunning box (Strappini et al., 2013). If bruises still occur after the operator has been trained not to strike the cattle with the door, there is likely to be a mechanical problem. The door controls must be designed so that downward motion of the door will quickly stop when the operator pushes either a control valve or a button. On a poorly designed pneumatic system, the door may continue to move downward for 30 to 60 cm after the operator attempts to stop it. Pneumatic systems should be designed so that air pressure is used to raise the door but the cylinder is not pressurized to push the door downward to the closed position. If the air cylinder is powered down, the door may continue to descend after the operator attempts to stop it.

Other chapters of this book will discuss bruises that occur during transport, but the first step is to determine if bruises or injuries are occurring inside or outside the abattoir. If the bruises or injuries are happening inside the abattoir, they will occur on animals from many different farmers. If the bruises are occurring outside the abattoir, the bruises will be associated with either a certain livestock source (farm or market) or a specific transporter. The rear door of a stun box striking animals is one obvious cause of bruises. Less obvious causes are pinch points and sharp edges in a restraint device and protruding gate latches.

## 5.6 EQUIPMENT PROBLEMS THAT WILL REQUIRE MAJOR RENOVATIONS OR EQUIPMENT PURCHASES TO REMEDY

One of the worst and most expensive problems to fix is an undersized $CO_2$ stunning unit. In one slaughterhouse, to compensate for increased line speed, three pigs were forced into a gondola designed for two pigs. An electric goad was used to jump the third pig on top of the others. The only way to remedy this is to either slow down the slaughter line or buy a larger gas stunning unit to suit the throughput rate.

Problems with overloaded equipment may occur in both large and very small slaughterhouses. In another case, an abattoir that worked well at 26 cattle per hour had many problems when the speed was raised to 35 per hour. In small slaughterhouses, electric stunning of pigs on the floor can be low stress because lining up in a single file race is eliminated. At high speeds, the author has observed that floor stunning may become rough and sloppy. This is another situation where an equipment such as a conveyor restrainer would need to be installed.

## 5.7 HANDLING PROBLEMS THAT MUST BE CORRECTED AT THE FARM

The author has observed increasing problems with handling animals at a slaughterhouse that are caused by problems that must be corrected at the farm, feedlot, or ranch. These problems originate at the farm and they make good animal welfare at the abattoir extremely difficult to achieve.

### *5.7.1 Emaciated, weak animals*

Old breeding animals that have become weak, emaciated, and debilitated are extremely difficult to handle and maintaining an acceptable level of welfare is problematic. Producers should bring in sows, old dairy cows, and other breeding stock when they are still fit for travel. The author has observed that problems with animals that become non-ambulatory or have extreme difficulty walking are greatly reduced when the abattoir management charges an extra fee for handling non-ambulatory animals.

### *5.7.2 Cattle from very extensive systems are difficult and dangerous to handle*

Extensively raised cattle which have had little contact with people and are not used to being handled and restrained can be difficult and dangerous to manage during pre-slaughter handling. They will usually remain calmer if all races, stun boxes, and crowd pens have high, solid sides. Animals are specific about the things they habituate to. For example, if a horse becomes habituated and has little or no reaction to a blue and white umbrella, this does not transfer to a canvas sheet (Leiner and Fendt, 2011). The horse may become extremely agitated on seeing the flapping canvas and remain calm when seeing the umbrella. Cattle from extensive pastures that are accustomed to a person on a horse may become frightened the first time they see a person on foot at an abattoir. Cattle should be acclimated and accustomed to going in and out of pens with a person on foot before they come to an abattoir (Grandin, 2014). Cattle that are fattened in barns may be very tame and have a small flight zone when the farmer walks in the aisle. The flight zone may suddenly enlarge when the pens are entered for the first time. Farmers should get fattening cattle accustomed to people walking among them inside their pens. Cattle that are not habituated to people walking through their pen may have a huge flight zone when they are moved down an alley for truck loading. They may race back and forth in an attempt to get people out of their flight zone. At the abattoir, when they enter a lairage pen, they may suddenly "rebound" and suddenly run back out into the alley. These cattle are dangerous to handle and may knock over or injure a person. This is more likely to be a problem with animals with genetic predisposition for temperament and excitability (Grandin and Deesing, 2013). Animals learn using their sensory organs and their learning is specific to the context and circumstances (see Chapter 2 for detail). Memories are stored as specific pictures and sounds. Grandin and Johnson (2005) described a horse that feared black cowboy hats while white cowboy hats had no effect. The fear memory was very specific. Animal memories are stored as visual images (Judd and Collett, 1998; Hanggi, 2005). A person in the aisle is a different picture compared to a person inside the fattening pen. Animals need to be habituated to a person in both places before they arrive at the abattoir.

### *5.7.3 Pigs piling up during handling*

Research has shown that pigs will be easier to handle in the future if they have been handled during finishing (fattening) on the farm (Geverink et al., 1998; Abbott et al., 1997; Lewis et al., 2008). To improve ease of pig movement at the slaughterhouse, the farmer should walk through the fattening/finishing pens every day to get the pigs accustomed to quietly getting up and moving away when people walk through the pen (Grandin, 2014). Some of the most difficult to handle pigs at the abattoir were pigs that had never had a person enter their pen until the day they were loaded for market.

### *5.7.4 Lame, fatigued animals*

Intensively raised cattle and pigs are being grown to heavier weights. In the US, market weight for pigs can be as high as 140 kg. There has also been greater genetic selection for productivity. In both cattle and pigs fattened in feedlots, the author has observed mobility problems associated with poor leg conformation. In one slaughterhouse, approximately 10% of large fed feedlot cattle were lame (difficulty walking), even though they came from a dry feedlot. Another factor that has increased lameness and made animals sluggish and difficult to move is feeding animals with high doses of beta-agonists (Marchant-Forde et al., 2003). Pigs fed high doses of ractopomine were more likely to become stressed if they were handled in an aggressive manner (James et al., 2013).

### *5.7.5 Porcine stress syndrome (PSS) in pigs*

In most countries, breeding producers have selected against pigs which have porcine stress syndrome, a recessive genetic disorder. Unfortunately, there are some places where pigs exhibiting classic PSS symptoms are arriving at abattoirs. This problem must be corrected at the farm. Even though these pigs had been transported in a modern lorry with movable decks, their welfare was poor.

## 5.8 THE IMPORTANCE OF HANDLING SCORING

People manage things that are measured. Scoring of handling performance should be done with animal welfare outcome-based measures or indicators. Assessment of animal welfare is moving from specifying inputs (resources or management) to measuring welfare outcomes (Velarde and Dalmau, 2012). When handling practices are measured it makes it possible to determine if they are improving or becoming worse. This enables managers to continuously improve animal handling. The following handling variables should be measured (Grandin, 1998):

- Percentage of animals falling at any place in the facility. This includes the unloading ramp, lairage, crowd pen, leadup race, and stun box before stunning. A fall is scored if the body touches the floor and the incidence should be 1% or less (Grandin, 1998; OIE, 2014).
- Percentage of animals moved with an electric goad. Ideally, electric goads should not be used. However, certain circumstances may warrant the use of electric goads and, under this situation, the incidence should be 5% or less.
- Percentage of pigs or cattle vocalizing (bellow, moo, squeal) in the stun box or restrainer. Should be 5% or less (Grandin, 2001, 2012; Munoz et al., 2012). For systems where pigs are stunned in small groups on the floor, squeals are counted if they occur in the small stunning pen. Do not use vocalization scoring for sheep.

Many slaughterhouses are continuously scoring to determine if the pre-slaughter handling and stunning practices are improving or becoming worse. For example, measuring the percentage of animals stunned effectively with one application of the stunner and the percentage that remain insensible after hoisting. For electrical stunning, the percentages of animals with correct stunner placement is counted. The target is 99% correct placement on cattle, pigs, or sheep. All scores are per animal which makes scoring simpler. The animal is either moved with an electric goad or not moved with one; it is either silent or vocal.

### *5.8.1 Scoring as a trouble-shooting tool to show continuous improvement*

When handling or facility problems are being corrected, scoring can be used as a measuring and trouble-shooting tool. A baseline score can be determined and then the system can be re-scored after changes are made. For example, after handlers were trained to only use an electric goad on pigs that refused to move when there was space ahead for them to move, adding a light on a dark race entrance reduced electric goad use from 38 to 4% (Grandin, 2010).

Scoring can also be used to detect problems within a handling system. A good measurement for facility problems is the percentage of animals that back up in a race or the percentage of animals that refuse to enter a race. If a lower percentage of animals back up or more animals enter more easily, the modification is effective. Scoring can also be used to determine if an intervention such as training employees reduced serious carcass and meat quality problems such as bruising. Paranhos da Costa et al. (2014) found that training employees reduced bruising in beef cattle.

## 5.9 REFERENCES


Abbott, T.A., Hunter, E.J., Guise, J.H. and Penny, R.H.C. (1997) The effect of experience of handling on pig's willingness to move. *Applied Animal Behaviour Science* 54, 371–375.

Barbalho, P.C. (2007) Availacao de programas de trienamento em manejo racional de bovinos em frigorificos para melhoria do bem-estar animal. Dissertacao de mestrado. Programa de Pos-Graduacao em Zootecnia, Faculdade de Ciencias Agarias e Veterinarias, Universidade Estadual Paulista (UNESP) Jaboticabal-SP, Brazil.

Bates, L.S.W., Ford, E.A., Brown, S.N., Richards, C.J., Hadley, P.J., Wotton, S.B. and Knowles, T.G. (2014) A comparison of handling methods relevant to religious slaughter of sheep. *Animal Welfare* 23, 251–258.

Becker, B.G. and Lobato, J.F.P. (1997) Effect of gentle handling on the reactivity of Zebu crossed calves to humans. *Applied Animal Behaviour Science* 53, 219–224.

Binstead, M. (1977) Handling cattle. *Queensland Agriculture Journal* 103, 293–295.

Bourquet, C., Deiss, V., Golbert, M., Durand, D., Boissey, A., and Terlouw, E.M. (2010) Characterizing the emotional activity of cows to understand and predict their stress reactions to the slaughter procedure. *Applied Animal Behavior Science* 125, 9–21.

Bourquet, C., Deiss, V., Tannugi, C.C. and Terlouw, E.M.C. (2011) Behavioural and Physiological reactions of cattle in a commercial abattoir: Relationships with organization aspects of the abattoir and animal characteristics. *Meat Science* 88, 158–168.

Core, S., Miller, T., Widowski, T. and Mason, G. (2009) Eye white as a predictor of temperament in beef cattle. *Journal of Animal Science* 87, 2174–2178.

Deiss, V., Temple, D., Ligout, S., Racine, C., Bouix, J., Terlouw, C, and Boissey, A. (2009) Can emotional reactivity predict stress response at slaughter in sheep? *Applied Animal Behavior Sciences* 119, 193–202.

Dokmanovic, M., Velarde, A., Tomovic, V., Giamoclija, N., Markovic, R., Janjic, J., and Battie, M.Z. (2014) The effect of lairage time and handling procedures prior to slaughter on stress and meat quality parameters in pigs. *Meat Science* 98, 220–226.

Dunn, C.S. (1990) Stress reactions of cattle undergoing ritual slaughter using two methods of restraint. *Veterinary Record* 126, 522–525.

Dwyer, C.M. (2004) How has the risk of predation shaped behavior responses of sheep to fear and distress? *Animal Welfare* 13, 269–281.

Edwards, L.N., Engle, T.E., Correa, J.A., Paradis, M.A., Grandin, T. and Anderson, D.B. (2010a) The relationship between exsanguination blood lactate concentration and carcass quality in slaughter pigs. *Meat Science* 85, 435–440.

Edwards, L.N., Grandin, T., Engle, T.E., Porter, S.P., Ritter, M.J., Sosnicki, A.A. and Anderson, D.B. (2010b) Use of exsanguination blood lactate to assess the quality of pre-slaughter handling. *Meat Science* 86, 384–390.

Ewbank, R., Parker, M.J., and Mason, C.W. (1992) Reactions of cattle to head restraint at stunning: A practical dilemma. *Animal Welfare* 1, 55–63.

Ferguson, D.M. and Warner, R.D. (2008) Have we underestimated the impact of pre-slaughter stress on meat quality in ruminants? *Meat Science* 80, 12–19.

Fordyce, G. (1987) Weaner training. *Queensland Agricultural Journal* 113, 323–324.

Fordyce, G., Dot, R.M. and Wythes, J.R. (1988) Cattle temperaments in extensive herds in northern Queensland. *Australian Journal of Experimental Agriculture* 28, 683–688.

Geverink, N.A., Kappers, A., van de Burgwal, E., Lambooij, E., Blokhuis, J.H. and Wiegant, V.M. (1998) Effects of regular moving and handling on the behavioral and physiological responses of pigs to pre-slaughter treatment and consequences for meat quality. *Journal of Animal Science* 76, 2080–2085.

Grandin, T. (1980) Observations of cattle behavior applied to the design of cattle handling facilities. *Applied Animal Ethology* 6, 9–31.

Grandin, T. (1981) Bruises on southwestern feedlot cattle. *Journal of Animal Science* 53 (Supplement 1), 213.

Grandin, T. (1982) Pig behaviour studies applied to slaughter-plant design. *Applied Animal Ethology* 9, 141–151.

Grandin, T. (1992) Observation of cattle restraint devices for stunning and slaughtering. *Animal Welfare* 1, 85–91.

Grandin, T. (1993) Behavioral principles of handling cattle under extensive conditions, In: T. Grandin, (Editor) *Livestock Handling and Transport*, 1st Edition. CAB International, Wallingford, UK, pp. 43–57.

Grandin, T. (1996) Factors that impede animal movement at slaughter plants. *Journal of the American Veterinary Medical Association* 209, 757–759.

Grandin, T. (1997) The design and construction of handling facilities for cattle. *Livestock Production Science* 49, 103–119.

Grandin, T. (1998) Objective scoring of animal handling and stunning practices in slaughter plants. *Journal of the American Veterinary Medical Association* 212, 36–39.

Grandin, T. (2001) Cattle vocalizations are associated with handling at equipment problems in beef slaughter plants. *Applied Animal Behaviour Science* 71, 191–201.

Grandin, T. (2003) Transferring results of behavioral research to industry to improve animal welfare on the farm, ranch, and the slaughter plant. *Applied Animal Behavior Science* 81, 215–228.

Grandin, T. (2010) Improving livestock, poultry, and fish welfare in slaughter plants with auditing programs. In: T. Grandin (Editor) *Improving Animal Welfare: A Practical Approach*. CABI Publishing, Wallingford, Oxfordshire, UK, pp. 160–185.

Grandin, T. (2011) Auditing animal welfare and practical improvements in beef, pork and sheep slaughter plants, Centenary International Symposium, Humane Slaughter Association, Portsmouth, UK, June 30–July 1.

Grandin, T. (2012) Developing measures to audit welfare of cattle and pigs at slaughter. *Animal Welfare* 21, 351–356.

Grandin, T. (2014) Handling and welfare of livestock in slaughter plants. In: T. Grandin (Editor) *Livestock Handling and Transport*, 4th Edition. CAB International, Wallingford, UK, pp. 329–353.

Grandin, T. and Deesing, M. (2008) *Humane Livestock Handling*. Storey Publishing, North Adams, Massachusetts.

Grandin, T. and Deesing, M.J. (2013) Genetics and behavior during handling restraint and herding, In: T. Grandin and M.J. Deesing (Editors) *Genetics and the Behavior of Domestic Animals*. Academic Press (Elsevier), San Diego CA, pp. 115–158.

Grandin, T. and Johnson, C. (2005) *Animals in Translation*. Scribner (Simon and Schuster), New York.

Gruber, S.L., Tatum, J.D., Engle, T.E., Chapman, P.L., Belk, K.E. and Smith, G.C. (2010) Relationships of behavioral and physiological symptoms of preslaughter stress on beef longissimus muscle tenderness. *Journal of Animal Science* 88, 1148–1159.

Hanggi, E.B. (2005) The thinking horse: cognition and perception reviewed. In: T.D. Broken, (Editor) *Proceedings of the*

*51st Annual Convention of the American Association of Equine Practitioners*, Seattle, Washington, 3–7 December. American Association of Equine Practitioners, Lexington, Kentucky, pp. 246–255.

Hemsworth, P.H., Rice, M., Karlen, M.G., Calleja, L., Barnett, J.L., Nash, J. and Coleman, G.J. (2011) Human-animal interactions at abattoirs: Relationships between handling and animal stress in sheep and cattle. *Applied Animal Behaviour Science* 135, 24–33.

James, B.W., Tokach, M.D., Goodband, R.D., Nelssen, J.L., Dritz, S.S., Owen, K.Q., Woodworth, J.C. and Solabo, R.G. (2013) Effects of dietary L. carnitine and ractopomine HCl on the metabolic responses to handling in finishing pigs. *Journal of Animal Science* 91, 4426–4439.

Janczak, A.M., and Braastad, B.O. (2004) A short note on the effects of exposure to a novel stimulus (umbrella) on behaviour and percentage of eye white in cows. *Applied Animal Behaviour Science* 69, 309–314.

Judd, S.P.D. and Collett, T.S. (1998) Multiple stored views and landmark guidance in ants. *Nature* 392, 710–714.

Kilgour, R. and Dalton, C. (1984) *Livestock Behavior: A Practical Guide*. Westview Press, Boulder, Colorado.

Lanier, J.L., Grandin, T., Green, R., Avery, D. and McGee, K. (2000) The relationship between reaction to sudden intermittent movements and sounds to temperament. *Journal of Animal Science* 78, 1467–1474.

LeDoux, J.E. (1996) *The Emotional Brain*. Simon and Schuster, New York.

Leiner, L. and Fendt, M. (2011) Behavioral fear and heartrate responses in horses after exposure to novel objects: Effects of habituation. *Applied Animal Behavior Science* 13, 104–109.

Lemmon, W.B. and Patterson, G.H. (1964) Depth perception in sheep: Effects of interrupting the mother-neonate bond. *Science* 145, 835–836.

Lewis, C.R.G., Hulbert, C.E. and McGlone, J.J. (2008) Novelty causes elevated heart rate and immune changes in pigs exposed to handling alleys and ramps. *Livestock Science* 116, 338–341.

Ligout, S., Foriquie, D., Sebe, F., Bouix, T. and Boissey, A. (2011) Assessment of sociality in farm animals: The use of the arena test in lambs. *Applied Animal Behavior Science* 135, 57–62.

Marchant-Forde, J.N., Lay, D.C., Pajor, J.A., Richert, B.T. and Schinckel, A.P. (2003) The effects of ractopamine on the behavior and physiology of finishing pigs. *Journal of Animal Science* 81, 416–422.

Meischke, H.R.C. and Horder, J.C. (1976) A knocking box effect on bruising in cattle. *Food Technology in Australia* 28, 369–371.

Muller, R., Schwartzkopf-Genswein, K.S., Shah, M.A. and von Keyserlinkg, M.A.G. (2008) Effect of neck injection and handler visibility on behavioral reactivity of beef steers. *Journal of Animal Science* 86, 1215–1222.

Munoz, D., Strappini, A., and Gallo, C. (2012) Indicadores de bienstar animal para detector problemas en al cajon de insensibilizacion de bovinos (Animal welfare indicators to detect problems in the cattle stunning box). *Archivos de Medicine Veterinaria* 44, 297–302.

OIE (2014) Chapter 7.5. *Slaughter of Animals, Terrestrial Animal Health Code*. World Organization for Animal Health, Paris, France.

Pajor, E.A., Rushen, J. and dePaisille, A.M.B. (2003) Dairy cattle choice of handling treatments in a Y maze. *Applied Animal Behaviour Science* 80, 93–107.

Panksepp, J. (2011) The basic emotional circuits of mammalian brains: Do animals have affective lives? *Neuroscience and Biobehavioral Reviews* 35, 1791–1804.

Paranhos da Costa, M.J.R., Huertas, S.M., Strappini, A.C. and Gallo, C. (2014) Handling of transport of cattle and pigs in South America, In: T. Grandin (Editor) *Livestock*

*Handling and Transport*. CAB International, Wallingford, Oxfordshire, UK, pp. 174–192.

Probst, J.K., Hillman, E., Leiber, F., Kreuzer M. and Neff, A.S. (2013) Influence of gentle touching applied a few weeks before slaughter on avoidance distance and slaughter stress in finishing cattle. *Applied Animal Behavior Science* 144, 14–21.

Rogan, M.T. and LeDoux, J.E. (1996) Emotion, systems cells, and synaptic plasticity. *Cell* 85, 469–475.

Romeyer, A. and Bouissou, M.F. (1992) Assessment of fear reactions in domestic sheep and influence of breed and rearing conditions. *Applied Animal Behavior Science* 34, 93–119.

Rushen, J. (1986) Aversion of sheep for handling treatments: paired-choice studies. *Applied Animal Behaviour Science* 16, 363–370.

Stockman, C.A., McGilchrist, P., Collins, T., Barnes, A.L., Miller, D., Wickham, S.L. Greenwood, P.L., Café, L.M., Blache, D., Wemelsfelder, F. and Fleming, P.A. (2012) Qualitative behavioural assessment of Angus steers during pre-slaughter handling and relationship with temperament and physiological responses. *Applied Animal Behaviour Science* 142, 125–133.

Strappini, A.C., Metz, J.H.M., Gagllo, C., Frankena K., Vargas, R., de Freslon, I. and Kemp, B. (2013) Bruises in culled cows: when, where, and how they are inflicted. *Animal* 7, 485–491.

Talling, J.C., Waran, N.K., Wathes, C.M. and Lines, J.A. (1998) Sound avoidance in domestic pigs depends on characteristics of the signal. *Applied Animal Behaviour Science* 58, 255–266.

van Putten, G. and Elshof, W.J. (1978) Observations of the effects of transport on the well being and lean quality of slaughter pigs. *Animal Regulation Studies* 1, 247–271.

Velarde, A. and Dalmau, A. (2012) Animal welfare assessment at slaughter in Europe: Moving from inputs to outputs. *Meat Science* 92, 244–251.

Velarde, A., Rodriguez, P., Dalmau, A., Fuentes, C., Llonch, P., von Hollenben, K.V. et al. (2014) Religious slaughter: Evaluation of current practices in selected countries. *Meat Science* 96, 278–287.

Warner, R.D., Ferguson, D.M., Cottrell, J.J. and Knee, B.W. (2007) Acute stress induced by preslaughter use of electric prodders causes tougher meat. *Australian Journal of Experimental Agriculture* 47, 782–788.

Warriss, P.D., Brown, S.N. and Adams, S.J.M. (1994) Relationship between subjective and objective assessment of stress at slaughter and meat quality in pigs. *Meat Science* 38, 329–340.

Waynert, D.E., Stookey, J.M., Schwartzkopf-Gerwein, J.M., Watts, C.S. and Waltz, C.S. (1999) Response of beef cattle to noise during handling. *Applied Animal Behaviour Science* 62, 27–42.

Westervelt, R.G., Kinsman, D., Prince, R.P. and Giger, W. (1976) Physiological stress measurement during slaughter of calves and lambs. *Journal of Animal Science* 42, 833–834.

chapter six

# Mechanical stunning and killing methods

*Bert Lambooij*[1] *and Bo Algers*[2]

**Learning objectives**

- Basic neurological and physiological aspects of mechanical stunning methods.
- Description of the methods for mechanical stunning.
- Effective use of guns.
- Monitoring points for correct stunning.
- Advantages and disadvantages of stunning methods.

## CONTENT



## SUMMARY

Mechanical stunning methods can be divided into penetrating and non-penetrating applications of a missile into the brains of animals. Irrespective of the method, disruption of normal brain function is achieved principally due to

[1] Wageningen UR Livestock Research, P.O. Box 338, 6700 H Wageningen, The Netherlands
[2] Swedish University of Agricultural Sciences, Department of Animal Environment and Health, P.O. Box 234, 53223 Skara, Sweden

brain concussion by transmitting the energy from the missile (e.g. bolt) into the cranium upon impact and by haemorrhages due to rupture of the arteries to the brain. The penetrating mechanical stunning methods cause subsequent disruption of the brain tissue and help to prolong the duration of unconsciousness and insensibility.

Mechanical methods that principally refer to penetrative stunning or killing include use of captive bolt pistols and rifles. Cartridges with gunpowder, compressed air or a spring under tension are used to drive missiles through the skull of farm animals. Stunning and stun/kill methods are developed to induce, when applied correctly, pathological brain states that are incompatible with the persistence of consciousness and sensibility. The missile should create a large, deep, penetrating and well-defined haemorrhagic track which traverses almost the full thickness of the brain. The ideal shooting position is frontally on the head. The velocity of the penetrating bolt can be reduced by incorrect application or a damp cartridge or dirty gun. Successful penetrative stunning is followed by an immediate collapse, prompt and persistent absence of rhythmic breathing, loss of corneal reflex or spontaneous blinking, immediate onset of tonic seizure lasting several seconds and no righting reflex or vocalization. The standard operating procedure will involve sampling of all animals by slaughterhouse personnel for indicators of consciousness immediately after stunning, before sticking and during bleeding. In addition, the animal welfare officer will sample a fraction of all animals to monitor the effectiveness of the process, and will correct the operator or other aspects of the stunning process if necessary.

## 6.1 INTRODUCTION

In the past, stunning of farm animals was sometimes performed by a heavy blow to the head with a blunt instrument (Wijngaarden-Bakker, 1980). As early as the year 1336, in the town of Neurenberg, it was laid down that all animals should be "struck" before bleeding (Fahrbach, 1948). This was probably intended to facilitate slaughtering and to protect the slaughterman from struggling animals. During the 19$^{th}$ century, however, concern for animal welfare had grown and a blow on the head was considered as unsatisfactory. The first development was a spike which was driven into the brain with or without the help of a hammer, a mask being placed on the animal's head beforehand. These instruments were not used for small animals as placing the mask took too much time (Fahrbach, 1948). Thereafter, a firearm method was developed, whereby a bullet was shot into the head. Drawbacks were the loud noise and a bullet lodged somewhere in the neck of the slaughtered animal. An advantage was the immediate immobilization of the animal (Fahrbach, 1948). Finally, captive bolt stunning was developed for use in small animals, but came to be very useful for large animals (Fahrbach, 1948). At present, the penetrating captive bolt is legally accepted for use in slaughterhouses in most countries of the world and is nowadays the most common stunning method used for cattle.

The main mechanical methods accepted for stunning purposes by the OIE Terrestrial Animal Health Code (2014) and the Council Regulation (EC) No 1099/2009 of 24 September 2009 on the protection of animals at the time of killing are:

1. *Penetrative captive bolt device*: Severe and irreversible damage of the brain provoked by the shock and the penetration of a captive bolt. Used in all species during slaughter, depopulation and other situations.
2. *Non-penetrative captive bolt device*: Severe damage to the brain by the shock of a captive bolt without penetration. Used in ruminants, poultry, rabbits and hares during slaughter, depopulation and other situations for poultry, rabbits and hares.
3. *Firearm with free projectile*: Severe and irreversible damage to the brain provoked by the shock and the penetration of one or more projectiles. Used in all species during slaughter, depopulation and other situations.
4. *Maceration*: Immediate crushing of the entire animal. Used in chicks up to 72 hours and egg embryos during all situations other than slaughter.
5. *Cervical dislocation*: Manual or mechanical stretching and twist of the neck provoking cerebral ischaemia. Used in poultry up to 5 kg live weight during slaughter, depopulation and other situations.
6. *Percussive blow to the head*: Firm and accurate blow to the head provoking severe damage to the brain. Used in piglets, lambs, kids, rabbits, hares, fur animals and poultry up to 5 kg live weight during slaughter, depopulation and other situations.

## 6.2 NEUROLOGICAL AND PHYSIOLOGICAL ASPECTS

Mechanical methods can be divided into penetrating (penetrative captive bolt device, firearm with free projectile and maceration) and non-penetrating (non-penetrative captive bolt device, cervical dislocation and percussive blow to the head) applications and principally operate by causing concussion, brain haemorrhages and unconsciousness (see introduction). Other methods cause mechanical disruption of tissues which leads to unconsciousness by neural shock or anoxia by blood loss after a few seconds (puntilla), and therefore they are not acceptable. Mechanical methods that principally refer to captive bolt guns (CBG) also include use of free bullet and rifles. Cartridges with gunpowder, compressed air or a spring under tension are used to drive bolts (missiles) against or through the skull of farm animals as well as farmed fish. Consequently, there will be transfer of energy to the animal's head, causing concussion but also structural damage as the bolt travels through the brain. Immediate insensibility and unconsciousness is caused by rapid propagation of shock waves of kinetic energy through the brain and abrupt acceleration and deceleration of the relatively soft brain within the bony skull. If the bolt is too thin or it is fired through a trephined skull there will not be enough energy transfer to the head to induce effective stunning (Karger, 2009; EFSA, 2004; Raj and O'Callaghan, 2001; Daly and Whittington, 1989). Stunning and stun/kill methods, when applied correctly, are developed to induce pathological brain states that are incompatible with the persistence of consciousness and sensibility. Insensibility occurs immediately after its application by direct physical trauma (focal and diffuse injury) from the missile to the structures of the diencephalon and brainstem (Finnie et al., 2002). Pressure effects, in the form of flattening of the cerebrum contralateral to the shot, may be due to trauma from countercoup forces or associated with subdural haemorrhage after the shot. The magnitude of deviation from the normal brain electrical activity can be determined using electroencephalogram (EEG) or electrocorticogram (ECoG). Immediately after shooting, major changes occur on the EEG (*delta* and *theta* waves tending to an iso-electric line). It is assumed that the animal is unconscious by analogy to similar EEG changes described in man (Lopes da Silva, 1983; EFSA, 2004).

### *6.2.1 Concussive blows*

Cerebral concussion is provoked by a traumatically induced derangement of the nervous system, resulting in an instantaneous diminution or loss of consciousness without gross anatomical changes in the brain. Irrespective of the type of force which produces the traumatic depolarization of the cell membrane, there is evidence that powerful pressure waves are provoked within the cranial cavity by a blow on the head and that the frequency and force of the waves vary in different parts of the brain. It has been suggested that it is not the pressure as such developed by these waves that is the important factor but the rapid oscillations in this pressure (Nilsson and Nordström, 1977). The captive bolt should create a large, deep, penetrating and well-defined haemorrhagic track which traverses almost the full thickness of the brain (Figure 6.1; Algers and Atkinson, 2014).

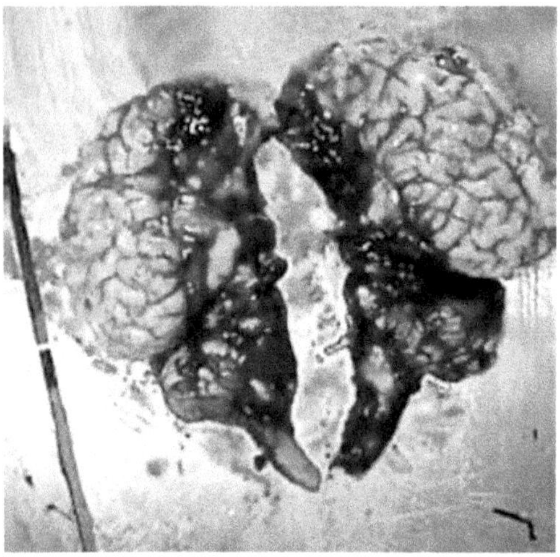

**Figure 6.1:** Brain haemorrhage in well-stunned cattle

### *6.2.2 Captive bolt and free bullet*

The aim of captive bolt stunning methods is to cause concussion and major arterial haemorrhages in the brain by transmitting the energy from the missile (bolt) into the cranium and brain (Figure 6.1). Immediately after stunning, animals express a tonic spasm for approximately 10 seconds prior to relaxation, however, excessive convulsions often follow (Lambooij et al., 1981). The ideal shooting position is frontally on the head (Grandin, 1981).

In general, penetration of a missile into the brain can cause injury in the following three ways, depending on its velocity and shape: by laceration and crushing (< 100 m/s), by shock waves (about 100 to 300 m/s) and by temporary cavitation effect (> 300 m/s; Hopkinson and Marshall, 1967). In fact, using the formula of $e = m Xv^2$, where e = energy, m = mass, and v = velocity, it has been shown that the delivered energy required for effective stunning is determined by the velocity of the missile. However, secondary tissue damage by penetration also prevents any chance of recovery. Free bullets have a lower mass than bolts of captive bolt

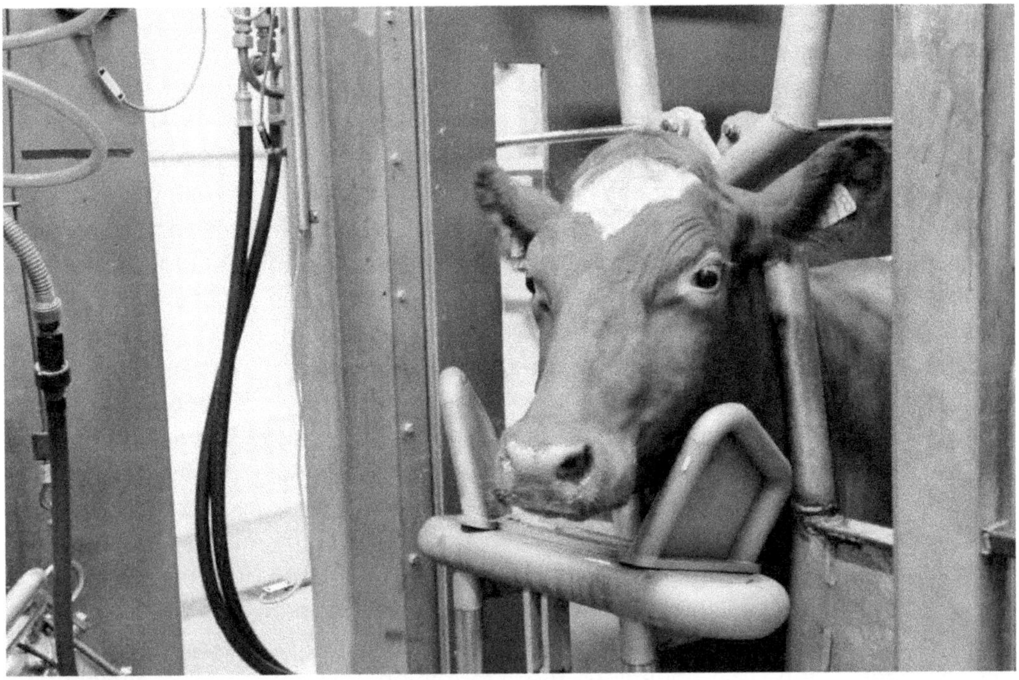

**Figure 6.2:** Stun box and restraint device with chin lift.
*(Photo: Audun Flåtten)*

stunners, and travel with higher velocity (typically > 300 m/s for rifles). To improve energy transfer, bullets are constructed to fragment and/or deform when hitting the target (Holleben et al., 2010; Schiffer et al., 2014). A stunning gun for large animals commonly produces a bolt velocity close to 75 m/s. It is very likely that significant pressure changes arise in all volumes of the brain during stunning, where the pressure change will be greater in volumes that are located in front of the stunning gun compared to the pressure changes in volumes that are located at the periphery (Davidson et al., 2014). In captive bolt stunning methods, the most important factor is to cause concussion of the brain, which can be determined in the laboratory using EEGs as indicated by the highly synchronized electrical activity, that is, low frequency and high amplitude slow waves. This kind of activity is not compatible with the persistence of consciousness and is physically manifested as tonic seizure, absence of breathing and fixed eyes, which can be used as animal-based indicators of effective CBS in slaughterhouse conditions. Shock waves and cavitation by free bullet cause the arteries to rupture and heavy bleeding. This leads to a levelling out of arterial pressure with intracranial pressure that substantially reduces cerebral perfusion pressure, and so cell function ceases.

Frontal shot placement is also considered to be optimal for captive bolt stunning of cattle. The recommended place is considered to be at the intersection of two imaginary lines drawn from the lateral canthus (outside corner) or middle of the eye to the base of the contralateral horn

(Figures 6.2 and 6.6; Anil and Lambooij, 2009; Gilliam et al., 2012) or a little higher in order to increase the probability of impacting the crucial area around the brainstem, and to decrease the risk of destroying only the frontal sinuses (Kohlen, 2011). In that case, the danger of regaining consciousness before bleeding can be reduced due to physical destruction of the brainstem. The maximum penetration depth is determined by the length of the bolt. Depending on the device (8-cm or 12-cm bolt lengths are common) and thickness of skin and hair, the brainstem might not even be inflicted at all (Kohlen, 2011). A large-bore projectile passing through the skull might destroy the brainstem anyway. Sufficient haemorrhages are associated with a "deep" stun. This underlines the close relationship between brain haemorrhages which might be explained by the high kinetic energy of the large bore employed in this shot (approximately 4000 Joule). In case of a failed shot by such a large-bore calibre, it might happen that the animal breaks down and appears to be properly stunned, due to the intensity of the shot (Schiffer et al., 2014). Regarding the question of a passing shot and the .22 Magnum projectiles, which remain in the cranial cavity, wound ballistics, as described by Di Maio (1999), are to be taken into consideration: when a live organism is shot, the severity of the wound is determined not only by the amount of brain tissue damage but also the so-called temporary cavity. The temporary cavity is caused by surrounding tissue flinging radially outward from the trajectory, which is crucial to the total damage of the brain.

A temporary cavity will be much bigger than the diameter of the projectile. It only lasts for 5 to 10 milliseconds, but it determines the permanent wound track, tearing tissues in the path. The size of the temporary cavity and, thus, the severity of brain damage is a result of the amount of energy lost during penetration through various tissues in the pathway rather than the total energy possessed by the bullet. The pressure propagates to a larger extent in the direction of the bolt travel (Davidsson et al., 2014) which may be of importance for obtaining basal brain haemorrhages (Nydahl, 2014). A projectile that does not pass through the cranial cavity expends its energy while fracturing the skull and then is captured in the cranium. This implies that the whole energy of the projectile remains within the cranial cavity and impacts wound formation, which may inflict rapidly fatal haemorrhages (Finnie, 1997; Di Maio, 1999). When the temporary cavitation effect is related to the velocity of the projectile, it occurs in case of > 300 m/s (Anil and Lambooij, 2009). This applies to the .22 Magnum projectile with a velocity of 580 m/s, but not to ordinary captive bolt stunning (< 100 m/s). In the case of a shot remaining within the skull such as the .22 Magnum projectiles, full penetration of the brain can also occur, with the projectile remaining at the base of the cranial cavity plus the temporary cavitation effect. The .22 Magnum is suggested to be an efficient calibre for immediately stunning or killing small- or medium-sized cattle by a precise frontal shot (Schiffer et al., 2014).

### *6.2.3 Mechanical disruption of tissues*

By cervical dislocation in small animals, the head is turned in the opposite direction to the body while stretching the neck and concomitantly turning and stretching; blood vessels are crushed and bleeding occurs. After dislocation or thrusting a knife, a tonic cramp occurs resulting in paralysis after 5 to 10 seconds. Removal or inhibition of the contact between brain and spinal cord causes apnoea and loss of sensory perception from the body and spinal shock, with the exception of the face innervated by the 5th cranial nerve. An alternative method for animals with a long neck involves stretching by hand.

Instantaneous fragmentation in high-speed rotating blades will kill small animals. Macerators with rotating blades with a speed of at least 2,800 revolutions/min at a power of 4 KW are effective, however, 6,000 revolutions/min are recommended. The method is used in hatcheries to kill unwanted day-old chicks and unhatched eggs. However, there is a risk that flying birds may keep high up in the macerator; this can be prevented by fast flow-through rates or by killing these animals another way.

## 6.3 DESCRIPTION OF THE METHOD

To achieve successful stunning, the captive bolt device must be correctly placed and a bolt of adequate length and diameter must be used. Captive bolt guns can be either trigger or contact fired. With contact fired guns, there is no possibility to correct the position of the gun once it touches the head of the animal (Figures 6.3 and 6.4). As a consequence there are a lot of failed shots with this type of gun in some plants, depending on the skill of the staff and adequacy of

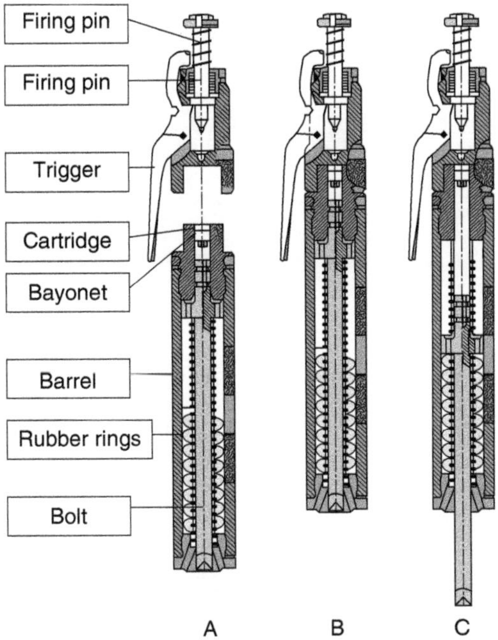


**Figure 6.3:** A penetrating captive bolt.
*(FAO, 2001)*

the restraining devices (Holleben et al., 2010).

### *6.3.1 Captive bolt and free bullet*

This method is commonly used for ruminants to stun them for slaughter and after stunning, death is caused through exsanguination before the animals regain consciousness. The ideal shooting position is frontally on the head. The commercially available captive bolt for ruminants can be used for the biggest animals and the "Goldhase Schusz Apparat" (EFSA, 2005) for smaller animals. Cartridges with gunpowder, compressed air or a spring under tension are used to drive bolts (missiles) against or through the skull of farm animals. Missiles used for stunning and killing of animals are a bullet or a bolt. Immediately after stunning, the animals express a tonic spasm

**Figure 6.4:** Penetrative captive bolt (Jarvis) stunner suspended in front of stun box.
*(Photo: Elisiv Tolo)*

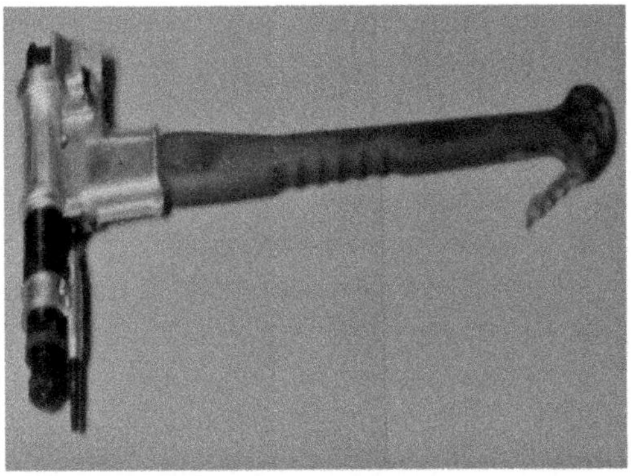

**Figure 6.5:** Mushroom headed concussion stunner with a trigger

for approximately 10 seconds prior to relaxation, however, excessive convulsions may occur (Eichbaum et al., 1975).

In poultry, mechanical devices which are penetrating as well as non-penetrating have been developed specifically to kill, rather than stun birds (Hewitt, 2000). Because the skull bones are not ossified in poultry, both penetrating and non-penetrative devices induce severe structural damage to the brain and immediate death, provided the bolt parameters are adequate, rather than concussion of the brain (EFSA, 2004).

### *6.3.2 Concussive blows*

The instrument used for concussion stunning in early times was a simple bludgeon (Fahrbach, 1948). At present a "non-penetrating mushroom headed bolt" which derives its power from a cartridge of appropriate charge or from compressed air is used in slaughterhouses (Figure 6.5). In some countries outside the EU, concussion stunning is allowed to be used for calves. Non-penetrative captive bolt devices are also used to stun and kill chicken and turkeys. They are fitted with a plastic or metal concussive head. The bolt head is fired with high velocity onto the head of the chicken or turkey and causes severe structural damage to the skull and brain. The device is used in small-scale slaughter plants, where prompt sticking is recommended and for casualty/emergency killing (Hewitt, 2000). Gregory and Wotton (1990) showed that stunning chickens with a non-penetrative captive bolt resulted in pronounced changes in their visual evoked responses. A blow to the head administered by hand with a blunt instrument, for example a heavy pipe or stick, leads to instantaneous unconsciousness in poultry up to 5 kg due to cerebral commotion, provided the blow is delivered with sufficient impact and right on the target. As efficiency depends largely on operators' skill and can hardly be standardized, this method is only suitable for small-scale slaughter (Schütt-Abraham, 1995).

### *6.3.3 Neck dislocation*

This method might be used in small animals such as rabbits and birds. One way is to place the thumb and index finger on either side of the neck at the base of the skull and to pull the hind limbs or tail away from the head. Another is to place a blunt instrument at the base of the neck before pulling on the hind legs. A rapid pull induces separation of the vertebrae, mainly the cervical, but sometimes the thoracic vertebrae are separated (Keller, 1982). In birds, the head may be twisted and extended dorsally while stretching. Due to turning and stretching, blood vessels may be damaged and bleeding may occur at the site of vertebral separation; sometimes decapitation may also occur.

### 6.3.4 *Maceration leading to fragmentation*

Maceration is used in small animals such as day-old chicks and unhatched eggs. However, there is a risk that for birds that can fly there will be a tendency to escape towards the higher parts in the macerator, which can be prevented by fast flow-through rates. This method is used in small animals such as day-old chicks and unhatched eggs.

## 6.4 EFFECTIVE USE OF THE METHOD

The velocity of the bolt of the captive bolt gun used for stunning farm animals is about 100 m/s in air. At this relatively low velocity, the shape of the bolt should crush the cortex and deeper parts of the brain either by itself or by the generation of forward shock waves. Studies have suggested that it is the kinetic energy imparted to the cranium by the bolt that produces insensibility (Daly and Whittington, 1989), while the actual physical damage by the bolt to specific brain structures is responsible for producing irrecoverable insensibility (Figure 6.6; Gibson, et al., 2012).

According to Algers and Atkinson (2014), bolt velocity is reduced if:

- incorrect cartridges for the species and size of animal are used;
- the cartridges are damp when kept in humid environments;
- the gun is dirty or has worn out or damaged parts.

The impact of a captive bolt can be seriously hindered if:

- the operator applies the gun at an incorrect angle, where correct is at a 90° angle;
- the gun is far away from the animal's forehead;
- animals are moving within the stun box, obtaining the incorrect position.

Effective captive bolt stunning is associated with immediate absence of evoked cortical responses in the brain (Daly and Whittington, 1989; Daly et al., 1987). Absence of primary cortical evoked responses indicates failure in neurotransmission at a level that occurs before conscious perception of a stimulus. After captive bolt stunning, severe changes are observed in the EEG (Lambooij et al., 1981). Unlike evoked responses, the spontaneous EEG is not as reliable as an indicator of brain disturbance following captive bolt stunning, and evoked responses are preferred (Gregory, 2007; Daly et al., 1988; Daly et al., 1987). The stun quality in commercial cattle slaughter has been monitored by Atkinson et al. (2013) who reported inadequate stunning occurring in 12.5% of the cattle (16.7% of bulls, compared with 6.5% other cattle). Bulls displayed symptoms rated the highest level for inferior stun quality 3 times more frequently than other cattle. Despite being shot accurately, 13.6% bulls were inadequately stunned compared with 3.8% other cattle. A total of 12% of cattle were re-shot, and 8% were inaccurately shot. Grandin (2001) reported percentages of cattle successfully stunned with one shot from a captive bolt stunner to be 100% in 12% of processing plants, 99% in 24%, 95 to 98% in 54% of the plants and more than 95% in 10% of plants in the USA in 1999. All cattle where the first shot missed

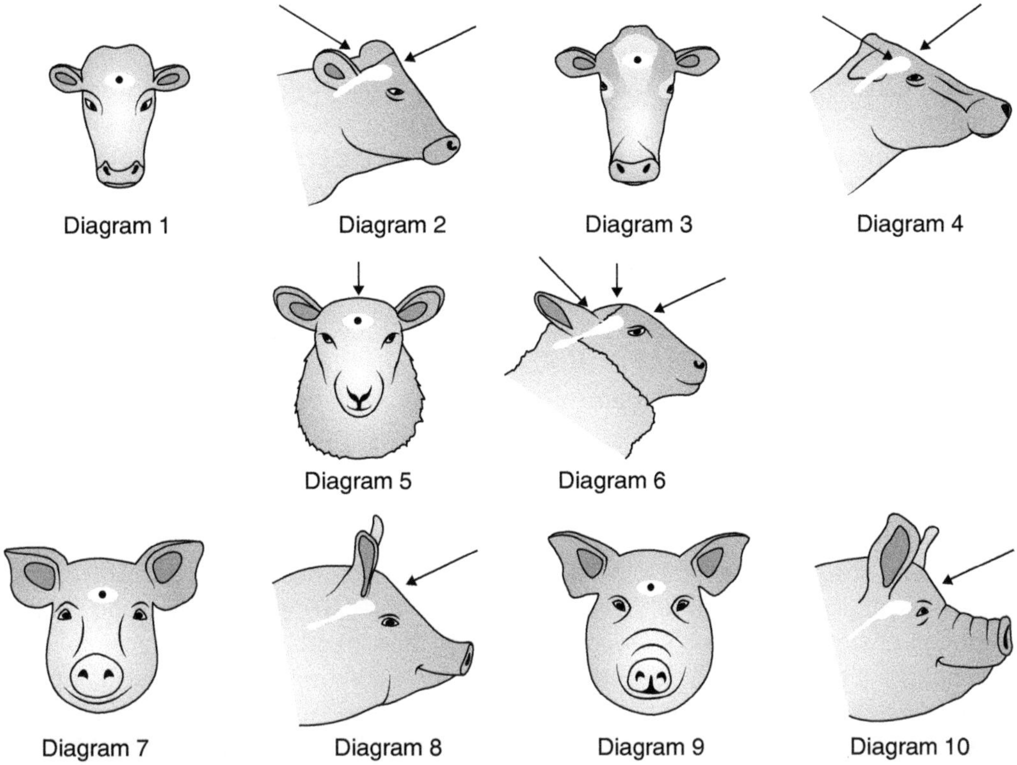


**Figure 6.6:** The shot must be aimed as shown in diagrams 2, 4 and 6. The poll position is recommended as an alternative (Lambooy and Spanjaard, 1981).

*(Source: http://www.grandin.com/humane/captive.bolt.html)*

were immediately re-stunned (Grandin, 2001). After these stunning procedures the animals may not die immediately depending on the degree of injury to the brain. Therefore, it is recommended to kill the animal by exsanguination, delivering compressed air into the cranium, or pithing to damage the deeper parts of the brain and to prevent convulsions. When concussion by a blow to the head is incorrectly performed the animal may be injured and neither stunned or killed. In cattle, to ensure efficient stunning, the captive bolt must be fired at the crossover point of imaginary lines drawn between the base of the horns and the contralateral eyes and certainly no further away than 2 centimetres radius from this point (EFSA, 2004; Finnie, 1993a,b; Lambooij, 1981; Lambooij et al., 1981). Shooting accuracy becomes more critical using low powered devices (Gregory, 2007). The shooting angle may also be of importance to ensure the laceration of brain arteries at the base of the brain (Nydahl, 2014). Deviations from the recommended shooting position and from the perpendicular shooting direction increases intensity of muscle spasms after shooting and this may impede further processing including hoisting and sticking (Marzin et al., 2008).

Using the compressed air-driven captive bolt gun requires a fixation of the cattle head as the weapon used is heavier and will not easily follow the move of the cattle head. Significant improvements in shot accuracy and greater efficiency of shooting have been shown but at the cost of increased stress imposed on the animal during restraint. More precise and humane stunning occurs when the pneumatic stunner and head restraint system is used. In a study by Atkinson and Algers (2009) the stun quality was far superior in the abattoirs using restraint and pneumatic stunning with 99.5% cattle properly stunned compared to 85% of cattle stunned properly when using gun powder-driven captive bolt.

In sheep and goats generally the same principles apply as for cattle. The ideal shooting position for polled sheep is the highest point of the head in the midline, pointing straight down to the throat. The ideal shooting position for horned sheep and for all goats is the position just behind the middle of the ridge that runs between the horns. Then the captive bolt should be aimed towards the mouth (HSA, 2006; EFSA, 2004).

It is generally known that the removal of inhibitory influences from higher centres of the brain, before the spinal cord becomes anoxic, results in convulsive activity and enhancement of some spinal reflexes. This may affect research on muscle and brain. Most investigations concerning the mechanism of concussion have been performed using laboratory animals (i.e. rats, cats, primates). It is evident from these investigations that concussion does not always cause an immediate loss of consciousness. In humans, amnesia after the blow occurs. Successive severe blows result in prolonged loss of reflex activity and cause almost complete disappearance of all frequencies, that is, an isoelectric line on the EEG (Nilsson and Nordstrom, 1977).

Depending on the amount of brain damage induced, non-penetrative captive bolt stunning can cause either permanent or temporary unconsciousness. In lambs and calves, the majority showed signs of recovery (Blackmore and Delany, 1988). These signs and the development of righting reflexes did not usually occur in less than 2 minutes (Blackmore, 1979). To ensure effective stunning in adult cattle, the non-penetrative captive bolt must be placed 2 centimetres above the cross-over point of imaginary lines drawn between the base of the horns and the contralateral eyes. This must be achieved very precisely using proper body and head restraint, because only slight variation in the ideal shooting position and angle decreases stunning efficiency (HSA, 2006; Grandin, 2002). A mass of hair on the forehead or moulding of foreheads may hinder successful contact of the concussive head to the bone and thus lead to decreased energy transfer (Endres, 2005). An optimal relation between dimension of the concussive head of the stunner and strength of the blow, to achieve sufficient effectiveness without fracturing the skull, was not found. This was because especially in the young bulls the shape and hairiness of the foreheads showed huge variations. Fracturing of the skull resulted in less effective stunning (Endres, 2005).

Free projectiles provide optimized energy transfer by deforming and/or fragmenting when hitting the target (Anil and Lambooij, 2009). Fragmentation of free bullets and the secondary wound tracks caused by the fragments of the bullet enhance brain damage and have greater efficiency compared to a penetrating captive bolt pistol (Finnie, 1997). Different types of projectiles have been tested to induce fragmentation or deformation, however, there was no perforation (Schiffer et al., 2014). It has been neither possible to distinguish between different styles of projectiles such as semi jacketed, homogeneous or lead-free nor to focus on the construction of projectiles such as soft point or hollow point. Generally, hollow point projectiles provide optimized transfer of energy without the danger of over penetration (AVMA, 2013).

Lateral shots at isolated cattle heads were not as successful as expected and, based on this data, frontal shots at cattle are recommended. All of the tested calibres can be preliminarily recommended for precise frontal shots at small cattle, because they caused "severe" brain damage, with the exception of the .22 Magnum calibre. This calibre seems to be most favourable for gunshot of relatively small cattle breeds such as the Galloways (Schiffer et al., 2014). Shots that failed to produce immediate insensibility either missed the brain or entered off midline superficially damaging the cerebellum and lobes of the cerebrum and missing the structures of the thalamus and brainstem. The top of the head (crown) position of the shot is similar to that found to produce irrecoverable insensibility in sheep shot with penetrating captive bolt without a secondary procedure (bleeding or pithing; Gibson et al., 2012).

In cattle and sheep, many studies report a gender difference in inadequate stunning, with male animals generally having a higher prevalence of being incompletely concussed (Atkinson et al., 2013; Gibson et al., 2012; Gregory et al., 2007). A study in 1,608 cattle shot with contact firing penetrating captive bolt showed that 51% of the shots were inaccurately placed (Gregory et al., 2007). Meanwhile, other studies have reported the incidence of incorrectly placed shots as 35% (Fries et al., 2012), 8% (Atkinson et al., 2013; von Wenzlawowicz et al., 2012) and 35% without head restraint (von Wenzlawowicz et al., 2012). Sheep are generally shot on the crown or top of the head, however, overall 19% of shots were incorrectly placed with horned rams being more challenging to shoot accurately (Gibson et al., 2012).

## 6.5 MONITORING POINTS FOR CORRECT STUNNING

Generally, scientific studies conclude that captive bolt stunning is an effective method of rendering animals insensible prior to the act of slaughter, provided that the animals are shot in the correct position, with the appropriate CBG/cartridge combination for the species and animal type and finally that the CBGs are regularly cleaned and serviced to minimize deterioration in performance (Grandin, 1994, 2002). The use of protocols can help standardize assessments and allow for benchmarking of stun quality at commercial slaughter. This could contribute to setting minimal standards as a safeguard to animal welfare at stunning (EFSA, 2013, Atkinson et al., 2013).

To achieve successful stunning, the captive bolt device must be correctly placed and a bolt of adequate length and diameter must be sufficiently accelerated. Consequently, there will be transfer of energy to the animal's head, causing concussion but also structural damage as the bolt travels through the brain. When using a non-penetrative captive bolt in sheep as well as cattle, unconsciousness should be induced with a single blow at the frontal position of the head. Subsequent shots may not be effective due to the swelling of the skin occurring from the first shot, and therefore, should not be allowed. If the first shot is unsuccessful, the animal should be stunned immediately using a penetrating captive bolt or electric current (EFSA, 2004).

Eyeball rotation following CBG stunning has been previously reported as the first sign of a shallow depth of concussion. In sheep and cattle shot with CBG the incidence of eyeball rotation has been reported as 4% and 3.8%, respectively (Gibson et al., 2012; Gregory et al., 2007).

Effective stunning is considered to occur when animals show (Atkinson et al., 2013; Algers and Atkinson, 2014):

- immediate collapse;
- no eye movements, that is, fixed stare, with a glazed appearance and unresponsive to light or touch;
- absence of rhythmic breathing, spontaneous blinking or corneal reflex;
- normal expected clonic and tonic seizures/spasms;
- no attempt to get up or raise the head (righting reflex);
- no vocalizations.

Ineffective stunning is considered to occur if animals show any of the following symptoms:

- nystagmus
- eyeball rotation
- vocalization
- rhythmic respirations
- righting reflex
- spontaneous blinking
- corneal reflex.

When cattle do not show clear signs of deep unconsciousness, it is difficult to ascertain if or when pain or fear perception ceases or returns under practical working conditions in an abattoir (Figure 6.7). Full eyeball rotation was the most frequently seen symptom (Atkinson et al., 2013).

The standard operating procedure for slaughter of cattle will involve a sampling fraction of 100% by slaughterhouse personnel, as the operators check each animal for indicators of consciousness immediately after stunning, before sticking and during bleeding. In addition, the animal welfare officer will sample a fraction of all animals to monitor the effectiveness of the process, and will correct the operator or other aspects of the stunning process if necessary (EFSA, 2013).

Both types of guns are normally fired on the forehead (usually frontal bone) of an animal, but other sites may be selected when there are horns or thick ridges on the skulls. Captive bolts must always be fired perpendicular (at right angle) to the skull bone surface (at the chosen site); otherwise bolts may skid and fail to fully impact the skull (Holleben et al., 2010). The animal's head must be suitably presented to the operator to facilitate accurate shooting. The animal should be rendered unconscious using a single shot and effective bleeding is required and needs to be performed immediately after stunning to ensure rapid brain death following exsanguination (EFSA, 2004, p. 48).

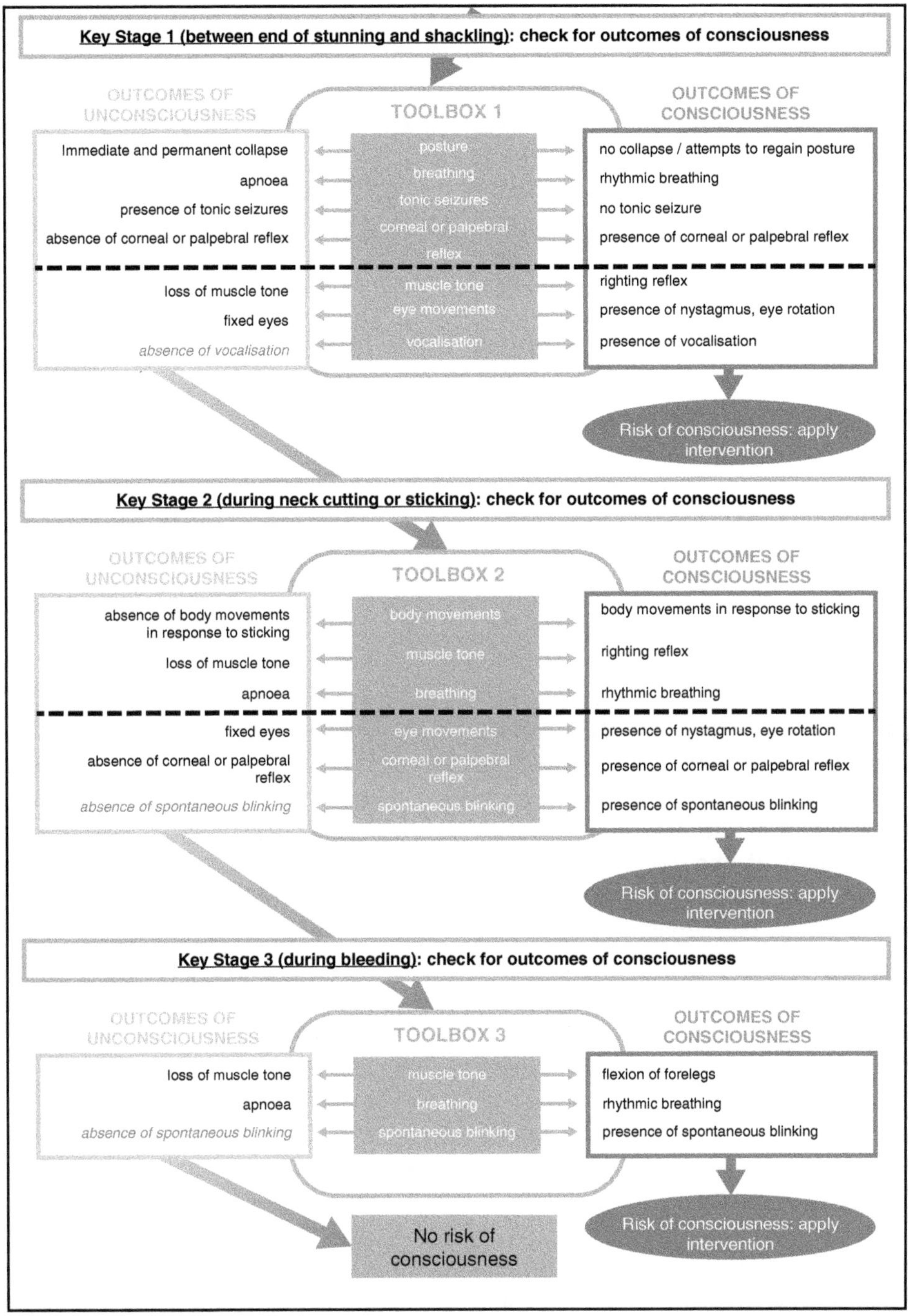


**Figure 6.7:** Indicators that are considered suitable to be used for detection of conscious animals at each key stage of the procedure of captive bolt stunning in bovines. The indicators listed above the dashed line are recommended to be used to recognize consciousness. The indicators below the dashed line can also be used to check for signs of consciousness, but their sensitivity is low and they should not be relied upon on their own.

*Source: (EFSA, 2013)*

Regarding safety concerns, the danger of injuring either people or conspecifics is minimized when a projectile is used that does not have sufficient momentum to pass through the head (Blackmore, 1984). Even if a frontal shot at live cattle implies that the bullet is likely to lodge in the animal (in case of a flight path pointing towards the *foramen magnum*), concerns might arise due to emission of chips of the bullet, or the bullet, or parts of it, remaining within edible tissue. The properties of suitable lead-free ammunition for slaughter purposes of cattle require further investigations.

Poor maintenance of guns is often the reason for stunning failures. Guns have to be cleaned and maintained regularly, otherwise the velocity of the bolt may be impaired. Worn rubber rings or springs used to retract the bolt from the skull have to be replaced at once; failure to do so can result in damage to the bolt and gun (e.g. prevent the pistol getting stuck in the skull and becoming damaged). If the tip of the bolt is protruding from the muzzle of the gun, this indicates bad maintenance; this results in enlargement of the gas expansion chamber within the gun and leads to underpowered shots. Deformed bolts will not achieve the necessary speed. Too short or narrow bolts can lead to decreased stunning effectiveness, through decreased kinetic energy transfer (Gregory et al., 2007; EFSA, 2004, p. 45; Holleben et al., 2002).

When using the techniques of dislocation, the necessary handling and restraint can be stressful for the animal and sedating the animal before it is killed can mitigate this stress and any subsequent pain.

## 6.6 ADVANTAGES AND DISADVANTAGES

- After neck dislocation, electrical activity of the brain may persist for as long as 13 seconds in some mammals and birds (Gregory and Wotton, 1990) during which time animals may feel pain due to afferent stimuli from the trigeminal nerve.
- After cervical dislocation, convulsions only occur when separation between the cranial and the fifth thoracic vertebra occurs, while severance caudal to this location results in paralysis of conscious animals (Eichbaum et al., 1975).
- If the macerator is overloaded, animals may be not be humanely killed.
- All these mechanical disruption techniques are aesthetically controversial. The interpretation of electrical activity in the brain after neck dislocation is controversial as to what feeling remains, and is still a matter of debate (Gregory and Wotton, 1990; Hollson, 1992). Sedating animals before cervical dislocation will minimize distress and any subsequent pain. This may be required in some cases of maceration where the animal may escape the blades.
- In special cases only, the use of a free bullet can be recommended due to the hazard of injuring other animals including the operator(s), as a free bullet has low controllability. The gunshot method must only be implemented by highly professional marksmen in order to guarantee high reliability but if so, in comparison to ordinary captive

bolt stunning, it has the potential to ensure a maximum degree of animal welfare (Schiffer et al., 2014).
- Brain particles are found in the blood, lungs, heart and muscle after penetrative stunning methods. Brain particles are visually observed in up to 33%, 12% and 1% of the carcasses after captive bolt stunning with air injection afterwards, bolt penetration using air pressure and cartridges with gunpowder, respectively (Anil et al., 1999; Schmidt et al., 1999).

## 6.7 REFERENCES


Algers, B. and Atkinson, S. (2014). Stunning: Mechanical Stunning. In: Devine, C. and Dikeman, M. (Editors-in-chief) *Encyclopedia of meat sciences 2e*, Vol. 3. Elsevier: Oxford. pp. 413–417.

American Veterinary Medical Association (AVMA) (Ed) (2013). AVMA Guidelines for the Euthanasia of Animals: 2013 Edition. Schaumburg, IL, USA. https://www.avma.org/KB/Policies/Documents/euthanasia.pdf

Anil, H. and Lambooij, B. (2009). Stunning and slaughter methods. In: Smulders, F.J.M. and Algers B. (Eds) *Welfare of production animals: assessment and management of risks*. 1st Edition. Wageningen Academic Publishers: Wageningen, The Netherlands.

Anil, M.H., Love, S., Williams, S., Shand, A., McKinstry, I.L., Helps, C.R., Waterman-Pearson, A., Seghatchian, J. and Harbour, D. (1999). Potential contamination of beef carcasses with brain tissue at slaughter. *Veterinary Record* 145, 460–462.

Atkinson, S. and Algers, B. (2009). Cattle welfare, stun quality and efficiency in 3 abattoirs using different designs of stun box loading, stun box restraint and weapons. Project Report Department of Animal Environment and Health, Swedish University of Agricultural Sciences (SLU): Skara, Sweden.

Atkinson, S., Velarde, A. and Algers, B. (2013). Assessment of stun quality at commercial slaughter in cattle shot with captive bolt. *Animal Welfare* 22, 473–481.

Blackmore, D.K. (1979). Non-penetrative percussion stunning of sheep and calves. *Veterinary Record* 105, 372–375.

Blackmore, D.K. (1984). Differences in behaviour between sheep and cattle during slaughter. *Research in Veterinary Science* 37, 223–226.

Blackmore, D.K., and Delany, M.W. (1988). *Slaughter of stock - a practical review and guide*. Veterinary Continuing Education, Massey University: Palmerston North, New Zealand. pp. 137.

Council Regulation (EC) No 1099/2009 of 24 September (2009) on the protection of animals at the time of killing. *Official Journal of the European Union* L303, 1–30.

Daly, C.C. and Whittington, P.E. (1989). Investigation of the principal determinants of effective captive bolt stunning of sheep. *Research Veterinary Science* 46, 406–408.

Daly, C.C., Gregory, N.G. and Wotton, S.B. (1987). Captive bolt stunning of cattle: effects on brain function and role of bolt velocity. *British Veterinary Journal* 143 (6), 574–580.

Daly, C. C. Kallweit, E. and Ellendorf, F. (1988). Cortical function in cattle during slaughter: conventional captive bolt stunning followed by exsanguination compared with shechita slaughter. *Veterinary Record*, 122 (14), 325–329

Davidsson, J., Algers, B. and Risling, M. (2014). Stunning physics – Studies into cavity formation in gel samples, pressure distribution in gel samples and in the brain during simulated stunning experiments. Project Report, Department of Animal Environment and Health, Swedish University of Agricultural Sciences (SLU): Skara, Sweden. 33 pp.

Di Maio, V.J.M. (1999). *Gunshot wounds.* 2nd Edition. CRC Press: Boca Raton, USA.

EFSA (Ed) (2004). Opinion of the Scientific Panel on Animal Health and Welfare on a request from the Commission related to welfare aspects of the main systems of stunning and killing the main commercial species of animals. *EFSA Journal* 45, 1–29.

EFSA (Ed) (2005). Aspects of the biology and welfare of animals used for experimental and other scientific purposes. Annex to the *EFSA Journal* 292, 1–136.

EFSA (Ed) (2013). Scientific Opinion on monitoring procedures at slaughterhouses for bovines. *EFSA Journal* 11(12), 3460, 1–65.

Eichbaum, F.W., Slewer, O. and Yasaka, W.J. (1975). Postdecapitation convulsions and their inhibition by drugs. *Experimental Neurology* 49, 802–812.

Endres, J.M. (2005). Effektivität der Schuss-Schlag-Betäubung im Vergleich zur Bolzenschussbetäubung von Rindern in der Routineschlachtung [Effectiveness of concussion stunning in comparison to captive bolt stunning in routine slaughtering of cattle]. Vet. med. Diss Ludwig-Maximilians-Universität: München. 210 pp.

Fahrbach, R. (1948). Die heute üblichen Betäubungsverfahren bei Schlachttieren und ihre historische Entwicklung. *Thesis*, Hannover.

FAO (2001). Guidelines for humane handling, transport and slaughter of livestock. (Compiled by P.G. Chambers and T. Grandin; edited by G. Heinz and T. Srisuvan). *RAP Publication* 2001/4. Food and Agricultural Organisation of the United Nations Regional Office for Asia and the Pacific. Humane Society International: Washington, DC, USA.

Finnie, J.W. (1993a). Brain damage caused by a captive bolt pistol. *Journal of Comparative Pathology* 109, 253–258.

Finnie, J.W. (1993b). Pathology of experimental traumatic craniocerebral missile injury. *Journal of Comparative Pathology* 108, 93–101.

Finnie, J.W. (1997). Traumatic head injury in ruminant livestock. *Australian Veterinary Journal* 75 (3), 204–208.

Finnie, J.W., Manavis, J., Blumbergs, P.C. and Summersides, G.E. (2002). Brain damage in sheep from penetrating captive bolt stunning. *Australian Veterinary Journal* 80 (1–2), 67–69.

Fries, R., Schrohe, K., Lotz, F. and Arndt, G. (2012). Application of captive bolt to cattle stunning: a survey of stunner placement under practical conditions. *Animal* 6, 1124–1128.

Gibson, T. J., Ridler, A. L., Lamb, C. R., Williams, A., Giles, S. and Gregory, N. G. (2012). Preliminary evaluation of the effectiveness of captive-bolt guns as a killing method without exsanguination for horned and unhorned sheep. *Animal Welfare* 21, 35–42.

Gilliam, J.N., Shearer, J.K., Woods, J, Hill, J., Reynolds, J. and Taylor, J.D. (2012). Captive-bolt euthanasia of cattle: determination of optimal-shot placement and evaluation of the Cas Special Euthanizer Kit for euthanasia of cattle. *Animal Welfare* 21, 99–102.

Grandin, T. (1994). Euthanasia and slaughter of livestock. *Journal of the American Veterinary Medical Association* 204, 1354–1360.

Grandin, T. (1981). Bruises on south western feedlot cattle. *Journal of Animal Science* (Suppl.1) 53, 213 (Abstract).

Grandin, T. (2001). Cattle vocalisations are associated with handling and equipment problems at beef slaughter plants. *Applied Animal Behaviour Science* 71, 191–201.

Grandin, T. (2002). Return-to-sensibility problems after penetrating captive bolt stunning of cattle in commercial beef slaughter plants. *Journal of the American Veterinary Medical Association* 221 (9), 1258–1261.

Gregory, N. G. (1998). Stunning and slaughter.

In: Gregory N.G. (Ed.) *Animal welfare and meat science*. CAB Int.: Wallingford, Oxon, UK. pp.223–240.

Gregory, N.G. and Wotton, S.B. (1990). Comparison of neck dislocation and percussion of the hand on visual evoked responses in the chicken's brain. *Veterinary Record* 126, 370–572.

Gregory, N.G., Lee, C.J. and Widdicombe, J.P. (2007). Depth of concussion in cattle shot by penetrating captive bolt. *Meat Science* 77 (4), 499–503.

Hewitt, L. (2000). The development of a novel device for humanely despatching casualty poultry. Thesis, University of Bristol, UK.

Holleben, K. v., Schütte, A., Wenzlawowicz, M. v., Bostelmann, N. (2002). Call for veterinary action in slaughterhouses. Deficient welfare at CO2-stunning of pigs and captive bolt stunning of cattle. *Fleischwirtschaft International* 3/02, 8–10.

Holleben, K., v.,von Wenzlawowic, M., Gregory N., Anil, H., Velarde, A., Rodríguez, P., Cenci Goga, B., Catanese, B. and Lambooij, B. (2010). Animal welfare concerns in relation to slaughter practices from the viewpoint of veterinary sciences. Dialrel Report. http://www.dialrel.eu/images/veterinaryconcerns.pdf

Hollson, R.R. (1992). Euthanasia by decapitation: evidence that this technique produces prompt, painless unconsciousness in laboratory rodents. *Neurotoxicology Teratology* 14, 253–257.

Hopkinson, D.A.W. and Marshall, T.K. (1967). Firearm injuries. *British Journal of Surgery* 54, 344–353.

HSA (Humane Slaughter Association) (2006). *Captive-bolt stunning of livestock.* Guidance Notes No. 2. 4th Edition. Humane Slaughter Association: Wheathampstead, Herts., UK. pp. 23.

Karger, B. (2009). Penetrating gun shots to the head and lack of immediate incapacitation I. Wound ballistics and mechanisms of incapacitation. *International Journal of Legal Medicine* 108, 53–61.

Keller, G.L. (1982). Physical euthanasia methods. *Laboratory Animal* 11, 20–26.

Kohlen, S. (2011).Untersuchungen zum korrekten Treffpunkt für den Bolzenschuss bei der 648 Betäubung von Rindern bei der Schlachtung. Thesis, Tierärztliche Fakultät der Ludwig-Maximilians-Universität München, Germany.

Lambooij, E. (1981). Mechanical aspects of skull penetration by captive bolt pistol in bulls, veal calves and pigs. *Fleischwirtschaft International* 61, 1865–1867.

Lambooy, E. and Spanjaard, W. (1981). Effect of the shooting position on the stunning of calves by captive bolt. *Veterinary Record* 109 (16), 359–361.

Lambooij, E., Spanjaard, W.J. and Eikelenboom, G. (1981). Concussion stunning of veal calves. *Fleischwirtschaft* 61, 98–100.

Lopes da Silva, H.F. (1983). The assessment of consciousness: General principles and practical aspects. In: Eikelenboom G. (Ed.) *Stunning of animals for slaughter.* Martinus Nijhoff: The Hague. pp. 3–12.

Marzin, V., Collobert, J. F., Jaunet, L. and Marrec, L. (2008). Critères pratiques de mesure de l'efficacité et de la qualité de l'étourdissement par tige perforante chez le bovin. *Revue Médicine Véterinaire* 159, 423–430.

Nilsson, B. and Nordström, C.H. (1977). Rate of cerebral energy consumption in concussive head injury in the rat. *Journal of Neurosurgery* 47, 274–281.

Nydahl, M. (2014). Simulation of pressure wave propagation in connection to brain injury mechanism. Chalmers University of Technology, Gothenburg, Sweden. Master's thesis 2014:59, 32 pp.

OIE (World Organization for Animal Health). (2014). Terrestrial Code. 23rd Edition. Paris, France.

Raj, A. B. M. and O'Callaghan, M. (2001). Evaluation of a pneumatically operated captive bolt for stunning/killing broiler chickens. *British Poultry Science* 42, 295–299.

Schiffer, K.J., Retz, S.K., Richter, U., Algers, B. and Hensel, O. (2014). Assessment of key parameters for gunshot used on cattle: a pilot study on shot placement and effects of diverse ammunition on isolated cattle heads. *Animal Welfare* 23, 479–489.

Schmidt, G.R., Hossner, K.L., Yemm, R.S. and Gould, D.H. (1999). Potential for disruption of central nervous system (CNS) tissue in beef cattle by different types of captive bolt stunners. *Journal Food Protection* 62, 390–393.

Schütt-Abraham, I. (1995). Stunning methods for poultry: influence on birds' welfare and prospects for future EC regulations. In: Ricardo Cepero Briz: Poultry Products Microbiology/European Regulations and Quality Assurance Systems, WPSA meeting, Zaragoza (Spain), 25–29 September. pp. 333–344.

Von Wenzlawowicz, M., von Holleben, K. and Eser, E. (2012). Identifying reasons for stun failures in slaughterhouses for cattle and pigs in a field study. *Animal Welfare* 21, 51–60.

Wijngaarden-Bakker, L.H. van (1980). An archeozoological study of the Beaker settlement at Newgrange, Ireland. Thesis, Amsterdam.

chapter seven

# Electrical stunning and killing methods

*Mohan Raj*[1] *and Antonio Velarde*[2]

## Learning objectives

- To understand the basic neurological and physiological bases of electrical stunning.
- To become familiar with electrical stunning methods and their effective use.
- To learn how to assess stunning effectiveness and be able to detect animals at risk of recovering consciousness.
- To identify the main risk factors of unsuccessful stunning.

## CONTENT



## SUMMARY

Electrical stunning is a commonly used method for pigs, sheep, goats, rabbits, poultry and, to a limited extent, for cattle, under slaughterhouse conditions. It involves application of an electric current of sufficient magnitude to the brain of an animal such that a generalized epileptiform activity is induced, which can be

[1] School of Veterinary Science, University of Bristol, Langford BS40 5DU, United Kingdom
[2] IRTA. Animal Welfare Subprogram, Veïnat de Sies, s/n, 17121 Monells, Spain

demonstrated using electroencephalogram (EEG) under laboratory conditions. Based on this criterion, species-specific minimum currents have been prescribed in the EC Slaughter Regulation 1099/2009.

Electrical stunning can be applied as head-only or head-to-body. Head-only stunning electrodes should span the brain; whereas, head-to-body stunning can involve a single cycle such that one electrode is placed on the forehead and the other on the body of the animal behind the position of the heart, or two separate current cycles such that a head-only current cycle is immediately followed by a transthoracic or head-to-body application of the current. Head-to-body stunning must always result in cardiac ventricular fibrillation, which can be demonstrated using an electrocardiogram (ECG) under laboratory conditions. Therefore, first (head-only) current cycle can be applied with high frequencies (Hz) but the second current cycle should be applied using a 50 Hz sine wave alternating current that is more effective in inducing cardiac ventricular fibrillation.

The efficacy of head-only electrical stunning can be ascertained in slaughterhouse conditions using animal-based indicators such as immediate collapse, tonic–clonic seizures, apnoea and fixed eyes (i.e. absence of palpebral and corneal reflexes). The efficacy of the induction of cardiac ventricular fibrillation can be ascertained in slaughterhouse conditions using animal-based indicators such as immediate collapse, tonic seizure leading to a brief clonic seizure or relaxation of carcass (loss of muscle tone) and dilated pupils.

In order to fulfil humane slaughter requirements, the duration of unconsciousness induced by head-only electrical stunning must be longer than the sum of time that lapses between the end of stun and the time to onset of death. Since the effect of head-only electrical stunning is momentary, the onus of preventing resumption of consciousness thereafter relies on the efficiency of the slaughter procedure, that is, the prompt and accurate severance of blood vessels supplying oxygenated blood to the brain.

Unlike red meat species, poultry are shackled prior to electrical stunning and conveyed into an electrified waterbath and several birds (e.g. 40) can be stunned at any one moment. Under the electrical waterbath stunning situation, the current flows through the whole body and the effectiveness of stunning depends upon the electrical waveforms (sine or square wave alternating current, pulsed direct current or combinations), frequencies (Hz) and the amounts of current (Ampere) applied to individual birds. The efficacy of electrical waterbath stunning can be ascertained in slaughterhouse conditions using animal-based indicators, which includes arched neck, wings tightly held close to the breast, fixed eyes and absence of breathing. In general, electrical waterbath stunning can lead to cardiac arrest in birds when applied using low frequencies (e.g. 50Hz mains current) or reversible when applied using high frequencies (200 Hz or more). Cardiac arrest induced at waterbath stunning can be recognized from the complete loss of muscle tone and dilated pupils.

## 7.1 INTRODUCTION

A requirement under most of the humane slaughter regulations is that electrical stunning must induce an immediate loss of consciousness and sensibility that lasts until death occurs through blood loss after slaughter or through the induction of cardiac arrest (e.g. EC Regulation 1099/2009). Regulation also requires that animals must be dead before any carcass dressing or scalding operations begin. Electrical stunning has been considered as a humane method on the basis of the scientific principle that stimulation of brain with an electric current of sufficient magnitude induces generalized epilepsy.

## 7.2 NEUROLOGICAL AND PHYSIOLOGICAL ASPECTS

The stimulation of the brain with an electric current of sufficient magnitude induces an epileptiform activity similar to that recorded in humans during *grand mal* epileptic seizures (Croft, 1952; Hoenderken, 1978). In humans, generalized epilepsy occurs as a pathological state of neurones involving both cerebral hemispheres, is considered to be incompatible with normal neuronal function and, hence, is invariably accompanied by unconsciousness. This seizure-like state, immediately followed by an exhausted state, is suggestive of a loss of consciousness and appears to be associated with a lack of sensory awareness, which only lasts a finite period of time (Gregory and Wotton, 1984). It is also known that spreading depression (SD) also occurs as a pathophysiological consequence of generalized epilepsy. In sheep, a minimum of 0.2-second application of 1.0 Amp was found to be necessary to induce epileptiform activity in the brain (Cook et al., 1995), and therefore electrical stunning is considered to cause immediate loss of consciousness.

Epileptiform activity in the brain is induced by an increase in the release of several excitatory neurotransmitters (e.g. glutamate and aspartate) into the extracellular space (Cook et al., 1992; Figure 7.1.). These changes are transient and if the animal is not slaughtered immediately and is allowed to recover from the seizure-like state, it will start regaining consciousness rapidly. (Blackmore and Newhook,1982). It has been suggested that a period of analgesia lasting between 5 and 15 minutes and caused by a release of gamma amino butiric acid (GABA) follows the end of the seizure-like state; the precise duration of this period appears to be related to the magnitude of the changes in the electrical activity of the brain caused by the stun (Cook et al., 1992).

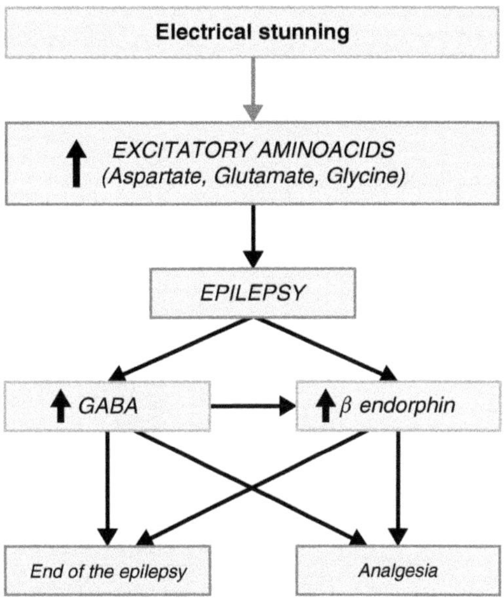


**Figure 7.1:** Relationship of the neurotransmitters during epilepsy and analgesia induced by electrical stunning.
*(Adapted from Cook et al., 1992)*

Under laboratory conditions, the induction and maintenance of unconsciousness and insensibility following electrical stunning can be ascertained by recording the brain activity using electroencephalograms (EEGs) or electrocorticograms (ECoGs). Scientific literature suggests that there are different kinds of partial or focal epilepsies involving one cerebral hemisphere or a group of neurones that are not always associated with loss of consciousness in humans. It is also known that the threshold current necessary to induce seizures is less than that required to induce generalized epileptiform activity in the brain. Therefore, electrical stunning parameters should be carefully selected on the basis of the induction of epileptiform activity in the brain recorded using EEGs under laboratory conditions. The epileptiform activity is followed by a period of quiescent or profoundly suppressed EEG, which is referred to as spreading depression and occurs due to hyperpolarization (Figure 7.2.). The power ($V^2$) content in the quiescent or isoelectric EEG is less than 10% of the pre-stun power content (Bager et al., 1990; Lukatch et al., 1997). Predominance of 8 to 13 Hz high amplitude (> 100 μV) EEG activity characterizes the occurrence of grand mal epilepsy in mammals; however, poultry species seem to differ from this norm. In this sense, some studies have reported low frequency (3–5Hz) polyspike activity or spike suppression in the EEG (Gregory, 1986).

When stunning-induced EEG or ECoG changes are ambiguous, abolition of auditory, somatosensory or visual evoked potentials (AEPs, SEPs, VEPs, respectively) in the brain can be used to ascertain brain responsiveness to these external stimuli.

Generalized epilepsy involving both cerebral hemispheres is always associated with tonic–clonic seizures. In sheep, the seizure activity occurring after electrical stunning includes one tonic and two clonic phases, with the convulsions being more intense and less coordinated in the second clonic phase than in the first clonic phase (Velarde et al., 2002).

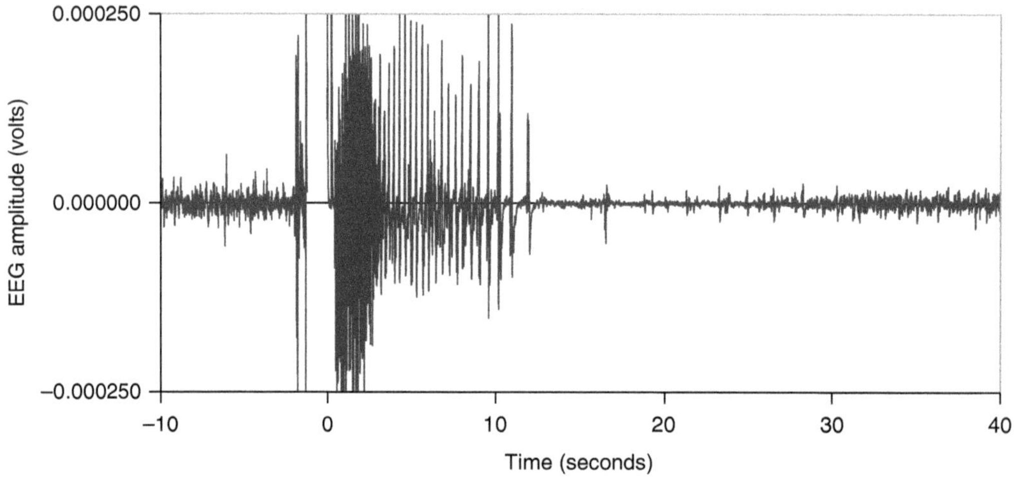


**Figure 7.2:** Illustration of electrical stunning induced EEG changes

*Note: duration of unconsciousness = duration of epileptiform activity plus duration of suppressed EEG activity.*

Although both phases are described as clonic convulsions, the pattern of physical activity, the parts of the body involved and the intensity of the convulsions are different, which suggests that the brain area responsible for each clonic phase may be different. According to the EEG, during the second clonic phase sheep are judged to have recovered consciousness but remain insensible to pain due to analgesia induced by the stun, as mentioned previously.

Head-to-body is an irreversible stunning method that consists of the application of an electrical current of sufficient magnitude across the brain and heart of an animal, leading to loss of consciousness as in head-only stunning and fibrillation of the heart. Cardiac ventricular fibrillation severely impairs cardiac output (reduced to less than 30% of normal) and normal blood circulation. Consequently, it induces hypoxia in the brain and myocardium, which either prolongs the period of unconsciousness and insensibility induced by the head-only electrical stun or leads to death under electrical stun/killing methods (Von Mickwitz et al., 1989). Under these circumstances, at the least, the ability of an animal to regain consciousness and sensibility is seriously impaired, even if it is not bled out. Therefore, stun-to-stick interval is not critical (irreversible stunning). Head-to-body stunning induces tonic seizure but clonic seizure may be brief or absent due to fibrillation of the heart, which can be observed on an electrocardiogram (ECG). Successful induction of cardiac arrest also leads to an isoelectric EEG, eliminating any chance of recovery in animals.

In poultry species, head-only electrical stunning leads to immediate onset of clonic convulsions, occurring as wing flapping (Raj, 2004; Raj and O'Callaghan, 2004a,b), and application of electrical stunning using a waterbath leads to tonic phase that can be recognized from the arched neck and wings held tightly against the breast (Raj et al., 2006a,b,c). Induction of cardiac ventricular fibrillation in a waterbath stunner leads to completely relaxed carcass manifested as drooping wings at the exit from the stunner (Raj, 2006).

## 7.3 DESCRIPTION OF THE METHOD

### *7.3.1 Head-only electrical stunning*

Head-only electrical stunning is normally applied to pigs, sheep, goat, rabbit and poultry species by placing the tongs on either side of the head (transcranial application). It may be achieved manually (Figure 7.3.) or automatically, by using purpose-built devices. In cattle, however, one (live) electrode is placed on the muzzle and the earth electrodes behind the ears.

If sufficient current is administered through the head (all brain parts are stimulated) a generalized epileptiform insult will occur. Ohm's law defines the

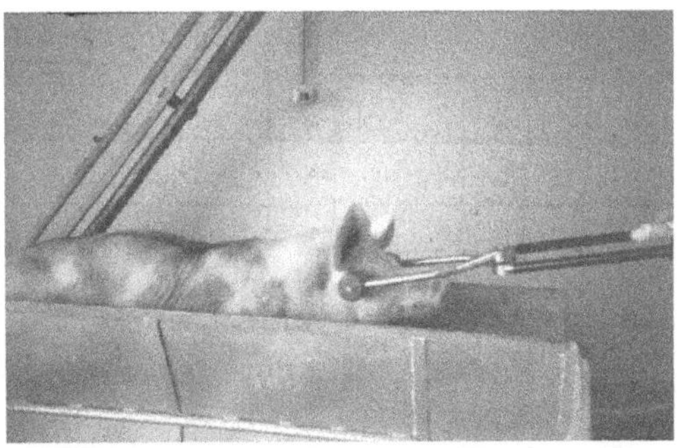

**Figure 7.3:** Head-only electrical stunning applied manually

relationship between voltage, resistance and current:

$$\text{Current (I)} = \frac{\text{Voltage (V)}}{\text{Resistance (R)}}$$

A minimum current (the threshold level), which is a function of the amount of voltage and inversely proportional to the electrical resistance of the stunning tongs and tissues (i.e. amount of fleece, dryness of skin, thickness of subcutaneous fat, and skull bone density and thickness) in the current pathway between the stunning electrodes, is necessary for the effectiveness of the electrical insult. Application of a voltage lower than that required to break down the electrical resistance in this pathway and hence deliver currents lower than the threshold necessary to induce grand mal epilepsy (either due to low voltage, high resistance or misplaced electrodes) or when the current does not pass through the brain due to shunting between the electrodes over the surface of the skull, will induce a potentially painful arousal or seizures rather than unconsciousness.

With constant voltage stunners, the amount of current delivered to the animal varies due to variation of the total impedance or resistance in the pathway between the electrodes. Research has shown that the minimum voltage required to deliver the required currents is determined by several factors, including:

- properties of the electrical stunner (e.g. constant current with variable voltage, constant current with constant voltage or constant voltage);
- material, design, construction and maintenance of stunning electrodes/tongs;
- size and shape of electrodes;
- position of electrodes on the animal;
- electrical resistance in the pathway;
- pressure applied during the application.

Electrodes need to be kept clean to reduce resistance to flow of current. If water or saline are not used, electrodes are required to be routinely cleaned with a powered wire brush (Gregory, 1998). Poor electrode maintenance or contact with the head can be recognized from burning of the skin, hair or feathers due to the development of heat, which normally occurs due to increased electrical resistance. In addition, maintaining good electrical contact during stunning and supply of voltage high enough to deliver a recommended current are also essential. In sheep, the position of the stun electrodes and parameters such as the presence or absence of wool and the wetness of the skin are essential determinants of a successful stun. The recommended tong position from a welfare perspective is between the eyes and the base of the ears on both sides of the head, preferably on wet skin (Velarde et al., 2000).

When a constant voltage stunner is used, the current starts to flow from zero to the maximum, which would take a certain time depending upon the voltage. The time taken to break down this resistance seems to be shorter when high voltages (250 V or more) are employed. By contrast, constant current stunners are designed and constructed in such a way that they anticipate high resistance in the pathway and hence start with the maximum available voltage, which is usually in excess of 250 V. Owing to this, the target current is reached within the first few current cycles (within milliseconds of the start of current application) and the applied voltage may also be modulated according to changes in resistance. Therefore, constant current stunners are preferred to constant voltage stunners on animal welfare grounds.

Electrical stunning devices involve varieties of waveforms and frequencies of currents. The generic waveforms of currents are pulsed direct currents (DC) and sine wave alternating currents (AC). The amount of current and voltage delivered using a sine wave AC is expressed as root mean square (RMS) or peak-to-peak because it flows in both positive and negative directions (bipolar). The most common electrical stunning methods use 50 Hz sine wave alternative current (AC) frequency. However, the frequency can be as high as 1800 Hz and, in comparison with low frequency currents, high frequencies require higher voltages to induce epilepsy and the duration of epilepsy is also shorter.

Stunning equipment should be fitted with devices displaying the delivered voltage and current during each stunning cycle and with an acoustic and/or optic signals, to indicate an interrupted stun, excessively short stun duration and/or increase in total electrical resistance in the pathway (due to dirt, fleece or carbonization), which could lead to insufficient current flow and inadequate stunning. Such devices would facilitate effective monitoring of electrical stunning. The details of electrical parameters, such as waveform, frequency and the output voltage and current in appropriate units (average or RMS) need to be readily available for inspection to verify that correct parameters are applied, ensuring that a current of sufficient magnitude beyond that needed to induce generalized epilepsy is applied. The equipment must be inspected at regular intervals, in order to ensure that it is operating correctly according to the specification and that it is in a good state of repair. A calibrated volt and/or current meter appropriate to the waveform of the current should be used to verify the output of the stunner.

Electrical stunning is widely used in rabbit slaughter plants. This method can be applied by using handheld or wall-mounted V-shaped electrodes (Anil et al., 1998; Maria et al., 2001). The head of the rabbit is placed in the V-shaped electrode in a way that allows for transcranial current flow. The electrodes are placed between the outer corners of the eyes and the base of the ears thus spanning the brain. To avoid pain as well as injury to the rabbit's back, the animal is restrained by one hand supporting its belly while the other is guiding the head by holding its ears. Few studies have investigated the use of electrical stunning of rabbits from a welfare point of view. Therefore, due to insufficient data, no recommendations can be given with regard to the magnitude, type and duration of the current required to ensure that all rabbits are adequately stunned and remain unconscious until death from bleeding supervenes (EFSA, 2006). Until further research has been carried out, the minimum recommended current for head-only stunning of rabbits is 400 mA. Head-only stunning is fully reversible and will give only a short temporary stun; the animals need to be promptly bled by severing both carotid arteries.

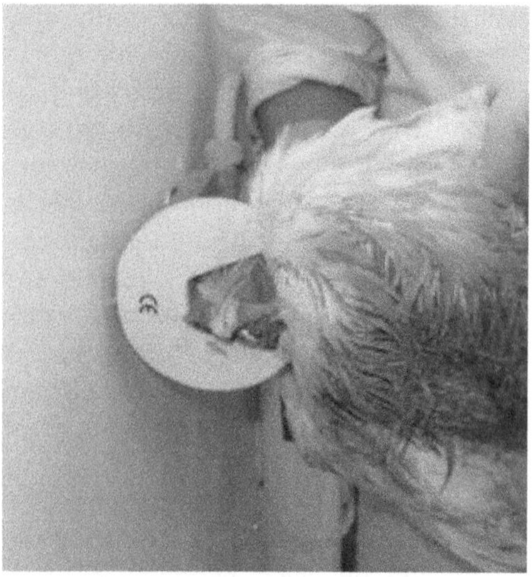

**Figure 7.4:** Illustration of a wall-mounted head-only electrical stunning system used for poultry in small-scale slaughterhouses

**Figure 7.5:** Head-only electrical stunning system used for poultry in large-scale slaughterhouses.

*(Courtesy of Bert Lambooij)*

Head-only electrical stunning of poultry is usually applied using a pair of handheld or wall-mounted stunning tongs (Figure 7.4; Berg and Raj, 2015). More recently, a completely automated head-only electrical stunning system has been developed (visit www.topkip.com) and used in large-scale poultry slaughterhouses (Figure 7.5; Berg and Raj, 2015).

### *7.3.2 Head-to-body electrical stunning*

Head-to-body stunning of red meat species involves induction of cardiac ventricular fibrillation (rapid and irregular beating of the heart) by passing an electric current across the heart in unconscious animals that have been stunned by head-only electrical stunning or simultaneous induction of unconsciousness and cardiac ventricular fibrillation. Under commercial conditions, cardiac ventricular fibrillation may be induced using a *single-cycle* or a *two-cycle* system. In a single-cycle system, induction of cardiac ventricular fibrillation involves application of an electric current by using a pair of tongs (or electrodes) placed on the head (in front of the brain) and body (behind the apex of the heart) or an electrified water-bath (in poultry species only), such that the electrical field spans the brain and heart (Figure 7.6). In a two-cycle system, two separate electric current cycles are used: a transcranial application is immediately followed by a second application of an electric current from head-to-body (behind the position of heart) or across the chest (transthoracic). The rear electrode on the body should be behind the position of the heart. In both cases, the magnitude of the current should be enough to induce cardiac ventricular fibrillation, leading to cardiac arrest. The amount of current delivered will depend upon the voltage and total impedance in the pathway (between the electrodes) that is affected by size of the electrodes, applied pressure, the phase of respiration during which the shock is applied, use of coupling gel and its salt content, and the distance between the electrodes, which

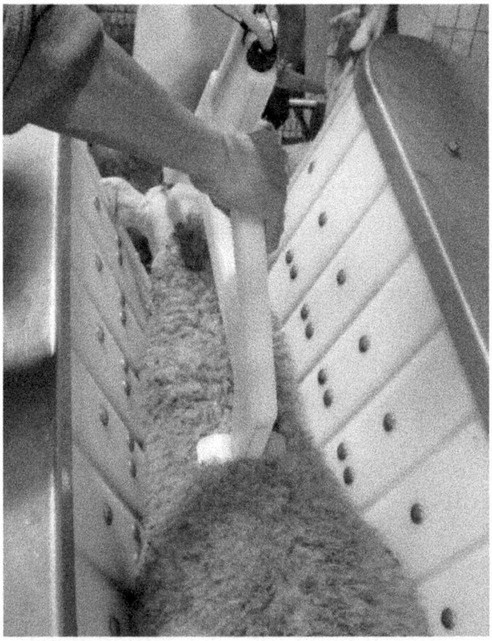

**Figure 7.6:** Head-to-body electrical stunning applied manually

is dependent upon the circumference of chest during transthoracic application (Niemann et al., 2003). Voltage should be at least 100 mV/cm at the level of the heart (Von Mickwitz et al., 1989). In small animals (piglets, young lambs, rabbits), it may be difficult to induce cardiac ventricular fibrillation because, due to the small size of the heart, the current passes through tissues surrounding the heart, rather than through the heart. The electrical resistance of various other tissues in the pathway may also play a role in this. High frequencies do not produce cardiac ventricular fibrillation, but they may reduce muscle spasms and convulsions. Therefore, first (head-only) current cycle can be applied at high frequencies but the second current cycle should be applied using a 50 Hz sine wave alternating current that is more effective in inducing cardiac ventricular fibrillation.

Several automatic electrical stunning methods are currently available. One device consists of a V-type restrainer that makes contact with the electrodes at the rear end of each pig and receives the stunning current. During stunning, the animals are turned out and fall onto a table. A second method also uses a conveyor belt system; at the end of the restrainer, the nose of the pigs interrupts a beam of light, which activates the electrodes. The electrodes are positioned between the eye and ear. After 1 second of stunning, a heart electrode is positioned behind the left shoulder and current delivered for 1.5 seconds. As a result of the body current, the animals do not show muscle contractions.

In cattle, head-only electrical stunning induces a short duration of the epileptiform insult. Induction of cardiac ventricular fibrillation resolves the problem of the short duration of unconsciousness. Chest sticking, by cutting the major blood vessels arising from the heart, should be preferred, to ensure efficient haemorrhaging and a rapid loss of brain activity (Anil et al., 1995). In currently used systems, a low voltage spinal discharge (electro-immobilization) is necessary to stop clonic convulsions and to make sticking possible. However, electro-immobilization may mask signs of consciousness.

Two types of automatic cattle stunners are commercially available. The first type consists of a modified cattle restraint where the head of the animal is restrained by a pair of metal neck yokes, a chin lift and a nose (muzzle) plate that can be electrified. A third and fourth electrode are placed on the brisket and on the tail base. It runs three programmes sequentially:

1. A 3-second head-only cycle (neck yoke/nose electrodes) induces loss of consciousness and sensibility.
2. A 15-second cardiac cycle (brisket/nose electrodes) induces ventricular fibrillation.
3. A 4-second spinal cycle (rear end/nose electrodes) reduces uncontrolled leg kicking after death.

The stunner delivers current at 550 V, 50 Hz sinusoidal AC, via a choke, which limits the current to a maximum of about 3.5 A.

The second type of automatic electrical beef stunner (Troeger, 2002) delivers 300 V, constant current while frequency can be selected from a 10–990 Hz range. It has only two cycles: a head-only cycle followed by a head–body cycle.

### *7.3.3 Waterbath stunning*

Multiple bird electrical waterbath stunning (Figure 7.7.) is the most common and cheapest method of rendering poultry unconscious prior to slaughter under commercial conditions, where high throughput rates are required.

Under this system, conscious birds are hung upside down on a moving metal shackle line and passed through an electrified waterbath, such that the current flows through the whole body towards the shackle, the earth. There are many welfare problems associated with the commercial electrical waterbath stunning systems. A good understanding of these problems would help to minimize suffering in birds.

*Pain and suffering associated with uncrating.* Three types of crates (containers) are mainly used for transporting poultry to high throughput slaughterhouses. From bird welfare and meat quality points of view, the open top drawer type crates are considered to be better than the old fashioned loose crates (which involve removing birds through a hole of

**Figure 7.7:** Electrical waterbath stunning system used for poultry in large-scale slaughterhouses

about A4-size paper) and modules (which involve tipping birds from a height of more than 2 meters).

*Pain and suffering associated with shackling.* Inevitably, the legs of poultry are compressed during shackling and the degree of compression has been reported to be as high as 20% (Sparrey, 1994). Based on the presence of nociceptors in the skin over the legs of poultry and the close similarities between birds and mammals in nociception and pain, shackling has been reported to be a painful procedure (Gentle, 1992; Gentle and Tilston, 2000). This pain and suffering would be worse in birds suffering from painful lameness due to diseases or abnormalities of the leg joint and/or bone (Butterworth, 1999; Danbury et al., 2000). This pain is also likely to be significant in birds suffering from dislocation of joints and/or fracture of bones induced by rough handling during catching, crating and uncrating. Owing to this, the maximum time for which chickens may be suspended from shackle lines before waterbath stunning is limited to 1 minute for chickens and 2 minutes for turkeys (EC Regulation 1099/2009).

*Pain and suffering associated with pre-stun electric shocks.* The complexity of commercial waterbath stunning systems and the physical contact between adjacent birds on the shackle line make it difficult, if not impossible, to control the current pathway and eliminate this potential bird welfare problem.

*Delivery of sufficient current to render each bird unconscious.* The commercial electrical waterbath stunner may contain many birds (e.g. 40) at any one moment and, as birds enter a stunner supplied with a constant voltage, they form a continuously changing parallel electrical circuit (Sparrey et al., 1993). Under this situation, according to Ohm's law, each bird will receive a current inversely proportional to the electrical resistance or impedance in the pathway. The effective electrical impedance can vary between birds, usually 1,000 to 2,600 Ohms in broilers and 1,900 and 7,000 Ohms in layer hens (Schutt-Abraham et al., 1987; Schutt-Abraham and Wormuth, 1991). Ironically, most of the electrical impedance in the pathway between the electrified waterbath and the earth is attributed to the poor contact between the legs and metal shackle. The implication of this is that tighter shackle-leg fitting will reduce the electrical impedance at the expense of increased suffering due to pain. However, owing to the variable electrical impedance, it will be impossible to deliver to each bird in a waterbath stunner a pre-set minimum current necessary to achieve humane stunning (Sparrey et al., 1992). On the other hand, it has been suggested that installation of constant current stunners under commercial conditions would ensure delivery of a pre-set current to each of the birds in a waterbath (Sparrey et al., 1993). During stunning with a variable voltage/constant current stunner, each bird is electrically isolated and the stunner modulates the voltage required to deliver a pre-set current by continuously monitoring the impedance in the pathway. However, considering that the birds are suspended on shackles 15 cm apart and the processing line is operating at a speed of up to 220 birds per minute, it has been argued that it will be extremely difficult to isolate each bird long enough to measure its resistance or impedance in the pathway and deliver the pre-set current (Bilgili, 1992). In addition, owing to the differences in the electrical resistance of various tissues in the pathway, it has been reported that only a small proportion of current (10 to 28%) applied in a waterbath may flow through the brain and the majority may flow through the carcass (Wooley et al., 1986a and 1986b), which seems to question the humanitarian advantages of a constant current waterbath stunning system.

*Live birds entering scald tanks.* Live birds can enter scald tanks under two scenarios. First, inadequately stunned birds and those that have missed the stunner, due to wing flapping or being runts, miss the neck cutter by holding their heads up. Occasionally, effectively stunned birds also miss the neck cutting machine due to the fact that they miss the rails that guide the neck towards the blade(s). Hence, if these birds were not slaughtered manually, they will enter the scald tank live and conscious. Second, adequately stunned birds could have a poor neck cut and hence enter the scald tank alive but unconscious. Although legislation requires that a manual backup should be present to cut necks of birds that missed the neck cutter, owing to fast throughput rates, manual backup alone is not sufficient to prevent this potential welfare problem.

The potential welfare problem of live birds entering scald tanks, recognized by the occurrence of "red-skin" carcass, was reported to be the result of poor slaughter procedures (Harris and Carter, 1977). In the 1980s, it was reported that almost one third of the birds processed under commercial conditions may be entering scald tanks alive (Heath et al., 1981; Griffiths and Purcell, 1984). Heath et al., (1983) suggested that red-skin carcasses are produced from poultry that are alive when they entered the scald tanks and this was later confirmed experimentally by Griffiths (1985) to be the consequence of an acute inflammatory reaction. In recent years, the potential for this problem to occur has increased due to the use of high-frequency currents (which do not induce cardiac arrest) in the waterbath and significant increases in throughput rates.

## 7.4 EFFECTIVE USE

### *7.4.1 Head-only and head-to-body electrical stunning*

The duration of unconsciousness induced by the electrical stunning method must be longer than the sum of time that lapses between the end of stun and slaughter (stun-to-stick interval) and the time to onset of death. Effective head-only electrical stunning produces a brief period of unconsciousness and is always accompanied by tonic–clonic seizures. As electrical stunning is reversible, it should be followed as soon as possible by a procedure ensuring death. The onus of preventing resumption of consciousness in effectively stunned animals relies on the efficiency of slaughter (bleeding) procedure, that is, the prompt and accurate severance of blood vessels supplying oxygenated blood to the brain. Severing the brachiocephalic artery in cattle and pigs and cutting two common carotid arteries in the necks of sheep, goat and poultry species would achieve this.

| Description | Conditions of use | Key parameters |
|---|---|---|
| Exposure of the brain to a current generating a **generalised epilptic form on the electro-encephalogram (EEG)**<br><br>Simple stunning | All species<br><br>Slaughter, depopulation and other situations | • Minimum **current** (A or mA)<br>• Minimum **voltage** (V)<br>• Maximum **frequency** (Hz)<br>• Minimum **time of exposure**<br>• Maximum stun-to-stick / kill intervals<br>• Frequency of calibration of the equipment<br>• Optimisation of the current flow<br>• Prevention of electrical shocks before stunning<br>• Position and contact surface area of electrodes |

**Figure 7.8:** Description, conditions of use and key parameters for head-only electrical stunning

The EC Slaughter Regulation 1099/2009 clearly states the description, conditions of use and key parameters, as presented in Figure 7.8.

The main determinant of the epileptiform activity that follows the stun is considered to be the amount of current flowing through the cerebral cortex (Hoenderken, 1978). The minimum currents required for head-only electrical stunning in Regulation 1099/2009 are listed in Table 7.1 below.

The EC Slaughter Regulation 1099/2009 clearly states the description, conditions of use and key parameters for head-to-body stunning, as presented in Figure 7.9.

The minimum currents required for head-to-body electrical stunning in Regulation 1099/2009 are listed in Table 7.2.

There are several factors that directly contribute to poor welfare for the following reasons:

- Poor stunning can occur through incorrect placement of the electrodes; inadequate electrical current

**Table 7.1** Minimum currents for head-only electrical stunning in Regulation 1099/2009

| Bovine – 6 months or older | Bovine – less than 6 months | Ovine and caprine | Porcine | Chicken | Turkeys |
|---|---|---|---|---|---|
| 1.28 A | 1.25 A | 1.0 A | 1.3 A | 240 mA | 400 mA |

| Description | Conditions of use | Key parameters |
|---|---|---|
| Exposure of the body to a current generating at the same time a **generalised epileptic form on the EEG and the fibrillation or the stopping of the heart**<br><br>Simple stunning in case of slaughter | All species<br><br>Slaughter, depopulation and other situations | • Minimum **current** (A or mA)<br>• Minimum **voltage** (V)<br>• Maximum **frequency** (Hz)<br>• Minimum **time of exposure**<br>• Maximum stun-to-stick / kill intervals<br>• Frequency of calibration of the equipment<br>• Optimisation of the current flow<br>• Prevention of electrical shocks before stunning<br>• Position and contact surface area of electrodes |

**Figure 7.9:** Description, conditions of use and key parameters for head-to-body electrical stunning

delivered to the brain due to wrong waveform (electrical) frequency, voltage or current. There is always some uncertainty in outcome due to equipment, and to varying resistances between animals.

- If the current is lower than recommended it may be that an animal is not rendered immediately unconscious and therefore might experience electric shocks. Electrical stimulation of muscle causes contractions that can be painful in a conscious animal (Croft, 1952).
- Cardiac arrest in a conscious animal may be painful.
- The time interval between rendering an animal unconscious and inducing death through bleeding out is critical in the head-only stunning situation, as the animal may recover consciousness before it is dead.
- Handling and restraint will cause stress as, at the outset, the animal will not be free to escape the pressure of the electrodes (which may be sharp). The change of environment, mixing in different groups

**Table 7.2** Minimum currents for head-to-body electrical stunning in Regulation 1099/2009

| Bovine – 6 months or older | Bovine – less than 6 months | Ovine and caprine | Porcine | Chicken | Turkeys |
|---|---|---|---|---|---|
| 1.28 A | 1.25 A | 1.0 A | 1.3 A | 240 mA | 400 mA |

| Description | Conditions of use | Key parameters |
|---|---|---|
| Exposure of the entire body to a current generating a **generalised epileptic form on the EEG and possibly the fibrillation or the stopping of the heart** through a waterbath<br><br>Simple stunning except where frequency is equal to less than 50 Hz | Poultry<br><br>Slaughter, depopulation and other situations | • Minimum **current** (A or mA)<br>• Minimum **voltage** (V)<br>• Maximum **frequency** (Hz)<br>• Frequency of calibration of the equipment<br>• Prevention of electrical shocks before stunning<br>• Minimising pain at shackling<br>• Optimisation of the current flow<br>• Maximum shackle duration before the waterbath<br>• Minimum **time of exposure** for each animal<br>• Immersion of the birds up to the base of the wings<br>• Maximum stun-to-stick / kill intervals for frequency over 50Hz(s) |

**Figure 7.10:** Description, conditions of use and key parameters for waterbath electrical stunning of poultry

and social isolation from cage/pen mates also have to be taken into consideration.

- Electrical currents through the brain will alter the biochemistry of the brain and brain neuro-peptide levels, and also may affect the biochemistry of other tissues such as muscle. Extravascular haemorrhages may occur in muscle, connective tissue and fat, and may cause muscle fibre ruptures and broken bones (Gregory and Wotton, 1991).

### *7.4.2 Waterbath stunning*

The EC Slaughter Regulation 1099/2009 clearly states the description, conditions of use and key parameters, as presented in Figure 7.10.

Since poultry are stunned using multiple bird waterbath stunners, the minimum average currents required for head-to-body electrical stunning in Regulation 1099/2009 are listed in Table 7.3.

## 7.5 MONITORING UNCONSCIOUSNESS

According to Council Regulation (EC) No 1099/2009, animals must be rendered unconscious and insensible by the stunning method and they must remain so until death occurs through bleeding. One

**Table 7.3** Electrical requirements for waterbath stunning equipment in Regulation 1099/2009

| Frequency (Hz) | Chickens | Turkeys | Ducks and geese | Quail |
|---|---|---|---|---|
| < 200 Hz | 100 mA | 250 mA | 130 mA | 45 mA |
| 200–400 Hz | 150 mA | 400 mA | Not permitted | Not permitted |
| 400–1,500 Hz | 200 mA | 400 mA | Not permitted | Not permitted |

way of achieving this animal welfare requirement is to monitor the state of consciousness and unconsciousness in animals at three key stages: (1) immediately after stunning; (2) at the time of neck cutting or sticking; and (3) during bleeding until death occurs. Efficient backup procedures are required where animals show signs of recovery of consciousness on the bleeding rail.

The effectiveness of stunning and sticking can also be recognized under the field conditions from characteristic changes in the behaviour of animals (e.g. loss of posture), physical signs (e.g. onset of seizures, cessation of breathing, fixed eye) and the presence or absence of response to physiological reflexes (e.g. response to an external stimulus such as blinking response to touching the cornea, known as corneal reflex, response to pain stimulus such as nose prick or toe pinching). In the scientific literature, these physical signs and reflexes have been referred to as indicators of unconsciousness or consciousness and used to monitor welfare at slaughter of animals (see, for example, EFSA, 2004, 2006). In red meat species, effective head-only electrical stunning induces immediate collapse of the animal and onset of tonic seizure, which is followed by two phases of clonic seizures. Research into head-only electrical stunning of sheep and pigs has shown that spontaneous breathing starts at the end of the first clonic phase and therefore animals should be ideally slaughtered during the tonic phase.

Nevertheless, the European Food Safety Authority (EFSA) have published scientific opinions on monitoring of animal welfare at slaughterhouses and, in these opinions, suggested toolboxes for monitoring welfare at key stages. Toolboxes for monitoring head-only electrical stunning of sheep and pigs are presented in Figures 7.11 and 7.12, respectively (EFSA, 2013a,b).

The head-only method induces tonic and clonic seizures, which are the outward symptoms of grand mal epilepsy. Following the stun, the hind legs are flexed under the abdomen and the forelegs fully extended. The body is tense and tonic (rigid), breathing is absent and the eyeballs may be rotated to a great extent that the pupils may not be visible. In some animals there will be running or paddling movements with the legs. A quiet phase can follow which is linked to exhaustion of the nervous system. The clonic (kicking) phase, which can be either a galloping, cantering or erratic kicking action, follows usually immediately after the tonic phase (Anil, 1991; Gregory, 1998). Electrically stunned pigs and sheep show two clonic phases and spontaneous breathing and signs of consciousness and sensibility at the end of

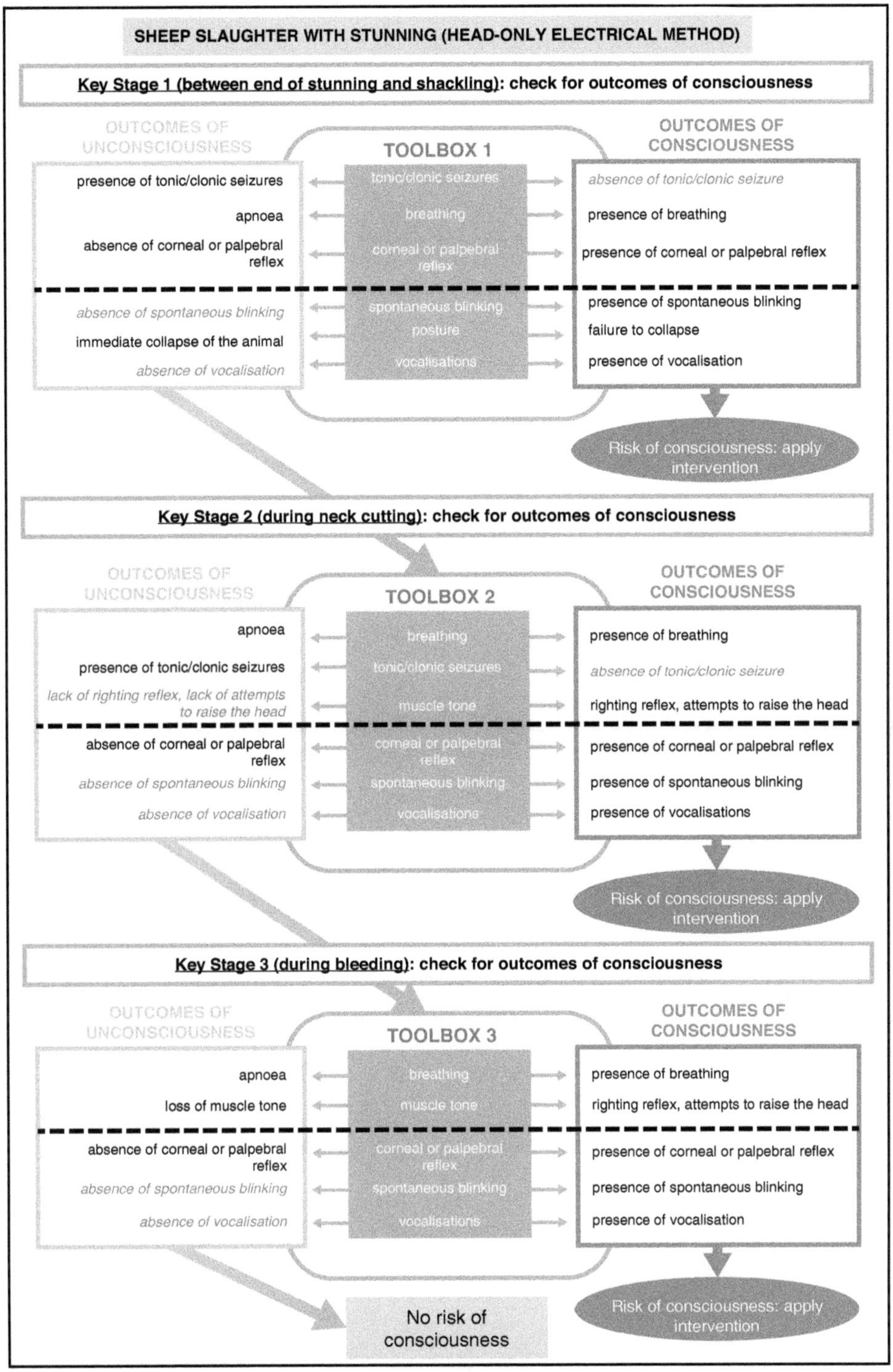


**Figure 7.11:** Toolbox for monitoring welfare at slaughter of sheep following head-only electrical stunning.

*(Source: EFSA, 2013a).*

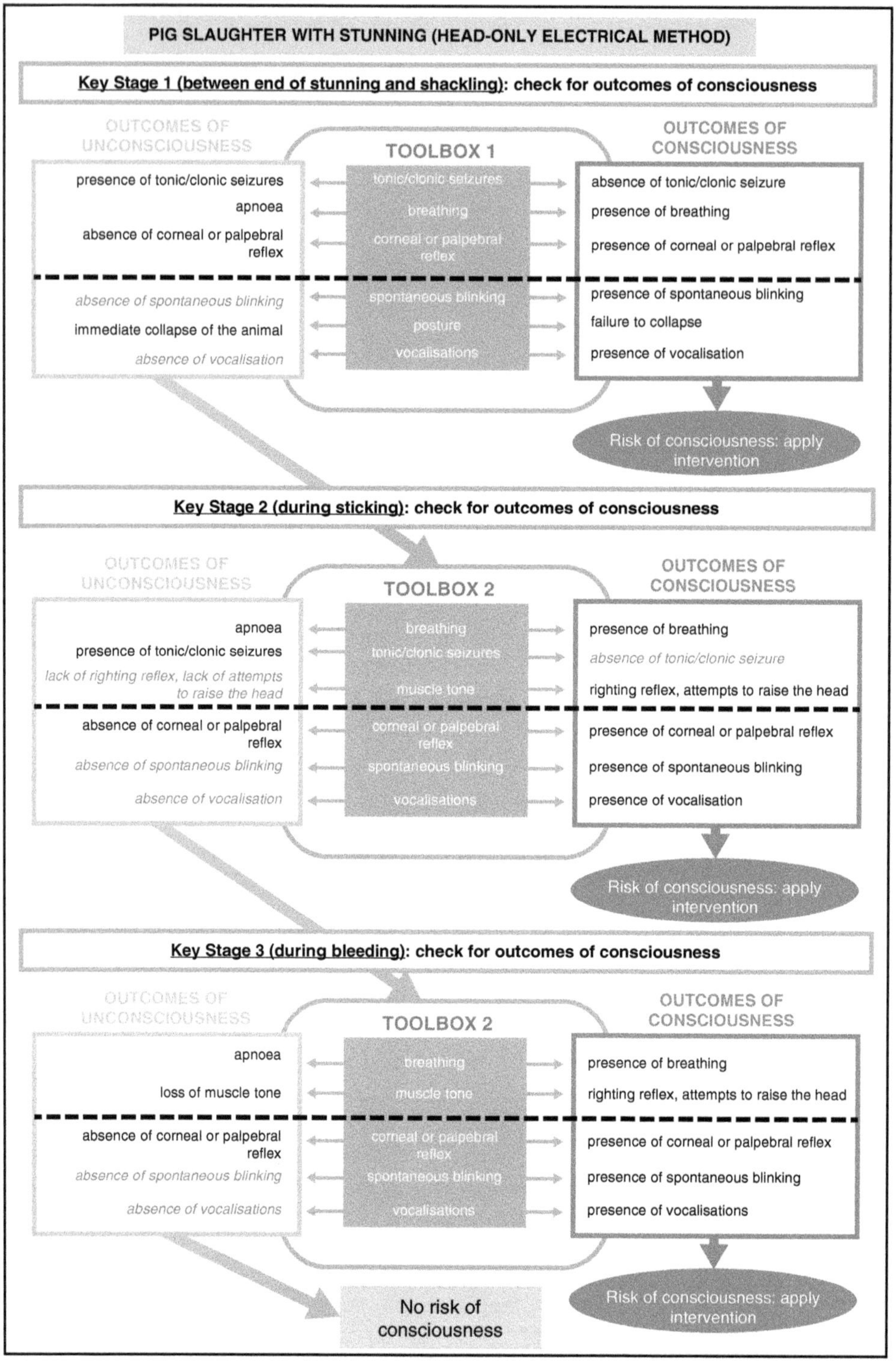


**Figure 7.12:** Toolbox for monitoring welfare at slaughter of pigs following head-only electrical stunning

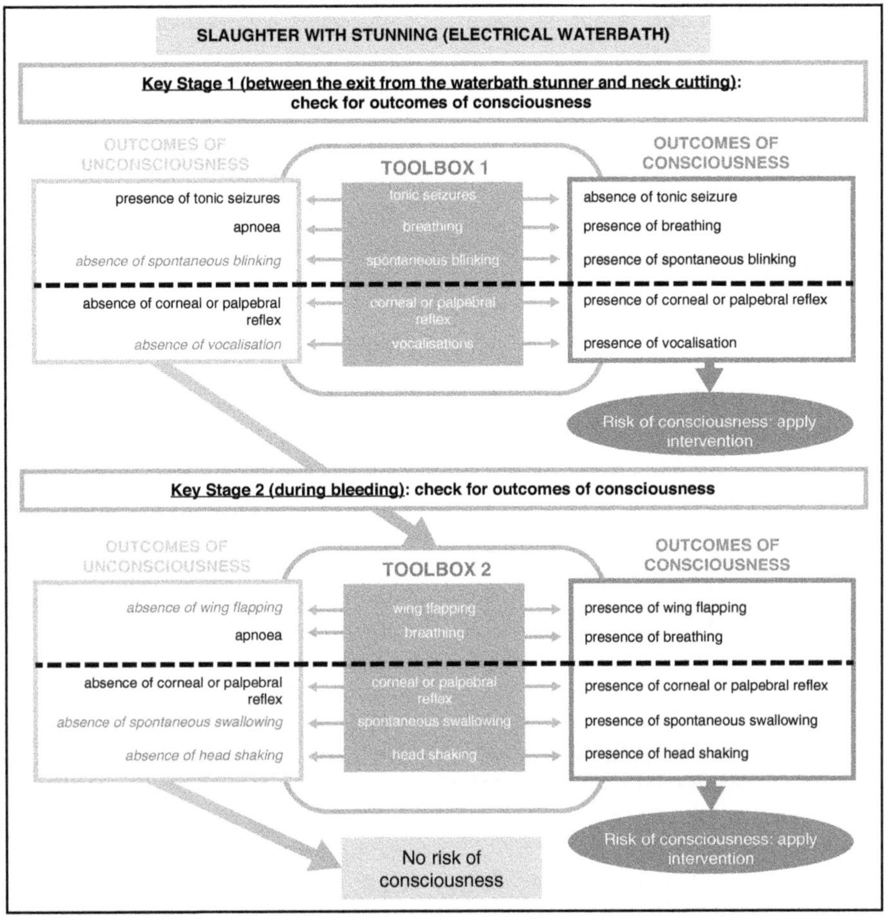


**Figure 7.13:** Toolbox for monitoring welfare at slaughter of poultry following waterbath electrical stunning

*(Source: EFSA, 2013c).*

the first clonic phase (Simmons, 1995; Velarde et al., 2002). Electrical stunning further induces pupillary dilatation, which gradually constricts as the animal returns to consciousness.

In sheep, spontaneous breathing returned before the end of the first clonic phase as already described by Velarde et al. (2000) and not simultaneously with the end of this phase as described by Gregory (1998). The corneal reflex returned simultaneously with the end of the first clonic phase.

The overall hazard is that the current does not span the brain and does not induce unconsciousness. In this case, animals fail to collapse and show absence of tonic–clonic seizures and presence of breathing (including laboured breathing). In extreme cases, animals may also vocalize.

Toolboxes for monitoring welfare during head-only stunning of cattle and poultry have not been developed.

In poultry species, however, effective head-only electrical stunning leads to immediate onset of wing flapping (clonic seizures) but the physiological responses are the same as in red meat species. Toolboxes for monitoring welfare at slaughter of poultry using waterbath stunning are presented in Figure 7.13 (EFSA, 2013c).

## 7.6 REFERENCES


Anil, M.H. (1991) Studies on the return of physical reflexes in pigs following electrical stunning. *Meat Science, 30: 13–21.*

Anil, M.H., McKinstry, J.L., Wotton, S.B., and Gregory, N.G. (1995) Welfare of calves 1. Investigations into some aspects of calf slaughter. *Meat Science, 41: 101–112.*

Anil, M.H., Raj, A.B.M., and McKinstry, J.L. (1998) Electrical stunning in commercial rabbits: Effective currents, spontaneous physical activity and reflex behaviour. *Meat Science, 48: 21–28.*

Bager, F., Shaw, F.D., Tavener, A., Loeffen, M.P.F., and Devine, C.E. (1990) Comparison of EEG and ECoG for detecting cerebrocortical activity during slaughter of calves. *Meat Science, 27: 211–225.*

Berg, L., and Raj, M. (2015) A review of different stunning methods for poultry—Animal welfare aspects (Stunning methods for poultry). *Animals, 5: 1207–1219.*

Bilgili, S.F. (1992) Electrical stunning of broilers – Basic concepts and carcase quality implications: A review. *Journal of Applied Poultry Research, 1: 135–146.*

Blackmore, D.K., and Newhook, J.C. (1982) Electroencephalographic studies of stunning and slaughter of sheep and calves. Part 3: The duration of insensibility induced by electrical stunning in sheep and calves. *Meat Science, 7: 19–28.*

Butterworth, A. (1999) Infectious components of broiler lameness: a review. *World's Poultry Science Journal, 55: 327–352.*

Cook, C.J., Devine, C.E., Tavener, A., and Gilbert, K.V. (1992) Contribution of amino acid transmitters to epileptiform activity and reflex suppression in electrically head stunned sheep. *Research in Veterinary Science, 52: 48–56.*

Cook, C.J., Devine, C.E., Gilbert, K.V., Smith, D.D., and Maasland, S.A. (1995) The effect of electrical head-only stun duration on electroencephalographic-measured seizure and brain amino acid neurotransmitter release. *Meat Science, 40: 137–147.*

Croft, P.G. (1952) Problem of electrical stunning. *The Veterinary Record, 64: 255–258.*

Danbury, T.C., Weeks, C.A., Chambers, J.P., Waterman-Pearson, A.E., and Kestin, S.C. (2000) Self-selection of the analgesic drug carprofen by lame broiler chickens. *Veterinary Record, 146: 307–311.*

EC (European Community). (2009) Council Regulation No 1099/2009 on the Protection of Animals at the Time of Killing. *Official Journal of the European Union, L303*: 1–30.

European Food Safety Authority. (2004) Welfare aspects of animal stunning and killing methods. *Scientific report of the Scientific Panel for Animal Health and Welfare on a request from the Commission.* Question. Adopted on the 15th of June 2004. Brussels. http://www.efsa.eu.int/ science/ahaw/ahaw_opinions/495/opinion_ahaw_02_ej45_stunning_report_v2_en1.pdf

European Food Safety Authority (EFSA). (2006) The welfare aspects of the main systems of stunning and killing applied to commercially farmed deer, goats, rabbits, ostriches, ducks, geese and quail. *EFSA Journal, 326: 1–18.*

European Food Safety Authority (EFSA). (2013a). Scientific opinion on monitoring procedures at slaughterhouses for pigs. *EFSA Journal, 11(12): 3523.*

European Food Safety Authority (EFSA). (2013b). Scientific opinion on monitoring procedures at slaughterhouses for sheep and goats. *EFSA Journal, 11(12): 3522.*

European Food Safety Authority (EFSA). (2013c) Scientific opinion on monitoring procedures at slaughterhouses for poultry. EFSA Journal, 11(12):3521.

Gentle, M.J. (1992) Ankle joint (Artc. Intertarsalis) receptors in the domestic fowl. *Neuroscience, 49: 991–1000.*

Gentle, M.J., and Tilston, V.L. (2000) Nociceptors in the legs of poultry: Implications for potential pain in pre-slaughter shackling. *Animal Welfare, 9: 227–236.*

Gregory, N.G. (1986) The physiology of electrical stunning and slaughter. In: *Humane Slaughter of Animals for Food.* Universities Federation for Animal Welfare: St Albans, UK. pp. 3–12.

Gregory, N.G. (1998) Stunning and slaughter. In: *Animal Welfare and Meat Science.* CABI Publishing. Wallingford, Oxon.

Gregory, N. G., and Wotton, S. B. (1984) Sheep slaughtering procedures. 1. Survey of abattoir practice. *British Veterinary Journal, 140: 281–286.*

Gregory, N.G., and Wotton, S.B. (1991) Effect of a 350 Hz DC stunning current on evoked responses in the chicken's brain. *Research in Veterinary Science, 50: 250–251.*

Griffiths, G.L. (1985) The occurrence of red-skin chicken carcasses. *British Veterinary Journal, 141: 312–314.*

Griffiths, G.L., and Purcell, D.A. (1984) A survey of slaughter procedures used in chicken processing plants. *Australian Veterinary Journal, 61: 399–401.*

Harris, C.E., and Carter, T.A. (1977) Broiler blood losses with manual and mechanical killers. *Poultry Science, 56: 1827–1831.*

Heath, G.B.S., Watt, D.J., Waite, P.R., and Ormand, J.M. (1981) Observation on poultry slaughter. *Veterinary Record, 108: 97–99.*

Heath, G.B.S, Watt, D.J., Waite, P.R., and Meakins, P.A. (1983) Further observations on the slaughter of poultry. *British Veterinary Journal, 139: 285–290.*

Hoenderken, R. (1978) Electrical stunning of pigs for slaughter. Thesis, State University, Utrecht.

Lukatch, H.S., Echon, R.M., MacIver, M.B., and Werchan, P.M. (1997) G-force induced alterations in rat EEG activity: a quantitative analysis. *Electroencephalography and Clinical Neurophysiology, 103: 563–573.*

Maria, G., Lopez, M., Lafuente, R., and Moce, M.L. (2001) Evaluation of electrical stunning methods using alternative frequencies in commercial rabbits. *Meat Science, 57: 139–143.*

Niemann, J.T., Garner, D., and Lewis, R.J. (2003) Transthoracic impedance does not decrease with rapidly repeated countershocks in a swine cardiac arrest model. *Resuscitation, 56: 91–95.*

Raj, A.B.M. (2004) Stunning and slaughter. In: *Welfare of Laying Hens,* G. C. Perry (Ed), CAB International (Pub), Oxon, UK, pp 375–389.

Raj, A.B.M. (2006) Recent developments in stunning and slaughter of poultry. *World's Poultry Science Journal, 62: 467–484.*

Raj, A.B.M., and O'Callaghan, M. (2004a) Effect of amount and frequency of head-only stunning currents on the electroencephalograms and somatosensory evoked potentials in broilers. *Animal Welfare, 13: 159–170.*

Raj, A.B.M., and O'Callaghan, M. (2004b) Effects of electrical water bath stunning current frequencies on the spontaneous electroencephalograms and somatosensory evoked potentials in hens. *British Poultry Science, 45: 230–236.*

Raj, A.B.M., O'Callaghan, M., and Knowles, T. G. (2006a) The effect of amount and frequency of alternating current used in water bath stunning and neck cutting methods on spontaneous electroencephalograms in broilers. *Animal Welfare, 15: 7–18.*

Raj, A.B.M., O'Callaghan, M., and Hughes, S. I. (2006b) The effect of amount and frequency of pulsed direct current used in water bath stunning and neck cutting methods on spontaneous electroencephalograms in broilers. *Animal Welfare, 15: 19–24.*

Raj, A.B.M., O'Callaghan, M., and Hughes, S. I. (2006c) The effects of pulse width of a pulsed direct current used in water bath stunning and neck cutting methods on spontaneous electroencephalograms in broilers. *Animal Welfare, 15: 25–30.*

Schutt-Abraham, I., and Wormuth, H.J. (1991) Anforderungen an eine tierschutzgerechte elektrische betaubung von schlachtegeflugel. *Rundeschau für Fleischhygiene und Lebensmitteluberwachung, 43: 7–8.*

Schutt-Abraham, I., Wormuth, H.J., and Fessel, J. (1987) Vergleichende untersuchungen zur tierchutzgerechten elektrobetaubung verschi edener schlachtgeflugelarten. *Berliner und Munchener Tierarztliche Worchenschrift, 100: 332–340.*

Simmons, N.J. (1995) The use of high frequency currents for the electrical stunning of pigs. PhD thesis, University of Bristol, UK.

Sparrey, J. (1994) Aspects in the design and operation of shackle lines for the slaughter of poultry. Unpublished M.Phil. thesis, University of Newcastle upon Tyne, Newcastle upon Tyne, U.K.

Sparrey, J.M., Kettlewell, P.J., and Paice, M.E. (1992) A model of current pathways in electrical waterbath stunners used for poultry. *British Poultry Science, 33: 907–916.*

Sparrey, J.M., Kettlewell, P.J., Paice, M.E., and Whetlor, W.C. (1993) Development of a constant current waterbath stunner for poultry processing. *Journal of Agricultural Engineering Research, 56: 267–274.*

Troeger, K. (2002) Blutentzug sofort nach Stromfluss-Ende. Fleischgewinnung. *Fleischwirtschaft, 7: 22–25.*

Velarde, A., Ruiz-de- la-Torre, L., Stub, C., Diestre, A., and Manteca, X. (2000) Factors affecting the effectiveness of head-only electrical stunning in sheep. *Veterinary Record, 147: 40–43.*

Velarde, A., Ruiz-de- la-Torre, J.L., Rosello, C., Fabrega, E., Diestre, A., and Manteca, X. (2002) Assessment of return to consciousness after electrical stunning in lambs. *Animal Welfare, 11: 333–341.*

Von Mickwitz, G., Heer, A., Demmler, T., Rehder, H., and Seidler, M. (1989) Slaughter of cattle, swine and sheep according to the regulations on animal welfare and disease control using an electric stunning facility. (SCHERMER, type EC) *Deutsche tierärztliche Wochenschrift, 96: 127–133.*

Wooley, S.A., Brothwick, F.J.W., and Gentle, M.J. (1986a) Flow routes of electric currents in domestic hens during pre-slaughter stunning. *British Poultry Science, 27: 403–408.*

Wooley, S.A., Brothwick, F.J.W., and Gentle, M.J. (1986b) Tissue resistivities and current pathways and their importance in pre-slaughter stunning of chickens. *British Poultry Science, 27: 301–306.*

chapter eight

# Gas stunning and killing methods

*Antonio Velarde[1] and Mohan Raj[2]*

**Learning objectives**

- To understand the basic neurological and physiological bases of gas stunning.
- To become familiar with gas stunning methods and their effective use in pigs and poultry.
- To learn how to assess stunning effectiveness and be able to detect animals at risk of recovering consciousness.
- To identify the main risk factors of unsuccessful stunning.

## CONTENT



## 8.1 INTRODUCTION

Exposure to gas mixtures is applied for the stunning or kill of pigs and poultry. The most commonly used are:

- Concentrations of carbon dioxide ($CO_2$) higher than 40% by volume in air (hypercapnia).
- Mixtures containing a minimum of 30% by volume of $CO_2$ and 20–30%

[1] IRTA. Animal Welfare Subprogram, Veïnat de Sies, s/n, 17121 Monells, Spain
[2] School of Veterinary Science. University of Bristol, Langford BS40 5DU, United Kingdom

by volume of added oxygen in air (hypercapnic hyperoxia).

- Mixtures of argon (Ar) and nitrogen ($N_2$) with less than 2% by volume of residual oxygen in air (anoxia).
- Mixtures of less than 30% by volume of $CO_2$ in argon or nitrogen or both with up to 5% by volume of residual oxygen (hypercapnic anoxia).

The design of the stunning system is determined by the gas mixture to be used and its specific gravity (the relative density with respect to air, at the same pressure and temperature). Gases with a specific gravity less than one will be "lighter than air", whilst those with a specific gravity greater than one are "heavier than air" (Table 8.1). Carbon dioxide (1.50) and argon (1.38) are higher than air and could therefore be contained within a pit. Carbon dioxide is cheaply and readily available as a by-product of the industries. On the other hand, argon has a low presence in the atmosphere (0.9 % by volume) and its availability for commercial stunning practices might be limited.

The presence of nitrogen in the atmosphere is around 79% by volume and might be a more suitable gas to be used for stunning pigs. It can be separated from atmospheric air with minimum cost and impact on the environment. However, the relative density of nitrogen (0.97) is slightly lower than air concentrations of $N_2$ and cannot be sustained within a pit at a concentration higher than 94% by volume (Dalmau et al., 2010). Nevertheless, this stability could be improved when nitrogen and $CO_2$ are combined (Dalmau et al., 2010). Gas mixtures of $N_2$ and up to 30% $CO_2$ have high stability and uniform concentrations along the pit. However, the higher the concentration of nitrogen in the gas mixture with $CO_2$, the lower the relative density of the mixture and, therefore, the more difficult it is to displace the oxygen from the pit. The use of carbon monoxide (CO) and nitrous oxide, either alone or with other gases, is not recommended in slaughterhouses as it is highly toxic and considered very dangerous.

In pigs, the most widely used gas stunning method is exposure to high concentration of $CO_2$ (Velarde et al., 2000). In commercial conditions, animals are loaded in groups into a cage or cradle and immersed into a concentration gradient of the gas, such that, as the cage is lowered into the well, the $CO_2$ concentration continues to rise until it reaches 80–90% at the bottom of the well. These systems have some animal welfare advantages compared with electrical stunning, as animals are stunned in groups with the minimum amount of restraint and handling stress (EFSA, 2004; Velarde et al., 2000). Gas systems can also be operated with mechanical push gates that separate large group of pigs into groups of five or six and gently push them into the stun box, abolishing the use of electric prodders. This,

**Table 8.1** Specific gravity of gases used for stunning or stun/killing (Kettlewell, 1986)

| Gas | Specific gravity at 300 K (27°C) at 1atm |
|---|---|
| Air | 1.00 |
| Argon | 1.38 |
| Carbon dioxide | 1.50 |
| Nitrogen | 0.97 |

together with the lower intensity of muscular contractions compared to electrical stunning, reduces the incidence of pale, soft and exudative (PSE) meat and haemorrhage (Velarde et al., 2001) and improves meat quality. However, $CO_2$ above 30% by volume in atmospheric air, causes aversion in pigs (Raj and Gregory, 1996). Raj et al. (1997) and Raj and Gregory (1995) showed argon in air or in association with low concentrations of $CO_2$ (up to 30% by volume) to be better on animal welfare grounds than high concentrations of carbon dioxide.

Poultry (mainly chickens and turkeys) are stunned using several gas mixtures:

- exposure to increasing concentrations of carbon dioxide in air;
- exposure to less than 40% by volume of carbon dioxide until onset of unconsciousness and then exposure to high concentrations of carbon dioxide in air;
- exposure to a mixture of carbon dioxide, oxygen and nitrogen until onset of unconsciousness and then exposure to high concentration of carbon dioxide in air;
- exposure to less than 2% by volume of residual oxygen in argon or nitrogen;
- exposure to a mixture containing up to 30% carbon dioxide in argon or nitrogen.

The main welfare advantage of using gas stunning for poultry is to eliminate pain and suffering associated with the inversion and shackling of live poultry and other known welfare issues associated with the waterbath electrical stunning systems. Gas stunning of poultry in their transport containers, as they arrive in the slaughterhouse, will eliminate the need for live bird handling, that is, uncarting and shackling, at the processing plant. Gas stunning poultry on a conveyor would eliminate the welfare problems associated with live bird shackling. Under both scenarios, birds will be stunned in large numbers and they will all have to be shackled and their necks cut. The interval between the end of exposure to gas mixture and neck cutting would be longer, for example several minutes after the end of exposure to gas mixtures. Owing to this, the duration of unconsciousness induced during gas stunning of poultry will have to be longer than that required under the conventional electrical stunning situations to prevent return of consciousness either prior to neck cutting or during bleeding. This is achieved by using appropriate gas concentrations and exposure times.

## 8.2 NEUROLOGICAL AND PHYSIOLOGICAL ASPECTS

Exposure of animals to gas mixtures leads to inhibition of neurones leading to progressive loss of brain function, and hence, gradual loss of consciousness (EFSA, 2004). Gas mixtures containing carbon dioxide induce hypercapnic hypoxia and inhibit neurones through acidosis. During $CO_2$ inhalation, the $O_2$ of erythrocytes becomes displaced by $CO_2$ and, as a direct consequence, $pO_2$ and $SatO_2$ decrease progressively. The respiratory and metabolic acidosis induced, reduces the pH of cerebrospinal fluid (CSF), which bathes the brain and spinal cord, and neurons thereby exerting its neuronal inhibitory and anaesthetic effects (Woodbury and Karler,

1960). Consequently, the animal loses consciousness (Gregory, 1987). In this regard, normal pH of CSF is 7.4 and unconsciousness begins when the CSF pH falls below 7.1 and reaches a maximum at pH 6.8.

Whereas, hypoxia or anoxia occurring as a result of the inhalation of argon or nitrogen induces unconsciousness by depriving the brain of oxygen. It has been established that cerebral dysfunction occurs in mammals when the partial pressure of oxygen in cerebral venous blood falls below 19 mmHg. The depletion of $O_2$ causes neuronal depolarization and intracellular metabolic crisis leading to cellular death in neurons (Rosen and Morris, 1991; Huang et al., 1994). Actually, brain oxygen deprivation leads to accumulation of extracellular potassium and a metabolic crisis, as indicated by the depletion of energy substrates and accumulation of lactic acid in the neurons (EFSA, 2004). The mechanism of induction of unconsciousness by hypoxia is due to the inhibition of N-methyl-D-aspartate (NMDA) receptor channels in the brain, which is essential for maintaining neuronal arousal during conscious state (EFSA, 2004). These effects can occur within a few seconds of inhalation of the anoxic agent. However, it is noteworthy that the survival times of various parts of the brain may differ according to the regional oxygen consumption rate. For example, the survival time of the cerebral cortex is considerably shorter than that of the medulla, in which the respiratory centre is located. Normal brain activity may be restored in anoxia-stunned animals if oxygen is administered or they are allowed to breathe atmospheric air. Inevitably, the recovery of consciousness in these animals is rapid.

The extent of neuron disruption in the thalamus and cerebral cortex caused by exposure to gas can be measured using electroencephalogram (EEG), which is normally recorded from the surface of the cerebral cortex. An example of changes occurring in the EEG of a chicken is presented in Figure 8.1 (Raj et al., 2008). In this chicken, rapid loss of spontaneous electrical activity occurred during the first 10 seconds of exposure leading to a profoundly suppressed EEG at the time of occurrence of loss of posture (i.e. earliest behavioural sign of onset of unconsciousness) at 12 seconds. The magnitude of EEG suppression occurring prior to the onset of convulsions at 17 seconds is considered to be indicative of a pathological brain state that is incompatible with the persistence of consciousness and sensibility. Changes occurring in EEGs can be quantified in laboratories using established procedures.

Laboratory studies have also used abolition of evoked electrical activity in the brain to determine the state of consciousness (EFSA, 2004). For example, a clicking noise produces an auditory evoked activity that can be recorded in the EEG as auditory evoked potential (AEP). The part of the EEG signal occurring in the interval, 10 to 100 ms after auditory stimulus presentation, is the middle latency auditory evoked potential (MLAEP) and has been used to evaluate changes in neural activity and assess the depth of anaesthesia (Martoft et al., 2002). From the MLAEP, the A-line ARX index (AAI) and the burst suppression index (BS%) can be estimated to assess unconsciousness (Rodriguez et al., 2008). The AAI is a numerical index, ranging from 0 to 99 that quantifies MLAEP variations of amplitude and latency. Higher values are

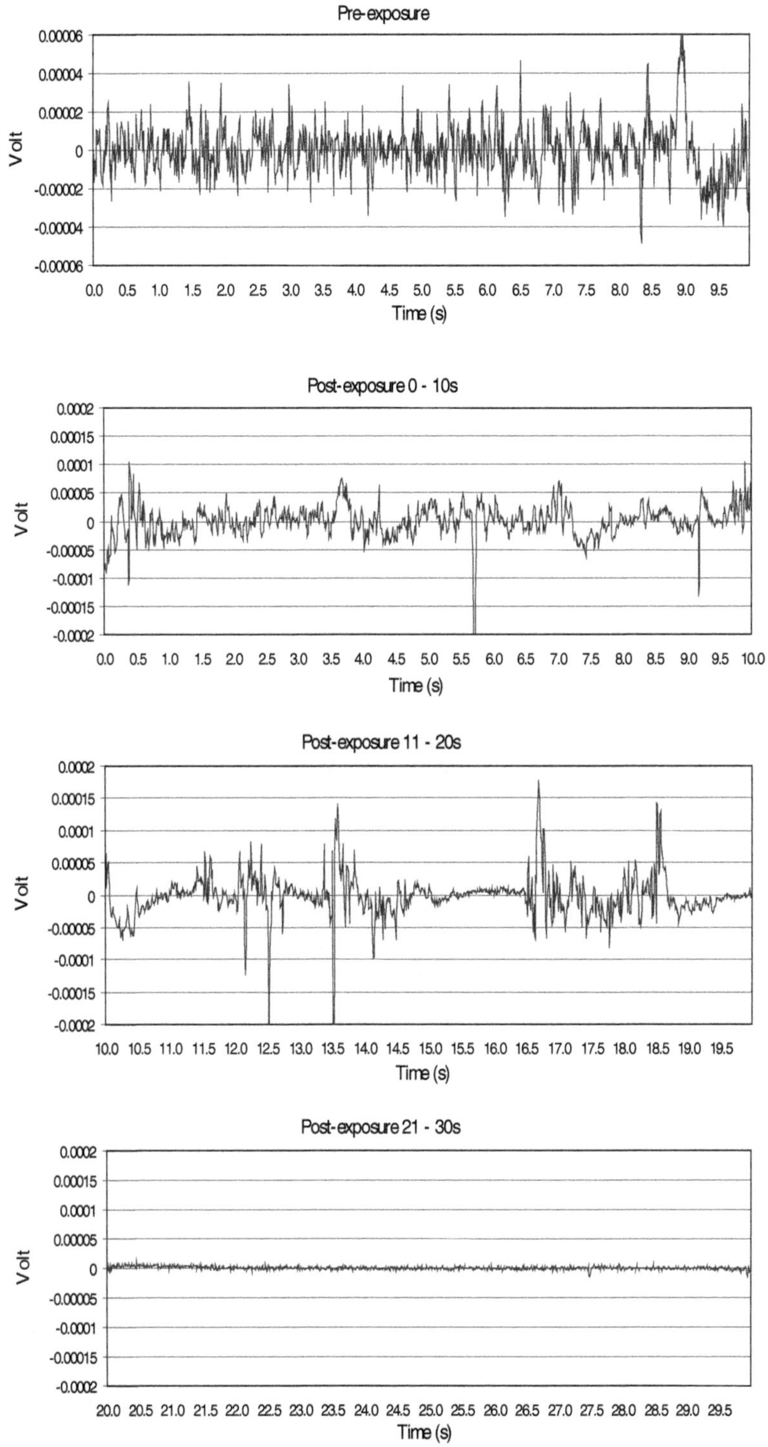


**Figure 8.1:** Changes occurring in the EEG of a chicken during gas stunning.

*(Source: Raj et al., 2008)*

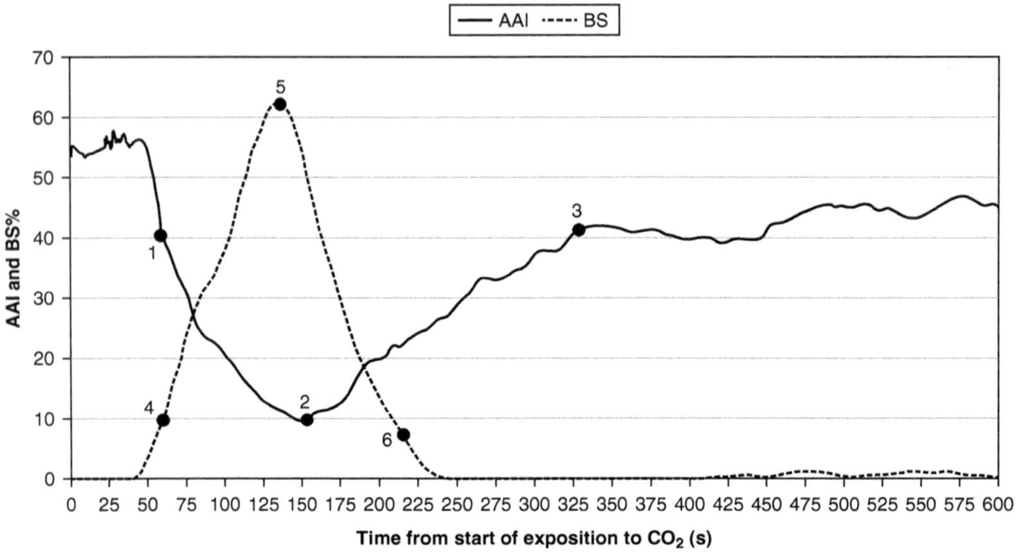


**Figure 8.2:** AAI and BS% mean values during exposure and recovery in pigs immersed in 90% CO2.

*(Source: Rodriguez et al., 2008)*

related to awareness, while decreases in AAI indicate a gradual loss of consciousness (Jensen, 1999). The BS% indicates the percentage of isoelectric activity during the preceding 30 seconds and also ranges from 0 to 100 (Litvan et al., 2002; Figure 8.2).

Similarly, flashing light can be used to record visually evoked potential (VEP) and low voltage electric pulses applied to nerves as a pain stimuli can be used to record somatosensory evoked potential (SEP) in the brain (Tables 8.2 and 8.3). Abolition of EPs in the brain has been used as an indicator of loss of consciousness and sensibility in animals during stunning and slaughter. However, since these evoked potentials can be present during general anaesthesia, their abolition would indicate more profound brain dysfunction than mere loss of consciousness and sensibility. The time to loss or abolition of EPs has been studied during gas stunning of poultry (e.g. Raj and Gregory, 1994; Raj et al., 1998).

The earliest behavioural indicator of unconsciousness during exposure to gas mixtures is loss of posture, which is followed by convulsions manifested as wing flapping in unconscious poultry (Raj et al., 1990, 1991). Convulsions occur due to spinal reflexes as a consequence of the loss of control of the brain over the spinal cord. Prolonged exposure to gas mixtures may lead to cessation of respiration. However, irregular cardiac activity (dysrhythmia) as recorded using electrocardiogram (ECG) can be elicited for several minutes after animals exit gas mixtures.

Entering the cradle and being lowered into the pit or moved within the tunnel do not cause aversion in pigs and poultry if the system contains atmospheric air (Velarde et al., 2007; Dalmau et al., 2010). However, since $CO_2$ does not

**Table 8.2** The average times (seconds) to onset of changes in the EEG and loss of SEPs during exposure of chickens to gas mixtures. (Source: Raj and Gregory, 1990; Raj et al., 1998)

| | Argon induced anoxia (2% residual $O_2$) | Mixture of 60% argon and 30% $CO_2$ (2% residual $O_2$) | 40% $CO_2$ + 30% $O_2$ + 30% $N_2$ | 45% $CO_2$ in air |
|---|---|---|---|---|
| EEG suppression | 17 | 19 | 40 | 21 |
| Loss of SEPs | 32 | 24 | 47 | 30 |
| Isoelectric EEG | 58 | 41 | Did not occur | 101 |

**Table 8.3** The average times (seconds) to onset of changes in the EEG and loss of SEPs during exposure of turkeys to gas mixtures. (Source: Raj and Gregory, 1994)

| | Argon induced anoxia (2% residual $O_2$) | Mixture of 60% argon and 30% $CO_2$ (2% residual $O_2$) | 49% $CO_2$ in air | 65% $CO_2$ in air | 86% $CO_2$ in air |
|---|---|---|---|---|---|
| EEG suppression | 41 | 16 | 21 | 15 | 13 |
| Loss of SEPs | 44 | 22 | 20 | 15 | 21 |
| Isoelectric EEG | 101 | 35 | 88 | 67 | 42 |

induce immediate loss of consciousness, inhalation of concentrations greater than 30% of carbon dioxide ($CO_2$) by volume in atmospheric air causes aversion, irritation of the mucous membranes (that can be painful) and respiratory distress during the induction phase in pigs and poultry (Raj and Gregory, 1995; Velarde et al., 2007). The presence of $CO_2$ in the blood is sensed by specific $CO_2$-sensitive chemoceptors that stimulate respiration (hyperventilation), heart rate and blood pressure. During $CO_2$ exposure, pigs show signs of aversion such as retreat attempts, headshaking, sneezing, breathlessness, freezing, escape attempts, gasping (a very deep breath through a gaping open mouth, indicative of breathlessness; Raj and Gregory, 1996) and vocalizations (Holst, 2001; Velarde et al., 2007). In poultry species, however, high concentrations of carbon dioxide induce severe head shaking, gasping, sneezing and vocalizations, which can be considered as indicators of distress. This interpretation is based on the fact that these behaviours also occur during respiratory disease.

The main causes of aversion and distress are irritation of the nasal mucosal membranes and lungs (Manning and Schwartzstein, 1995). Respiratory distress (Raj and Gregory, 1996) causes hyperventilation and a sense of breathlessness during the induction phase prior to loss of consciousness. The magnitude of the aversiveness may depend on several factors. The higher the $CO_2$

concentration, the more pronounced the aversion (Velarde et al., 2007). Conversely, a decrease in the concentration of $CO_2$ increased the time to loss of posture and, therefore, lengthened the perception of the aversive stimulus till the animal lost consciousness.

In contrast to hypercapnia, anoxia does not cause aversion prior to loss of consciousness. The time to induce unconsciousness when exposed to anoxia is longer than when exposed to hypercapnia (Raj et al., 1997). Raj and Gregory (1995) reported that the addition of $CO_2$ to a hypoxic atmosphere reduces the time needed to induce unconsciousness. However, according to Raj and Gregory (1995), the $CO_2$ concentration of the gas mixture should be up to 30% in the atmosphere in order to avoid aversion. The time to loss of consciousness occurring during exposure of animals, including poultry, to gas mixture depends upon the composition of the gas mixtures (Raj, Gregory & Wotton, 1990, 1991; Raj et al., 1992a,b; Raj and Gregory, 1993, 1994; Raj et al., 1998). The exposure of pigs to either argon-induced anoxia or the carbon dioxide–argon mixture for 3 minutes resulted in satisfactory stunning. However, bleeding should commence within 15 seconds to avoid resumption of consciousness. A 5-minute exposure to these gas mixtures followed by bleeding within 45 seconds prevented carcass convulsions during bleeding. The exposure of pigs to argon-induced anoxia or the carbon dioxide–argon mixture for 7 minutes resulted in death in the majority of pigs. Owing to the prolonged exposure time required to kill pigs with anoxia, it is not used under commercial conditions. However, further research and development is needed to evaluate the feasibility of inducing unconsciousness with anoxia and then killing pigs by other means (e.g. induction of cardiac arrest in unconscious pigs using an electric current).

## 8.3 PIGS

### *8.3.1 Description of the method*

In commercial conditions, pigs are loaded into a crate and lowered into a pit pre-filled with a high concentration of $CO_2$. Pigs are immersed into the gas following a concentration gradient, so that, the deeper the cage is lowered, the higher the $CO_2$ concentration, until it reaches 80-90% $CO_2$ concentration in atmospheric air. Two main systems exist, the dip-lift system and the paternoster system (Figure 8.3.). Dip-lift designs have only one box in the system that can be loaded with a nominal capacity of six pigs. In this system, groups of pigs are lowered directly into maximum concentrations of carbon dioxide at the bottom of the pit (EFSA, 2004). The paternoster designs have up to seven boxes (a nominal capacity of two to six pigs per box), rotating through the $CO_2$ gradient in a 3–8-metre deep pit, stopping at various intervals for loading of live pigs on one side and unloading unconscious pigs on the other side for sticking (EFSA, 2004). The number of pigs per group, the time taken to reach maximum $CO_2$ concentrations, and total exposure times are manipulated by the individual abattoirs according to their own discretion (Atkinson et al., 2012). The space allowance in the box should be enough to allow the animals to lie down without being stacked, even at maximum permitted throughput. Overloading increases the risk of unnecessary excitement or

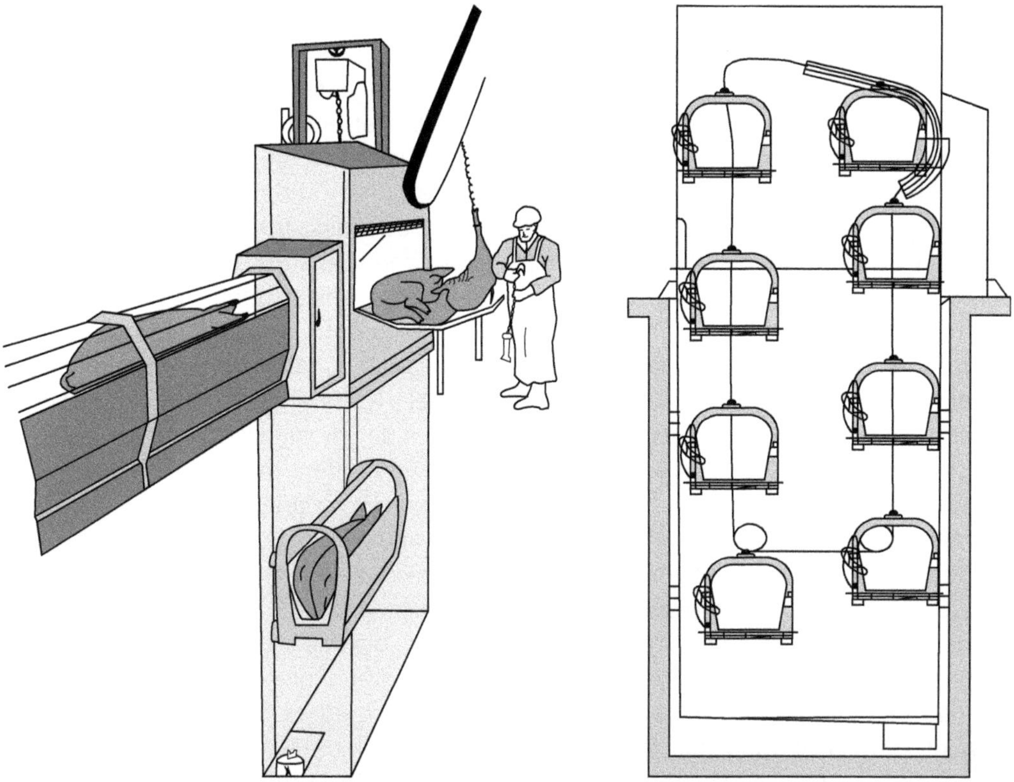

**Figure 8.3:** The two main systems in Europe. a) Dip-lift system; b) Paternoster

insufficient stunning effectiveness (e.g. hiding heads under other animals) and may lead to bruising increases.

Gas equipment should be designed and built to optimize the application of the stunning by gas, prevent injuries or contusions to the animal and minimize struggle and vocalization when animals are restrained. Stunning equipment should be fitted with devices displaying and recording the gas concentration and the time of exposure and giving alarms in case of insufficient gas flow. Gases should enter into the chamber or the location where animals are to be stunned and killed in a way that it does not provoke burns or excitement by freezing or lack of humidity. Moreover, the control of temperature and humidity of the gas mixture could improve the welfare of the animals. Inhalation of warm and humidified air helps to alleviate physical discomfort and distress. Moreover, because animals exposed to carbon dioxide gas also show gasping (oral breathing), it is thought that administration of a warm and humidified gas mixture helps to reduce the severity of distress.

### *8.3.2 Effective use*

Under normal conditions in the slaughterhouse, pig stunning with a high concentration of $CO_2$ can be reversible (or simple). The depth and duration of unconsciousness achieved with $CO_2$ gas stunning depends upon the animal, the $CO_2$ concentration, the speed at which animals are lowered towards the bottom of the pit, where the highest $CO_2$ concentration is achieved, and the duration of exposure (Raj and Gregory, 1996; Troeger and Woltersdorf, 1991). Due to individual biological variation, some pigs may regain consciousness while others do not, even if stunned in the same group (Forslid, 1987; Holst, 2001). Exposure to high concentrations of $CO_2$ shortens the period of time to unconsciousness and helps reduce the duration of hyperventilation and potential distress (Troeger and Woltersdorf, 1991; Raj and Gregory 1996; Barton-Gade, 1999). However, the degree of aversion depends on the carbon dioxide concentration. Velarde et al. (2007) reported that the aversion was higher when the stunning system contained 90% as opposed to 70% carbon dioxide due possibly to increased irritation of the nasal mucosal membranes and more severe hyperventilation. Conversely, a decrease in the concentration of carbon dioxide increased the time to onset of unconsciousness as determined using time to loss of posture and, therefore, lengthened the perception of the aversive stimulus till the animal lost consciousness. Immersion of pigs into 80 to 90% $CO_2$ usually leads to the induction of unconsciousness within 30 seconds.

To ensure good animal welfare the stun should ensure unconsciousness is induced for a sufficient duration to include not only the stun-to-stick interval but also the time taken for brain death to occur due to sticking. Because the effect of a stunning method is momentary, the onus of preventing resumption of consciousness following stunning relies on the efficiency of the slaughter procedure, that is, the prompt and accurate severance of blood vessels supplying oxygenated blood to the brain. Therefore, sticking must start as soon as possible after stunning. If not possible, the stun–stick interval can be increased proportionally without animals recovering consciousness, through increased exposure time to the gas (Holst, 2001). As the duration of unconsciousness determines the maximum acceptable stun-to-stick interval, it is therefore imperative for animal welfare that unconsciousness is closely monitored, and animals re-stunned when necessary. Prolonged exposure may result in irreversible stunning, and eliminate the chances of recovery of consciousness.

When exposed to a minimum of 70% carbon dioxide for 90 seconds, sticking (bleeding or exsanguination) should be performed as soon as possible (e.g. ideally within 15 seconds of exiting the gas) to prevent resumption of consciousness. When the duration of exposure to this level of carbon dioxide is increased, the incidence of death also increases. Exposure of pigs to a minimum of 90% by volume of carbon dioxide in air for 3–5 minutes results in death in the majority of pigs, which can be recognized from the presence of dilated pupils and absence of gagging (rudimentary respiratory activity) at the exit from the gas.

### *8.3.3 Monitoring points for correct stunning*

Animals should be monitored regularly during the entire process, from stunning to bleeding, and it should be ascertained that they do not show any signs of consciousness and also that death occurs before further carcass dressing operations or scalding begin. The EFSA Panel on Animal Health and Welfare (AHAW) set out to develop toolboxes of welfare indicators for developing monitoring procedures at slaughterhouses for pigs stunned with carbon dioxide at high concentration (EFSA, 2013a). It proposed welfare indicators together with their corresponding outcomes of consciousness, unconsciousness or death, to be used at three key stages of monitoring: (a) after stunning and during shackling and hoisting; (b) during sticking; and (c) during bleeding. The report concluded that, although it is traditional to look for outcomes of unconsciousness in pigs following stunning, the risk of poor welfare can be better detected if pig welfare monitoring is focused on detecting consciousness, that is, ineffective stunning or recovery of consciousness.

The recommended animal-based indicators for monitoring after stunning are muscle tone, breathing and the corneal or palpebral reflexes (Figure 8.4.). Effective gas stunning induces collapse and loss of posture, apnoea (absence of breathing) and abolition of corneal or palpebral reflex. If the exposure to carbon dioxide is ineffective and/or inadequate, the animal will show attempts to regain posture and/or sustained or presence of breathing, including laboured breathing. Pigs that are not effectively stunned or those recovering consciousness will also show positive corneal or palpebral reflex. Additionally, response to nose prick or ear pinch and vocalizations may be used. Ineffective stunning and recovery of consciousness due to poor stunning can be recognized from the response to nose prick or ear pinch. Vocalization is expected only in conscious animals. However, not all conscious animals will vocalize, and hence absence of vocalization does not always mean that the animal is unconscious. Since unconscious animals will not vocalize, this indicator is not applicable to monitoring unconsciousness.

For monitoring at sticking after carbon dioxide stunning, the recommended indicators to be used are muscle tone, breathing and vocalizations. An unconscious pig at this stage will show loss of muscle tone, hanging flaccidly on the overhead shackle or lying relaxed on the conveyor and is therefore not expected to show any changes in its posture. Loss of muscle tone can be recognized from floppy ears and relaxed jaw and completely relaxed body. In contrast, a pig recovering consciousness will attempt to regain posture, which will be manifested as righting reflex (e.g. severe kicking, head lifting, body arching), arching of the neck, body stiff (upright) ears and jaws and/or convulsions. These signs are more visible when the animals are hanging from the overhead rail. During sticking, unconscious pigs will continue to manifest apnoea. Pigs may show gagging, which is considered a rudimentary brainstem function indicative of a dying brain in pigs. However, in association with resuscitation efforts that might be provided during handling, shackling and hoisting, and gasping could lead to recovery of consciousness, especially if exsanguination is delayed. Pigs recovering consciousness whilst hanging on the overhead shackle will attempt to breathe, which may begin as regular gagging before leading to resumption of breathing. Additionally, the corneal or palpebral reflex and response to nose prick or ear pinch may be used.

For monitoring during bleeding after carbon dioxide stunning, the

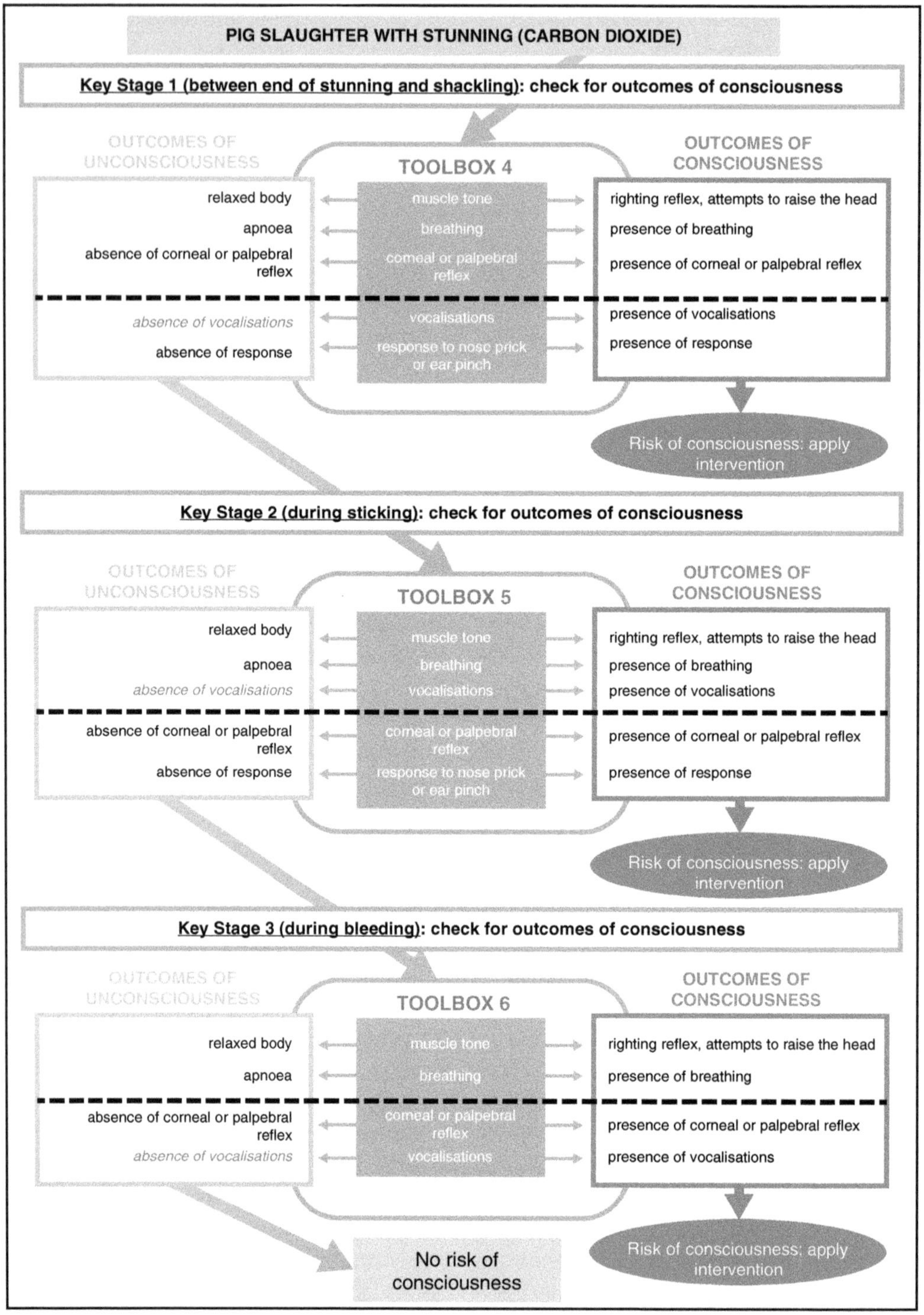


**Figure 8.4:** Indicators that are considered suitable to be used for detection of conscious animals at each key stage of the procedure of carbon dioxide stunning in pigs

*(Source: EFSA, 2013a)*

recommended indicators are muscle tone and breathing. The presence of righting reflex would indicate recovery of consciousness. Additionally, the corneal or palpebral reflex and vocalizations may be used. Pigs showing these signs of ineffective stunning will require immediate re-stun using a backup method.

The personnel performing stunning, shackling, hoisting and/or bleeding, will have to check all the animals and confirm that they are not conscious following stunning. For the animal welfare officer, who has the overall responsibility for animal welfare, a mathematical model for the sampling protocols was proposed, taking into account the throughput rate (total number of animals slaughtered in the slaughterhouses) and tolerance level (amount of potential failures – animals that are conscious after stunning; animals that are not unconscious or not dead after slaughter without stunning).

The EUWELNET project (www.euwelnet.eu) developed standard operating procedures (SOPs) for the assessment of effective unconsciousness following stunning at commercial slaughterhouses. The aim of these improved standard operating procedures (SOP) is to enable food business operators (FBO) and animal welfare officers (AWOs) to better implement welfare at slaughter and to provide the competent authorities (CAs) and official veterinarians (OVs) with a method to assess compliance with the Regulation. The SOPs include recommendations on the objectives, responsibilities, control measures, monitoring procedures, corrective actions and records.

## 8.4 POULTRY

### *8.4.1 Description of the method*

All the major poultry transport container manufacturers in Europe also produce gas stunning equipment. In principle, these systems use denser than air gas mixtures contained in a tunnel. Drawer-type modular poultry transport systems involves automatic removal of individual drawers or crates full of birds from the metal frames and then passing the drawers through a tunnel containing one of the desired gas mixtures (Figure 8.5a,b). Another system involves passing drawers or crates through a pit containing gas mixtures, very similar to the paternoster system used for pigs. The number of poultry contained in a drawer (stocking density) may vary for several reasons, including size of the drawer, live weight of poultry, environmental conditions, transport distance, and so on. Some other modular transport systems involve tipping live birds on to a conveyor and then carrying the birds through a tunnel containing a mixture of 40% carbon dioxide, 30% oxygen and 30% nitrogen in the first phase to induce unconsciousness and a high concentration of carbon dioxide in the second phase to irreversibly stun them (Figure 8.6).

Ideally, birds in crates or conveyors should be examined before they enter the gaseous atmospheres to remove dead birds.

Research has shown that, given a free choice, chickens and turkey will avoid an atmosphere containing high concentrations of carbon dioxide, that is, 40% or more by volume in air, but they do not avoid an atmosphere containing a low oxygen level created using argon or nitrogen (Raj and Gregory, 1991; Raj, 1996).

**Figure 8.5: a, b** Multistage $CO_2$ stunning system.
*(Meyn Food Processing Technology b.v., The Netherlands)*

The inference is that, under slaughterhouse conditions, poultry should be rendered unconscious with low concentrations (< 40% by volume) of carbon dioxide before exposing them to high concentrations of this gas. Unlike argon, exposure of poultry to carbon dioxide also causes head shaking and gasping (oral breathing with neck extension) in poultry and the amount of gasping occurring prior to loss of posture depends upon the concentration of this gas. Research

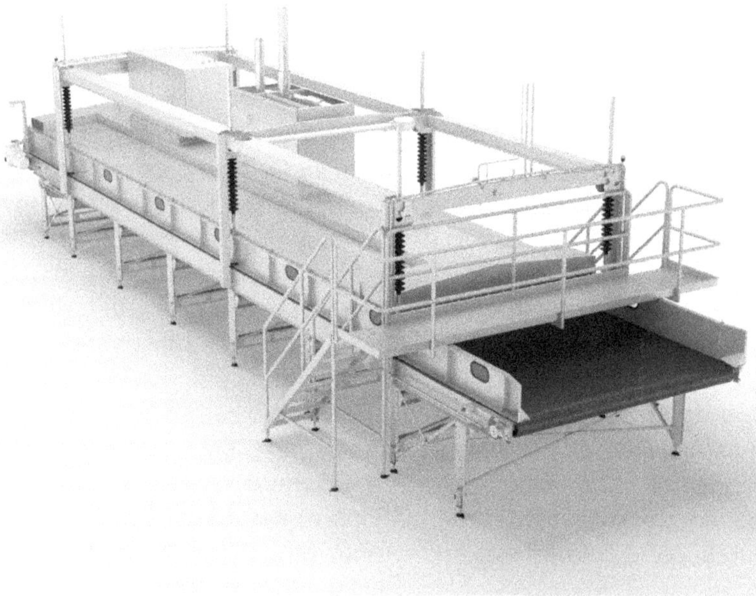

**Figure 8.6:** CAS Smooth Flow.
*(Marel Stork Poultry Processing b.v., The Netherlands)*

**Table 8.4** Amount of gasping occurring prior to loss of posture during exposure to gas mixtures in broilers. (Source: Lambooij and Pieterse, 1997)

| Gas mixture | Number of head shakes | Amount of gasping |
|---|---|---|
| 90% Ar in air | None | None |
| 30% $CO_2$ + 60% Ar in air | 3 | 3 |
| 30% $CO_2$ + 30% $O_2$ in air | 5 | 12 |
| 40% $CO_2$ + 30% $O_2$ in air | 4 | 9 |
| 40% $CO_2$ in air | 3 | 8 |

has shown that warming and humidification of carbon dioxide in the stunner reduced the severity of gasping and head shaking. Nevertheless, it can be argued that the cumulative stress associated with live bird handling, shackling, pre-stun shocks and poor welfare due to the complexity of the existing multiple bird waterbath stunning would be more than the stress associated with the induction of unconsciousness with carbon dioxide gas (Table 8.4).

### *8.4.2 Effective use*

Several critical control points have been identified to ensure welfare of poultry during gas stunning. Irrespective of whether birds are passed through the gas stunners whilst they are in their transport containers or on a conveyor, the passage should be smooth without tilting. Bunching or overcrowding should be avoided to ensure all the birds are exposed to the gas mixture. Gas concentrations should be continuously monitored at the bird's head level and maintained throughout the stunner. Devices used for monitoring gas concentrations should be calibrated regularly and maintained in good working condition. The duration of exposure to gas mixture should be adequate to prevent recovery of consciousness in birds. There should be backup stunners (e.g. captive bolts) to deal with birds showing signs of recovery of consciousness. Both the carotid arteries should be cut as soon as possible.

### *8.4.3 Monitoring points for correct stunning*

As for pigs, poultry should also be monitored regularly during the entire process, from stunning to bleeding, and it should be ascertained that they do not show any signs of consciousness and also that death occurs before birds enter the scalding tank. The EFSA Panel on Animal Health and Welfare (AHAW) set out to develop toolboxes of welfare indicators for developing monitoring procedures at slaughterhouses for poultry stunned with gas mixtures (EFSA, 2013b). This scientific opinion has proposed toolboxes of welfare indicators, and their corresponding outcomes of consciousness, unconsciousness or death, for developing monitoring procedures at slaughterhouses for poultry stunned using gas mixtures. It was suggested that, after stunning of the birds with gas mixtures prior to slaughter, the indicators should be repeatedly checked to detect signs of consciousness through the two key stages of monitoring during

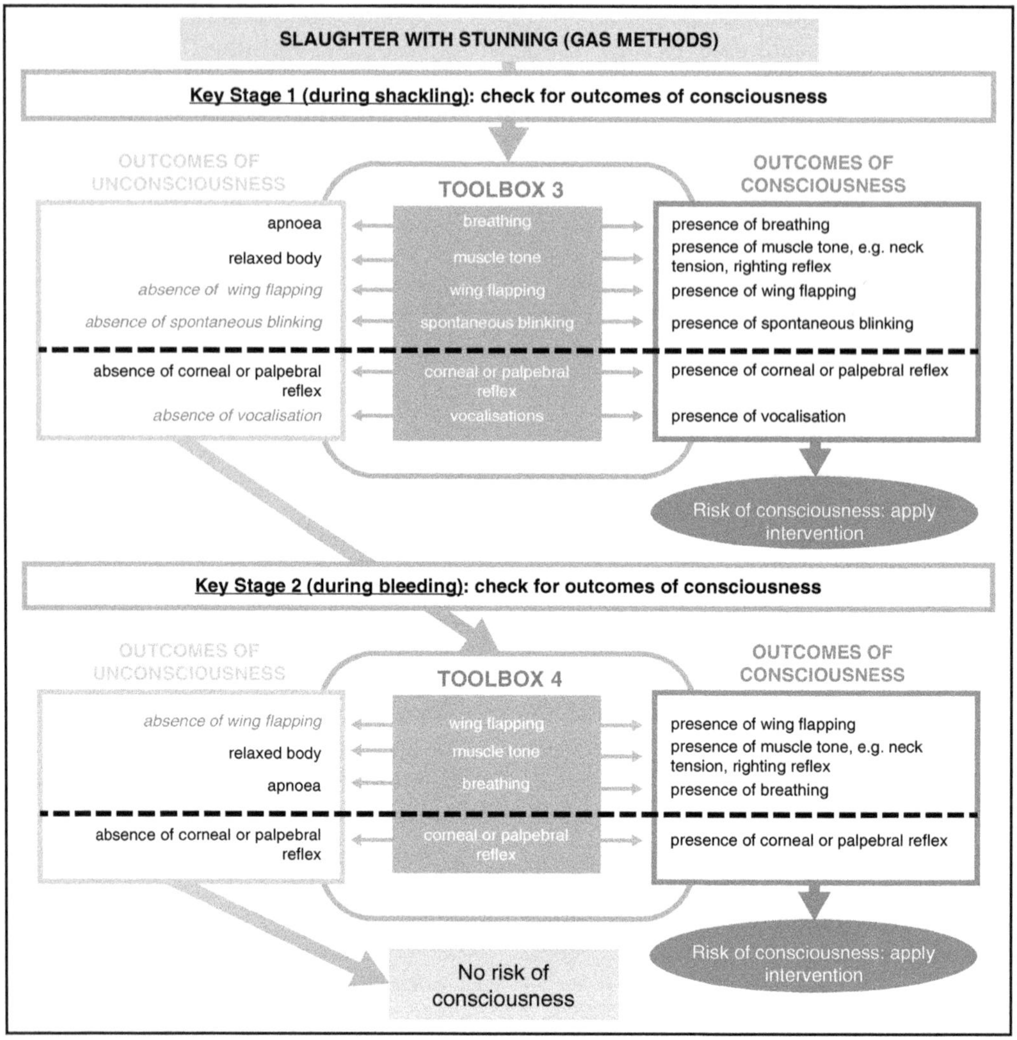


**Figure 8.7:** Indicators that are considered suitable to be used for detection of conscious animals at each key stage of the procedure of gas stunning in poultry

*(Source: EFSA, 2013b)*

the slaughter process: between the exit from the gas stunner and the entrance to the scalding tank, especially during shackling (key stage 1) and during bleeding (key stage 2). The recommended indicators for monitoring at key stage 1 are breathing, muscle tone, wing flapping and spontaneous blinking (Figure 8.7). Additionally, positive corneal or palpebral reflex and vocalizations may be used to recognize consciousness. The recommended indicators for monitoring at key stage 2 (during bleeding) are presence of wing flapping, muscle tone and breathing. In addition, the presence of corneal or palpebral reflex may also be used at this stage to recognize consciousness. In this EFSA Opinion, the most common risk factors involved in the welfare of animals during slaughter were linked to two categories: those risk factors that affect the quality of the stun and those that affect the quality of the assessment.

## 8.5 REFERENCES

Atkinson, S., Velarde, A., Llonch, P. and Algers, B. (2012) Assessing pig welfare at stunning in Swedish commercial abattoirs using CO2 group-stun methods. *Animal Welfare* 21: 487–495.

Barton-Gade, P.A. (1999) Preliminary investigations on the effect of immersion of pigs in carbon dioxide gas. Danish Meat Research Institute. Internal report Ref. 02.703. Unpublished data.

Dalmau, A., Llonch, P., Rodríguez, P., Ruíz-de-la-Torre, J.L., Manteca, X. and Velarde, A. (2010) Stunning pigs with different gas mixtures. Part 1: Gas stability. *Animal Welfare* 19: 315–323.

EFSA (2004) Welfare aspects of animal stunning and killing methods. Scientific report of the Scientific Panel for Animal Health and Welfare on a request from the Commission. Question. Adopted on the 15th of June 2004. Brussels. http://www.efsa.eu.int/science/ahaw/ahaw_opinions/495/opinion_ahaw_02_ej45_stunning_report_v2_en1.pdf.

EFSA AHAW Panel (EFSA Panel on Animal Health and Welfare) (2013a) Scientific Opinion on monitoring procedures at slaughterhouses for pigs. *EFSA Journal* 11(12): 3523, 62 pp. doi:10.2903/j.efsa.2013.3523.

EFSA AHAW Panel (EFSA Panel on Animal Health and Welfare) (2013b) Scientific Opinion on monitoring procedures at slaughterhouses for poultry. EFSA *Journal* 11(12): 3521, 65 pp. doi:10.2903/j.efsa.2013.3521.

Forslid, A., (1987) Transient neocortical, hippocampal and amygdaloid EEG silence induced by one-minute inhalation of high concentration carbon dioxide in swine. *Acta Physiologica Scandinavica* 130: 1–10.

Gregory, N.G. (1987) The physiology of electrical stunning and slaughter. In *Humane Slaughter of Animals for Food.* Universities Federation for Animal Welfare: St Albans, UK, 1986; pp. 3–12.

Holst, S. (2001) Carbon dioxide stunning of pigs for slaughter – Practical guidelines for good animal welfare. Proceedings of the 47th International Congress of Meat Science and Technology, Krakow, Poland. Vol. I, pp. 48–54.

Huang, Q.F., Gebrewold, A., Zhang, A., Altura, B.T. and Altura, B.M. (1994) Role of excitatory amino acid in regulation of rat pial microvasculature. *American Journal of Physiology* 266: R158–R163.

Jensen, E.W. (1999) Monitoring depth of anaesthesia by auditory evoked potentials. PhD Thesis, Faculty of Health Sciences, University of Southern Denmark, Odense, Denmark.

Kettlewell, P.J. (1986) Engineering aspects of humane killing of poultry. In: Cotract Report (no. CR/173/86/8333) of the National Institute of Agricultural Engineering, Bedford, UK.

Lambooij, E. and Pieterse, C. (1997) Alternative stunning methods for poultry. In: Lambooij, E. (Ed.) Proceedings of satellite symposium on developments of new humane stunning and related processing methods for poultry to improve product quality and consumer acceptability. European Poultry Meat and Egg Quality Symposia, Poznan, Poland, pp. 7–14.

Litvan, H., Jensen, E.W., Revuelta, M., Henneberg, S.W., Paniagua, P., Campos, J.M., Martínez, P., Caminal, P. and Villar Landeira, J.M. (2002) Comparison of auditory evoked potentials and the A-line ARX index for monitoring the hypnotic level during sevoflurane and propofol induction. *Acta Anaesthetic Scandinavica* 46: 245–252.

Manning, H.L. and Schwartzstein, R.M. (1995)

Pathophysiology of dyspnea. *New England Journal of Medicine* 333 (23): 1547–1553.

Martoft, L., Lomholt, L., Kolthoff, C., Rodríguez, B.E., Jensen, E.W., Jorgensen, P.F., Pedersen, H.D. and Forslid, A. (2002) Effects of CO2 anaesthesia on central nervous system activity in swine. *Laboratory Animals* 36(2): 115–126.

Raj, M. (1996) Aversive reactions of turkeys to argon, carbon dioxide, and a mixture of carbon dioxide and argon. *Veterinary Record* 138: 592–593.

Raj, A.B.M. and Gregory, N.G. (1990) Investigation into the batch stunning/killing of chickens using carbon dioxide or argon-induced hypoxia. *Research in Veterinary Science* 49: 364–366.

Raj, A.B.M. and Gregory, N.G. (1991) Preferential feeding behaviour of hens in different gaseous atmospheres. *British Poultry Science* 32: 57–65.

Raj, A.B.M. and Gregory, N.G. (1993) Time to loss of somatosensory evoked potentials and onset of changes in the spontaneous electroencephalogram of turkeys during gas stunning. *The Veterinary Record* 133: 318–320.

Raj, A.B.M. and Gregory, N.G. (1994) An evaluation of humane gas stunning methods for turkeys. *The Veterinary Record* 135: 222–223.

Raj, A.B.M. and Gregory, N.G. (1995) Welfare implications of the gas stunning of pigs 1. Determination of aversion to the initial inhalation of carbon dioxide or argon. *Animal Welfare* 4: 273–280.

Raj, A.B.M. and Gregory, N.G. (1996) Welfare implications of gas stunning of pigs 2. Stress of induction of anaesthesia. *Animal Welfare* 5: 71–78.

Raj, A.B.M., Gregory, N.G. and Wotton, S.B. (1991) Changes in the somatosensory evoked potentials and spontaneous electroencephalogram of hens during stunning in argon-induced hypoxia. *British Veterinary Journal* 147: 322–330.

Raj, A.B.M., Gregory, N.G. and Wotton, S.B. (1990) Effect of carbon dioxide stunning on somatosensory evoked potentials in hens. *Research in Veterinary Science* 49: 355–359.

Raj, A.B.M., Wotton, S.B. and Gregory, N.G. (1992a) Changes in the somatosensory evoked potentials and spontaneous electroencephalogram of hens during stunning with a carbon dioxide and argon mixture. *British Veterinary Journal* 148: 147–156.

Raj, A.B.M., Wotton, S.B. and Whittington, P.E. (1992b) Changes in the spontaneous and evoked electrical activity in the brain of hens during stunning with 30% carbon dioxide in argon with 5% residual oxygen. *Research in Veterinary Science* 53: 126–129.

Raj, A.B.M., Johnson, S.P., Wotton, S.B. and McInstry, J.L. (1997) Welfare implications of gas stunning of pigs 3. Time to loss of somatosensory evoked potentials and spontaneous electrocorticogram of pigs during exposure to gases. *British Veterinary Journal* 153: 329–340.

Raj, A.B.M., O'Callaghan, M.C., Thompson, K., Becket, D., Morrish, I., Love, A., Hickman, G. and Howson, S. (2008) Large-scale killing of poultry species on farm during outbreaks of diseases: evaluation and development of a humane containerised gas killing system. *World's Poultry Science Journal* 64: 227–243.

Raj, A.B.M., Wotton, S.B., McKinstry, J.L. Hillebrand, S.J.W. and Pieterse, C. (1998) Changes in the somatosensory evoked potentials and spontaneous electroencephalogram of broiler chickens during exposure to gas mixtures. *British Poultry Science* 39: 686–695.

Rodriguez, P., Dalmau, A., Ruiz-de-la-Torre, J.L., Manteca, X., Jensen, E.W., Rodriguez, B., Litvan, H. and Velarde, A. (2008) Assessment of unconsciousness during carbon dioxide stunning in pigs. *Animal Welfare* 17: 341–349.

Rosen, A.S. and Morris, M.E. (1991) Depolarising effects of anoxia on pyramidal cells of rat neocortex. *Neuroscience Letters* 124 (2): 169–173.

Troeger, K. and Woltersdorf, W. (1991) Gas anaesthesia of slaughter pigs. Stunning experiments under laboratory conditions with fat pigs of known halothane reaction type: meat quality and animal protection. *Fleischwirtschaft* 71: 1063–1068.

Velarde, A., Cruz, J., Gispert; M., Carrión, D., Ruiz-de-la-Torre, J.L., Diestre, A., Manteca, X. (2007) Aversion to carbon dioxide stunning in pigs: effect of the carbon dioxide concentration and the halothane genotype. *Animal Welfare* 16: 513–522.

Velarde, A., Gispert, M., Faucitano, L., Alonso, P., Manteca, X., Diestre, A. (2001) Effects of the stunning procedure and the halothane genotype on meat quality and incidence of haemorrhages in pigs. *Meat Science* 58: 313–319.

Velarde, A., Ruiz-de-la-Torre, J.L., Stub, C., Diestre, A. and Manteca, X. (2000) Factors affecting the effectiveness of head-only electrical stunning in sheep. *Veterinary Record* 147: 40–43.

Woodbury, D.M. and Karler, R. (1960) The role of carbon dioxide in the nervous system. *Anesthesiology* 21: 686–703.

chapter nine

# Fish stunning and killing

*Hans van de Vis and Bert Lambooij*

## Learning objectives

For farmed European eel, Atlantic salmon, trout, sea bream, sea bass, tuna and common carp, readers should:

- Obtain a brief overview of production of finfish in aquaculture worldwide.
- Understand that finfish represent a class of animals in which there is huge variety regarding evolutionary history, behaviours and habitats.
- Have knowledge of the steps carried out in harvest of farmed fish and know the meaning of the words "stunning" and "killing". Harvest is the process whereby a live fish is converted into an edible product.
- Comprehend the neurophysiological basis of stunning and killing, which is necessary to understand how stunning and killing methods can be assessed.
- Have an overview of stunning and killing methods that are used in practice and how the process of stunning and killing can be monitored in practice.
- Be aware of methods that are available for depopulation.
- Be aware of advantages and disadvantages of the implementation of stunning and killing methods in practice.

## CONTENT



Wageningen UR Livestock Research, P.O. Box 338, 6700 AH Wageningen, The Netherlands

## 9.1 INTRODUCTION

The world's aquaculture sector has been growing rapidly over the last few decades and is expected to do so until at least 2025 (Diana, 2009). About 354 finfish species were cultured in 2012 (FAO, 2014), and available statistics collected globally by FAO show that farmed finfish amounted to 44.2 million tons in 2012. This production volume refers to a range of 8.8–147 billion farmed finfish slaughtered in 2012. Within the class of finfish, there is diversity with respect to phylogeny (evolutionary history), behaviours and habitats and this is reflected in, for example, differences in anatomy and brain structure among fish species. Due to increasing societal awareness, especially in Europe, attention has been drawn to fish welfare in aquaculture and during catching, transport and slaughter. The World Organisation for Animal Health (known as the OIE), which includes 180 member countries, also adopted guidelines on the welfare of farmed fish, including welfare during transport and stunning and slaughter.

At present there is no legislation in the EU to specifically protect the welfare of farmed fish at slaughter. However, the European Regulation on the protection of animal welfare during slaughter and killing refers to all vertebrates (Council Regulation EC 1099/2009) and stunning and killing of fish is covered by the general provisions of this Regulation, that is, animals should be spared any avoidable excitement, pain or suffering during transport, lairage (a place where live animals are kept temporarily), restraining, stunning, slaughter or killing. In spite of such provisions, some commercial killing methods of farmed fish in the EU have been found to be stressful. For assessment of welfare aspects of stunning methods the general provision in the EU Regulation for vertebrates (Council Regulation EC 1099/2009) can be used as a general term of reference.

**Figure 9.1:** Arrival of Atlantic salmon transported by (a) well-boat, (b) lairage, (c) crowding and pumping of fish out of the cage, (d) a dewatering unit in front of a slaughter facility.

(c)

(d)

Specific definitions used in this chapter are presented below:

- *Harvest.* Harvest is the process whereby a live fish is converted into an edible product. This process can be categorized for farmed fish into the following steps, namely fasting, crowding, catching, transport to a slaughterhouse, keeping the fish temporarily (lairage) in cages near the plant quayside, in tanks at the slaughterhouse or pumping them directly into the processing line. In the processing line the fish are restrained, stunned and subsequently slaughtered (exsanguination or bleeding out). An overview of transport, lairage, crowding and pumping of Atlantic salmon is presented in Figure 9.1.
- *Teleost.* A large group containing most bony fishes. The fish species selected for this chapter are all teleosts.

### *9.1.1 Do fish feel pain?*

In spite of the commitment by the OIE to protect welfare of farmed fish, it is still argued whether fish can feel pain. To experience pain, fish should possess both a nociceptive system and a capacity for mental awareness (Sneddon, 2014). Studies indicate that pain sensory pathways exist in fish that are required for nociception (Sneddon 2002, 2003; Sneddon et al., 2014).

Given the absence of a cerebral cortex in fish, it might be argued that fish do not have a capacity for mental awareness, that is, to perceive pain and fear. However, feelings in humans may not depend exclusively on structures of the cerebral cortex (Damasio and Damasio, 2016). The issue whether fish may have experiences that relate to a negative affective state such as suffering is still debated in the scientific literature (Rose, 2002; Rose et al., 2014), as reviewed by Braithwaite and Ebbesson (2014). However, recent research (as reviewed by Braithwaite et al., 2013) shows that teleost fish have a capacity for mental awareness, as in studies on a limited number of teleost fish species relevant functional areas in the telencephalon have been found. These studies indicate that it is possible that teleost fish perceive pain and fear when they are not stunned before killing or slaughter. Accordingly, to protect welfare of farmed fish at slaughter, fish should be rendered unconscious and insensible by stunning without causing any avoidable excitement, pain or suffering prior to slaughter or killing.

It is obvious that for protection of welfare of fish at harvest all steps in the process have to be considered. However, within the scope of this chapter we will only focus on stunning methods and killing used for farmed European eel (*Anguilla Anguilla*), Atlantic salmon (*Salmo salar*), rainbow trout (*Oncorhynchus mykiss*), turbot (*Psetta maxima*), sea bream (*Sparus aurata*), sea bass (*Dicentrarchus labrax*), tuna (*Thunnus spp.*) and common carp (*Cyprinus carpio*).

## 9.2 ASSESSMENT OF STUNNING AND KILLING

We distinguish methods for stunning and stunning/killing. Both types of methods are presented in separate sections (9.3 and 9.4). These methods can be categorized into electrical, physical, mechanical and chemical methods.

It is important to note that in practice also methods are applied that boil down to killing of fish without stunning them first, e.g. asphyxia in an ice slurry. Information on these methods is provided in section 9.5.

For the methods presented in sections 9.3, 9.4 and 9.5, information is provided on the following aspects:

- neurophysiology and physiology;
- their effective use;
- monitoring points for effective application.

Regarding methods used for depopulation, information on these methods is scarce. As these methods are not intended for a regular harvest process, data are provided in a separate section, viz. 9.6.

In this section, methods for assessment of stunning and stunning/killing are presented. These methods should be used to establish ideal conditions for stunning or stunning/killing of fish without causing avoidable stress.

### *9.2.1 Neurological, physiological and behavioural aspects*

Published studies show that, in spite of the differences in the way stunning methods induce unconsciousness and insensibility, a fish can be judged to be unconscious, using electroencephalogram (EEG) recordings. For each method, typical changes in the EEG, including evoked responses or potentials in the brain, are necessary for an unequivocal assessment of the level of brain function in fish to determine whether or not the fish are effectively stunned. Visually evoked responses (VERs) or potentials (VEPs) and somatosensory evoked responses (SERs) on the EEG have been used in several fish species to determine the state of brain function following stunning.

Changes in the electrocardiogram (ECG) are useful to determine whether cardiac fibrillation or changes in heart rate occurs. In the case of fibrillation, the circulation of blood in the body is affected, which interrupts the supply of oxygen to the brain. Changes in heart rate may occur in fish that are subjected to live chilling or gas stunning. When these changes are observed prior to loss of consciousness they can be signs of distress in fish.

During and immediately after stunning, depending on the method and species involved, fish show typical behaviour patterns and physical reflexes, which can help to monitor the effectiveness of stunning under commercial conditions. Tonic and clonic cramps can be signs of an effective percussive stun. Immediately after an effective electrical or percussive stun, rhythmic breathing may be absent and the capacity of the fish to swim in a coordinated way is lost. In a stunned fish, the capacity to right itself is lost. When the application of live chilling with or without added $CO_2$ results in attempts of still conscious fish to escape, this is indicative for stress in the fish.

Stunning of fish by live chilling or exposure to gas does not induce immediate loss of consciousness. Analysis of the blood should be performed to determine whether the application of these methods is possible without causing avoidable stress in fish. Changes in levels of cortisol (a stress hormone), glucose, lactate and free fatty acids in the blood can be used as indicators.

For field observations, registration of EEGs and ECGs is not possible. In this case observation of behaviour has to be used. Behavioural measures have to be interpreted with caution (Van de Vis et al., 2014). Ineffective electrical stunning can be very painful and paralysis may occur without loss of consciousness. Also, as a result of exhaustion or low body temperature due to chilling, a conscious fish may not be able to show spontaneous behaviour and responses.

The toolbox in Table 9.1 shows tests that can be used for field observations to obtain an indication of whether fish are effectively stunned. Observed responses presented in italics and underlined are considered the most reliable indicators for the presence or absence of consciousness and sensibility in fish. Furthermore, a 3-point scoring system is defined for each test, where 0 represents no response, 1 an attenuated or abnormal response or behaviour and 2 a normal and clear response or behaviour.

Given the caution regarding interpretation of observed responses, we recommend to perform all tests in Table 9.1 and not limit the tests to one, for example the presence or absence of VORs.

### *9.2.2 Product quality*

It is known that each step in the harvest of fish, for example Atlantic salmon, can influence product quality. One of the challenges is to minimize handling stress in all steps prior to stunning and slaughter, as they can have a profound effect on the quality of the product. The relationship between stunning and product quality is described very briefly in sections 9.3, 9.4 and 9.5.

## 9.3 METHODS USED FOR STUNNING

### *9.3.1 Electrical stunning (Atlantic salmon, sea bass, rainbow trout and European eel)*

*Neurological and physiological aspects.* Electrical stunning can be performed by placing electrodes on either side of the head, or on the head and body so that current passes through the brains and heart. For fish, electrodes can also be placed in a water tank. An adequate voltage is necessary to drive sufficient current through the brains of an animal. This stimulation of the brain should induce a generalized epileptiform activity, as judged from EEG recordings and is indicative of unconsciousness and insensibility (Lambooij, 2014). In general, the expected outcome is either stunning when the electrodes are on the head only or electrocution (i.e. stunning and killing) when placed on head and body or the whole body is exposed to an electrical current in the water. Electrocution induces death by cardiac arrest. However, up to now reported data for, for example Atlantic salmon, sea bass, European eel, turbot and common carp, show that these animals cannot be killed by the use of electricity, as the fibrillation of the heart is not permanent (Daskalova et al., 2016; Lambooij et al., 2002b; Lambooij et al., 2010; Robb and Roth, 2003; Van de Vis et al., 2014). Hence, for fish, electrical stunning needs to be followed by a killing method.

**Table 9.1** Toolbox to assess *consciousness and sensibility* and *loss of consciousness and sensibility* in fish in the field: described responses in bold are the most reliable ones. The tests are based on Kestin et al., (2002), Morzel et al., (2003) and EFSA (2009c)

| | Self-initiated behaviour | | Response to stimuli | | | Clinical response | |
|---|---|---|---|---|---|---|---|
| *Name* | Equilibrium | Swimming | Tail pinch | Needle scratch | 6 V shock | Breathing | Vestibulo-ocular response (commonly called eye roll) |
| *Procedure* | Ability of the inverted fish to right itself in water. Observe this | Observe the fish undisturbed for coordinated movements in water. Bumps against the wall of tank should be absent. Observe behaviour | Catch the fish by the tail and administer tail pinch. Observe response | Apply needle along the imaginary middle line of the fish in water or air. Observe response | Apply 6 V dc to the mouth in air. Observe response | Observe the fish undisturbed. Breathing in water. Observe opercula for rhythmic movement | Vestibulo-ocular reflex (VOR) in air. Grasp the fish firmly. The movement of the eye is observed when the fish is rolled from side to side |
| Score | | | | | | | |
| 0 | *No attempts* | *Absent* | *No response* | *No response* | *No reaction* | *No rhythmic opercular movements* | *Eyes fixed relative to the head* |
| 1 | *Sluggish or delayed attempts* | *Sluggish movements* | *Sluggish attempts to escape* | *Muscular contraction felt under the hand or slow attempt to escape* | *Partial eye retraction into the sockets, slow head shake or escape attempt* | *Opercular movements either irregular, reduced, deeper or faster than normal* | *Slow and delayed or one eye shows VOR* |
| 2 | *Immediate response* | *Typical for the fish species concerned* | *Clear attempts to escape* | *Attempts to escape* | *Immediate and complete retraction of the eyes into the sockets, head shake or escape attempt* | *Regular opercular movements* | *Clear movements of the eyes* |

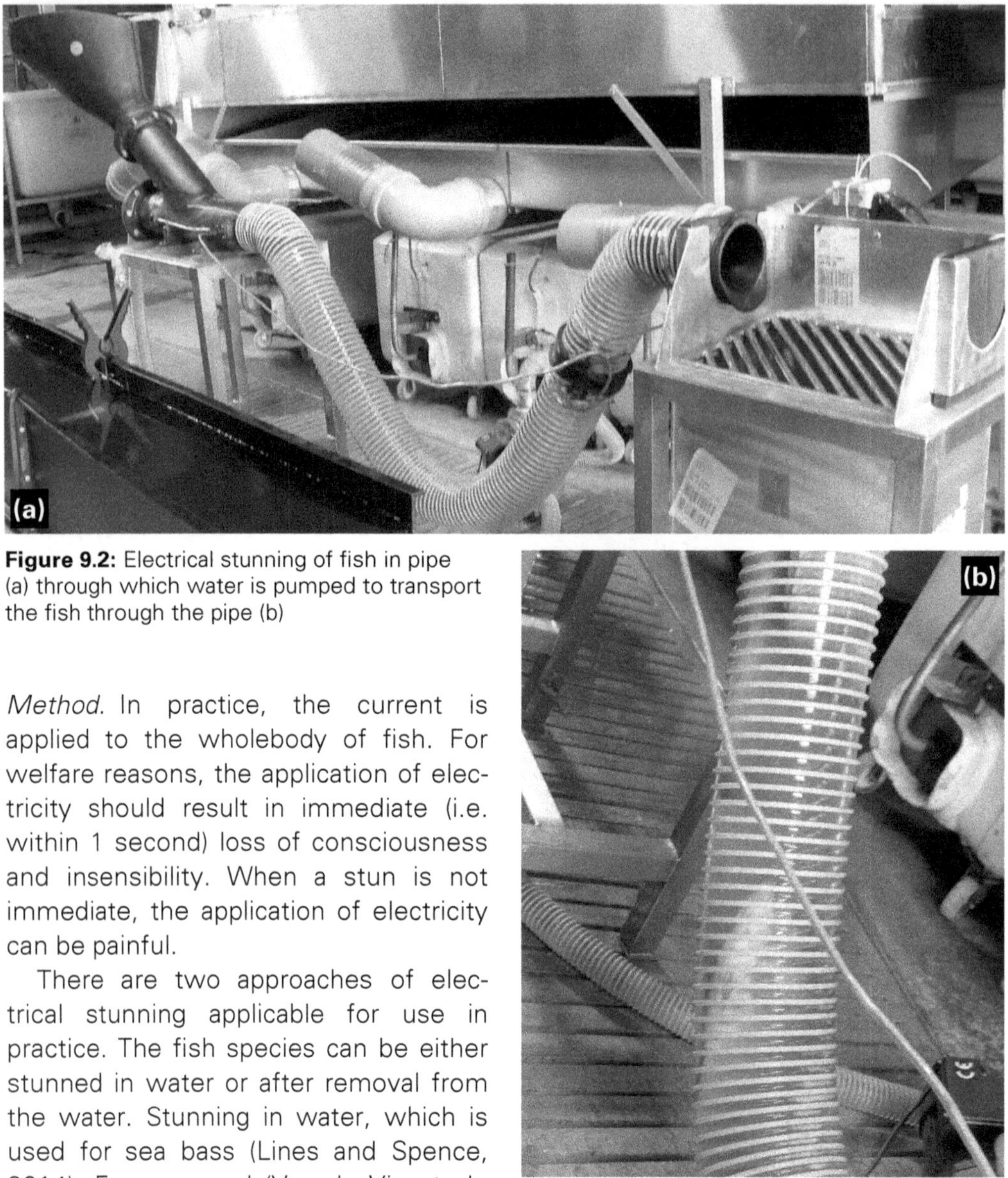


**Figure 9.2:** Electrical stunning of fish in pipe (a) through which water is pumped to transport the fish through the pipe (b)

*Method.* In practice, the current is applied to the wholebody of fish. For welfare reasons, the application of electricity should result in immediate (i.e. within 1 second) loss of consciousness and insensibility. When a stun is not immediate, the application of electricity can be painful.

There are two approaches of electrical stunning applicable for use in practice. The fish species can be either stunned in water or after removal from the water. Stunning in water, which is used for sea bass (Lines and Spence, 2014), European eel (Van de Vis et al., 2013) and rainbow trout (Lines and Spence, 2014; EFSA, 2009b), involves exposing the fish to an electrical field created using two plate electrodes in a water tank or ring or plate electrodes in a pipe (Figure 9.2) through which water is pumped. For stunning after removing out of water, which is used for Atlantic salmon (EFSA, 2009c) and European eel (Van de Vis et al., 2013), the fish is placed in a device that consists, for example, of a conveyer belt as negative electrode with steels flaps suspended (Figure 9.3). The rows of steel flaps are the positive electrodes. In principle, electrical stunning in water may be less stressful to a fish, as for

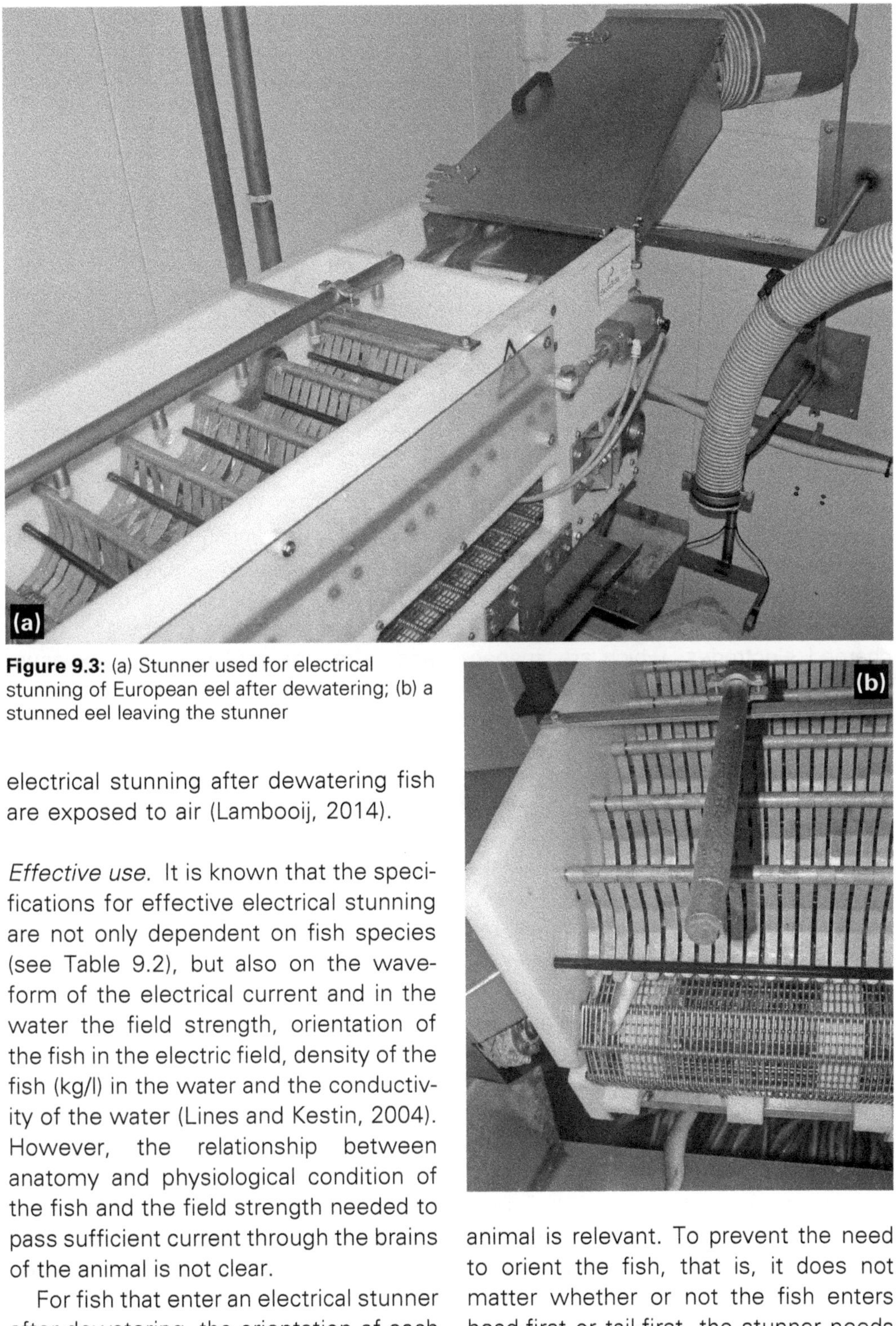


**Figure 9.3:** (a) Stunner used for electrical stunning of European eel after dewatering; (b) a stunned eel leaving the stunner

electrical stunning after dewatering fish are exposed to air (Lambooij, 2014).

*Effective use.* It is known that the specifications for effective electrical stunning are not only dependent on fish species (see Table 9.2), but also on the waveform of the electrical current and in the water the field strength, orientation of the fish in the electric field, density of the fish (kg/l) in the water and the conductivity of the water (Lines and Kestin, 2004). However, the relationship between anatomy and physiological condition of the fish and the field strength needed to pass sufficient current through the brains of the animal is not clear.

For fish that enter an electrical stunner after dewatering, the orientation of each animal is relevant. To prevent the need to orient the fish, that is, it does not matter whether or not the fish enters head-first or tail-first, the stunner needs

**Table 9.2** Conditions[a] to achieve an immediate stun by applying electricity in the water and after dewatering

| Species | Head-only[b] 50 Hz sinusoidal | Water bath 50 Hz sinusoidal | Stunning after dewatering[d] |
|---|---|---|---|
| European eel | 600 mA | 0.64 A/dm$^2$ | 1300 mA |
| Rainbow trout | 500 mA | | |
| Common carp | | 0.14 A/dm$^2$ | |
| Atlantic salmon | | | 670 mA |
| Turbot[c] | | 3.2 A/dm$^2$ | 1270/2390[e] |
| Sea bass[c] | | 3.4 A/dm$^2$ | |

a: More detailed information is presented in the papers included in the list of references; b: For head-only stunning, a pair of tongs was used to place the electrodes on the head of the fish; c: For stunning of turbot and sea bass, bipolar square waveform of 133 Hz, 43% duty cycle was used; d: For stunning, a direct current coupled with a 100 Hz alternating current at a voltage of 129 $V_{rms}$ for eel and 108 $V_{rms}$ for Atlantic salmon was used; e: For effective head-first and tail-first stunning of turbot, currents of 1270 and 2390 $mA_{rms}$ should be passed through an individual fish, respectively.

to be modified to ensure that the current is passed through the brains and not side-to-side through the tail. This can be achieved by using rows of negatively charged electrodes, which are suspended above the negatively charged conveyer belt. When the entire body of a fish is still under these negatively charged rows of suspended electrodes, no current is passed through the fish, as the conveyer belt is negatively charged too. Due to movement of the conveyer, the fish will be moved towards the first row of positively charged suspended electrodes. As result, a connection between the positive and the previous negative rows of suspended electrodes will be established. It should be noted that the current should pass through the tail of the fish towards the head and this distance is substantially longer than head-first stunning in which the current is passed side-to-side through the head of a fish. Due to this, longer distance and increase of the voltage by 20–40%, compared to head-first stunning is required. Further research is needed to establish immediate effectiveness for tail-first stunning in a commercial setting (Daskalova et al., 2016).

For both electrical stunning in water and after dewatering, the exposure time should be long enough to prolong the duration of unconsciousness and insensibility, which is needed to prevent recovery during the application of a killing method. A 1-second exposure to an electrical current is not sufficient, as most fish recover after 20–40 seconds from the electrical stun. Depending on the species, an exposure time of 5–20 seconds is needed to prolong the period of loss of consciousness and sensibility (Lines and Kestin, 2005), which is needed to prevent recovery during killing. After an exposure of 20 seconds to the electricity, European eels can be killed without recovery by immersion of the animals in a mixture of ice water and salt (ratio 0.08 kg NaCl/l ice water) for 15 minutes. The time interval post-stunning and immersion should not exceed 60 seconds for the eels (Van de Vis et al., 2013).

Exposing the Atlantic salmon for 5 seconds to electricity, as specified in Table 9.2, followed by a gill cut (EFSA, 2009c) resulted in one out of three fish recovering temporarily after 3 minutes (Lambooij et al., 2010). This shows that gill cutting does not bleed the fish effectively and, therefore, an alternative killing method needs to be developed to prevent recovery of the stunned fish. Common carp, sea bass, rainbow trout and turbot (Table 9.2) can be killed after electrical stunning by immersion in a slurry of ice and water without recovery (Daskalova et al., 2016; EFSA, 2009b,d,e; Lines and Spence, 2014; Van de Vis et al., 2013).

It is well known that chilling in ice water alters the homeostatic balance in fish. The issue is whether this alteration is of sufficient magnitude for stunned fish that are acclimatized to a water temperature of, for example, 8°C, which may occur in a temperate zone, as killing of stunned fish has only been studied by registration of EEG in fish that were acclimatized to temperatures of at least 13°C.

Poor welfare at stunning can be due to a voltage or field strength in the water that is too low to pass sufficient current through the brains of a fish or the waveform of the electrical current being incorrect. Other causes may be that a stunner is overloaded with fish resulting in a collapse of the power source that supplies the electricity for stunning. For both electrical stunning in water or after dewatering, exposure of fish to pre-shocks should be prevented. Another potential risk factor during head-only electrical stunning using dry electrodes, is that fish entering tail first will experience pain before the fish is rendered unconscious by the current.

There is some risk of poor welfare when applying electrical stunning in water (batch) systems and after dewatering, mainly due to mis-stuns which may result from the varying resistance between animals in the stunner.

Electrical stunning after dewatering may lead to skin burns, as the electrodes touch the skin of fish. Passing an electrical current through the skin may result in the production of heat. This is a hazard when an electrical stun does not result in immediate loss of consciousness and sensibility. In case of an effective stun, it is an economical hazard for a slaughterhouse. Spraying eels with fresh water during stunning could prevent the occurrence of skin burns (Van de Vis et al., 2013). For other species such as Atlantic salmon and African catfish this problem was not observed. Sound data on the overall incidence of skin burns in fish are lacking.

*Monitoring points.* For effective electrical stunning, it is essential that the conditions for an immediate stun are used. Exposure of fish to pre-shocks has to be avoided. These conditions are based on EEG and ECG recordings in combination with behavioural observations in a laboratory setting. To control electrical stunning in practice, the applied voltage (in water, the field strength), the strength of current and its waveform, and duration of exposure of fish to the electricity need to be monitored, as well as the time interval between fish leaving the stunner and the application of an effective killing method. It appears that gill-cutting after electrical stunning can lead to recovery in Atlantic salmon (Lambooij et al., 2010). Hence, an alternative method for killing the stunned fish should be developed. It is known for African catfish that decapitation prevents recovery from electrical stunning. It is essential that a sharp knife is used to cut all vessels.

For monitoring of electrical stunning followed by killing in practice, the animal-based indicators of unconsciousness and insensibility presented in the toolbox in Table 9.1 should be used.

It should be noted that there is always some uncertainty in the effectiveness of stunning due to extrapolation from experiments in the laboratory to implementation in a commercial setting, and to varying electrical resistances between fish. Attempts to electrocute the fish are not recommended, since data that show the fish can be killed by electricity are not available.

Electrical stunning with low frequency currents can cause spinal injuries and hematomas within the fillet, which can be reduced by using higher frequencies of 500 to 1000 Hz (Roth et al., 2004). An alternative waveform for an electrical current, which results in a very low incidence of injuries in Atlantic salmon, consists of a coupled direct and alternating electrical current (not fully sinusoidal with a frequency of 100 Hz) for stunning after dewatering (Roth et al., 2009). Longer exposure times (> 10 s) to electricity will eventually stimulate anaerobe glycolysis to such a degree that early onset of rigor mortis is observed in Atlantic salmon. Thus, for pre-rigor filleting of Atlantic salmon it is important to keep the electrical stimulation at a certain minimum.

### 9.3.2 Physical stunning

#### 9.3.2.1 Live chilling (Atlantic salmon)

*Neurological and physiological aspects.* In general, rapid live chilling can reduce body temperatures to the lower limit of an organism's thermal range, which can result in severe sublethal disturbances and mortality. Rapid chilling can induce loss of consciousness and sensibility, as observed in European eel that was exposed to a slurry of ice and water (see section 9.5). However, ECG recordings showed that for European eel this method is stressful (Lambooij et al., 2002a). For assessment of live chilling we recommend to use EEG and ECG recordings in fish in combination with behaviour observations and plasma analysis of cortisol, glucose, lactate and free fatty acids.

For Atlantic salmon, however, live chilling is used to calm the salmon down in behaviour prior to stunning (EFSA, 2009c). Subsequently, the conscious fish (EFSA, 2009c) are stunned and killed. A decrease in temperature from 16 to 4°C over 1 hour and 16 to 0°C over 5 hours did not result in a significant increase in plasma cortisol levels in Atlantic salmon (Foss et al., 2012).

*Method.* Live chilling of Atlantic salmon has been developed to achieve product quality control, efficiency and processor safety. For Atlantic salmon, cooling the fish is aimed to postpone the onset of rigor mortis and preserve the quality of this species. Live chilling of Atlantic salmon is performed in refrigerated seawater (RSW) tanks. The water is re-used and oxygen is supplemented to ensure sufficient levels (DO > 70%).

*Effective use.* Atlantic salmon seem to tolerate relatively steep temperature drops without compromising welfare (Foss et al., 2012). However, the most severe salmon temperature change (16 to 0 °C over 1 h) resulted in the death of fish (Foss et al., 2012). Deterioration of quality of the re-used water should be prevented.

*Monitoring points.* A study by Foss et al. (2012) suggested that Atlantic salmon are capable of coping with controlled changes in temperature of the water. A temperature drop from 16 to 4°C over 1 hour or from 16 to 0°C over 5 hours does not seem to affect the welfare of the fish in the short term (Foss et al., 2012). Hence, temperature, time and behaviour of the fish during and after the drop in temperature are monitoring points. As live chilling does not render Atlantic salmon unconscious and insensible, the fish should show normal or sluggish swimming behaviour with complete equilibrium. Excessive struggling (tail flaps) should be absent. The salmon should show clearly evident and regular respiration.

In practice, a gradual accumulation of waste products and carbon dioxide, which is excreted by the fish, can occur as large numbers of fish pass through the RSW tanks during the day. This deterioration of water quality causes severe stress in Atlantic salmon (EFSA, 2009c). The quality of the water that is re-used should be monitored, so that corrective actions can be taken in case it deteriorates.

#### *9.3.2.2 Live chilling with carbon dioxide (Atlantic salmon)*

*Neurological and physiological aspects.* Exposure of Atlantic salmon to live chilling with carbon dioxide, which is a recently developed method, has not been assessed by registration of EEGs and ECGs. Behavioural observations showed that exposure is aversive to these fish. It is likely that consciousness was not lost (Erikson, 2011).

*Method.* When the method is applied to farmed Atlantic salmon, the fish are exposed to cold water (temperature range -0.5 to 3 °C) containing low and moderate levels (65-257 mg/l) of added carbon dioxide and 70-100% saturation of added oxygen levels (Erikson et al., 2006). The water is re-used. Subsequently, the fish are killed by gill-cutting and bled in chilled seawater. An evaluation of live chilling with carbon dioxide showed that the use of carbon dioxide in combination with live chilling was found to be stressful (Erikson, 2011).

Live chilling of fish has been reported to produce contradictory results. Live chilled fish had an early onset and resolution of rigor mortis, as compared to percussive or electrically-stunned salmon observed by Roth et al. (2002; 2004), while Erikson et al. (2006) reported no difference for live chilling with moderate concentrations (37–80 mg/l) of carbon dioxide in the seawater and percussive stunning. The reason seems to lie in different live chilling practices, whereas oxygen supplement to refrigerated seawater prevents the fish undergoing hypoxia. Delaying the onset of rigor mortis facilitates pre-rigor filleting. Pre-rigor filleting is preferred, as the gaping score in the fillets is significantly lower, compared to post-rigor filleting.

*Effective use.* Atlantic salmon appear to find the method aversive; it is recommended not to use it.

*Monitoring.* The quality of the water that is re-used should be monitored, so that corrective actions can be taken in case it deteriorates. Temperature of and oxgyen level in the water, and duration of the exposure and behaviour of the fish are monitoring points. Excessive struggling (tail flaps), which is a sign of stress, should be absent. The fish should show clearly evident and regular respiration.

Erikson et al. (2006) also compared percussive stunning and live chilling with 37–80 mg/l carbon dioxide in refrigerated seawater with respect to colour of the fillet and textural properties. No significant differences were observed for these product quality parameters.

### 9.3.3 Chemical stunning

#### 9.3.3.1 Clove oil and Aqui-S™

*Neurophysiological and physiological aspects.* EEG recording revealed that Aqui-S™ renders fish unconscious, as judged from the appearance of theta and delta waves on the EEG. Post-stun behavioural observation of the stunned fish in a tank showed that recovery may not occur (Erikson et al., 2012).

*Method.* The main components in clove oil are eugenol (4-allyl-2-methoxyphenol) and isoeugenol (4-propenyl-2-methoxyphenol), which comprise 90–95% of clove oil by weight. Clove oil has been used in fish at concentrations of 25–100 mg/L in the water, depending on species and degree of anaesthesia needed. The positive features reported for clove oil as anaesthetic led to the development of a new anaesthetic in New Zealand, called Aqui-S™. Isoeugenol (4-propenyl-2-methoxyphenol) is the active compound in Aqui-S™.

*Effective use.* In the European Union and Norway it is not allowed for stunning/slaughtering of fish. Barriers to its use in the EU include the cost of overcoming the legislative requirements for introducing isoeugenol as anesthetic for food fish (Van de Vis et al., 2014). Recommendations for its effective use, are therefore, not presented.

*Monitoring points.* When the use of Aqui-S™ is allowed, temperature, time and behaviour of the fish during exposure are monitoring points. In case of water being re-used, the water quality should be monitored to avoid deterioration. Exposure of fish to Aqui-S™ should not lead to excessive struggling (tail flaps). Instead, gradual loss of equilibrium, sporadic swimming movements and weak or occasional opercular movement should occur, finally leading to loss of equilibrium, ability to swim, cessation of opercular movement and absence of responses to administered stimuli (Table 9.1).

## 9.4 METHODS USED FOR STUNNING/KILLING

### 9.4.1 Mechanical methods

#### 9.4.1.1 Percussion (Atlantic salmon and common carp)

*Neurological and physiological aspects.* Non-penetrative percussion is a traumatically-induced derangement of the nervous system, which results in an immediate diminution or loss of consciousness and sensibility (EFSA, 2004). Rapid oscillations of brain due to pressure waves created by the impact of the bolt have been suggested as the important factor, and not the pressure as such developed by these waves (Nilsson and Nordström, 1977). The appearance of theta, delta waves and spikes, followed by an iso-electric EEG in fish indicate unconsciousness and insensibility without recovery (Lambooij et al., 2007). It has been reported that percussed Atlantic salmon die of cerebral haemorrhage (Lambooij et al., 2010).

*Method.* Percussive stunning is the application of a blow to the head manually or by using a device. At commercial slaughterhouses for salmon, an automatic device is commonly used to apply a blow to the head. Prior to its application the fish are removed from the water. The main hazard for automated percussive stunning is variation in the size of fish within the population, causing a misstun in some fish. Machines for stunning and killing salmon should not be used if fish may be injured, but not stunned immediately because of their size or orientation in the machine. For percussive machines, size adjustment of the machines should be performed by skilled personnel, as this is crucial for stunning efficiency. For carp, the current stunning method for preparation for consumption consists of a blow or repeated blows on the head with a heavy wooden baton known as a priest (Lambooij et al., 2007). Under field conditions, carp are exposed to air for 30 minutes before the application of the blow to the head by hand (EFSA, 2009d; Van de Vis et al., 2006). It should be noted that air exposure for 30 minutes is stressful for carp, as the animal is lying on its side in a dry tank possibly covered by conspecifics that are also attempting to escape asphyxia. A blow to the head is painful when the stun is not achieved immediately. Percussive stunning should, therefore, result in an immediate stun (within 1 second) and recovery prior to death should not occur. Percussion cannot be applied with all species. For European eel it was found not to be effective.

*Effective use.* The method is suitable for immediate stunning without recovery in Atlantic salmon (Lambooij et al., 2010). A study revealed that brain function ceased 0.3 minutes after application of the blow (Robb et al., 2000). The stunning efficiency is related to applied force and adjustment to size of the fish (Roth et al., 2007a). EEG recordings showed that at an air pressure higher than 8.1 bar to drive the bolt, consciousness and sensibility are lost immediately in Atlantic salmon without recovery, under conditions used. However, at an air pressure higher than 8.1 bar, carcass damage occurred (Lambooij et al., 2010).

*Monitoring points.* It is essential that the blow is delivered correctly to ensure that consciousness and sensibility are lost immediately. Convulsions of the muscles indicate an effective stun. Percussive systems should have a separate air supply or alternatively have security valves to block the system if the pressure is reduced below a certain threshold. Especially for Atlantic salmon, asphyxia caused by exposure to air that takes too long should be avoided.

Regarding product quality of Atlantic salmon, it was reported that carcass damage may occur, as described previously.

#### *9.4.1.2 Spiking/coring (iki jime) and shooting of tuna*

*Neurophysiological and physiological aspects.* For experimental stunning and killing of Atlantic salmon the method used for coring of tuna was adapted (Robb et al., 2000). For this purpose a hollow bolt was constructed and investigation by EEG recordings showed that consciousness was not lost immediately (Robb et al., 2000).

Spiking/coring under water and shooting from the surface are both stressful, as the fish have to be crowded to the surface (EFSA, 2009a). Hosting or gaffing, which are performed prior to spiking or coring on board, impairs welfare of tuna severely (EFSA, 2009a). Both lupura and shooting tuna outside the water have not been assessed by registration of EEGs. For smaller tuna, spiking or coring the brain is used, as described above.

*Method.* Iki jime is a traditional Japanese technique used to kill tuna by brain ablation (EFSA, 2009a). For this killing method, the fish needs to be well restrained to achieve an accurate application. Spiking/coring is performed under water and on board. For spiking/coring under water, the fish are crowded close to the surface of the water (EFSA, 2009a). Fish that are spiked/cored in the water are bled there or on board. When the spiking or coring is performed on board, the fish are gaffed or hoisted and bled after the application of the method.

Spiking or coring is similar to penetrating captive bolt stunning of farm animals. For an effective application of this method, the spike or core is rapidly inserted into the brain via the pineal window. Subsequently, pithing is undertaken immediately after coring or spiking. For pithing, a wire is inserted into the hole in the skull, resulting from coring, and pushed into the spinal cord to destroy both brain and spinal cord (EFSA, 2009a). The abolition of VERs is indicative of the loss of consciousness and sensibility (Robb et al., 2000).

Two other methods evaluated for killing farmed tuna are (EFSA, 2009a): 1) shooting underwater to the head by using a power-head (lupara) that resembles a short gun barrel with a single-shot cartridge inside; 2) an alternative to lupura is shooting tuna from the surface of the water. For the latter method, the fish are brought to the surface rapidly and marksmen standing on the service boat or a platform next to the slaughter cage, shoot the fish on the head using a shotgun loaded with single-bullet cartridges. These two methods are used for tuna weighing more than 80 kg.

*Effective use.* Based on the risk assessment performed by EFSA (2009a), underwater shooting (lupara) is preferred, compared with shooting the fish from the surface. Applying gaffing or hoisting to bring tuna on board results in severely impaired welfare. For small tuna, EFSA (2009c) recommends performing spiking or coring while the fish are in shallow water, compared with the application on board.

However, whether or not consciousness and sensibility are lost immediately as a result of spiking/coring or the use of lupara has not been validated by registration of EEGs in these fish species.

*Monitoring points.* To assess whether lupara, and spiking/coring result in an effective stun in practice, it is recommended (EFSA, 2009a) to monitor the fish for loss of body movement and absence of reaction during handling. For spiking/coring is it crucial that the core or spike is inserted into the brain via the pineal window. To control stunning/killing of tuna in practice, EFSA (2009a) recommends setting up standard operation procedures which include valid, robust and feasible indicators to evaluate harvest of tuna in practice.

When lupara is used, a backup diver is required in case a second shot is needed.

## 9.5 KILLING METHODS WITHOUT STUNNING

### *9.5.1 Asphyxia in ice/a slurry of ice and water (sea bass, sea bream, turbot, European eel and trout) or air*

*Neurophysiological and physiological aspects.* Indices for the induction of unconsciousness and insensibility are the appearance of theta and delta waves and no response on pain stimuli on the EEG (Lambooij et al., 2002a). Abolition of VERs in the EEG recordings can also be used as an indicator for loss of consciousness and sensibility in fish. Changes in heart rate, as recorded on ECGs, can be signs of stress.

Chilling on ice or in air results in hypoxia in fish. Some fish species, however, are tolerant of hypoxia/anoxia. When fish are exposed to hypoxic or anoxic conditions, they will depend on anaerobic metabolism and have potential acid-base consequences. A sharp decrease in blood and tissue pH is a major limiting factor in hypoxic tolerance. In general, fish are not well-adapted to metabolic acidosis because of the relatively poor buffering capacity of their blood (Jackson, 2004). Crucian carp (*Carassius carassius*) is exceptional, as it can survive long periods of anoxia. This species is able to cope with the acid-base consequences of anaerobiosis. The fish is able to produce ethanol and $CO_2$ as major anaerobic end products (Shoubridge and Hochachka, 1980). Ethanol diffuses across the gills too.

In case of exposure to air, the gills collapse into a mass of tissue when the fish is out of the water. This collapse hinders breathing. However, it is known that, in general, fish do not die within 20–30 seconds after air exposure, as a variety of strategies, depending on the species, can occur in fish to cope with hypoxia and anoxia. A common strategy is hypometabolism (Jackson, 2004).

Sea bream and sea bass are commonly killed by removing them from water and leaving them to die in a slurry of ice and seawater (EFSA, 2009f). In most cases, violent attempts to escape are made and maximal stress responses are initiated (Robb and Kestin, 2002). When spontaneous movement has ceased the fish are processed.

EEG recordings in sea bream fish that were exposed to chilling on ice show that it may take 5 minutes before visually evoked responses in the brain (VERs) are abolished (Van de Vis et al., 2003). When the fish are left to die in air, the time taken to abolish VERs in sea bream is longer (Van de Vis et al., 2003). Sea bream removed from water at 23°C and let die by asphyxia in air, which is no longer performed in practice in Europe, showed an average time to loss of VERs of 5.5 minutes (van de Vis et al., 2003).

The time needed to kill sea bass by asphyxia in ice water varied between 20 minutes (Poli et al., 2004) and 34 minutes (Acerete et al., 2009).

Assessment of chilling of turbot

revealed a stress response during early immersion in a slurry of ice and seawater. Even an exposure of 75 minutes to the slurry may not induce unconsciousness in turbot, however, brain activity does decrease to a lower level (Lambooij et al., 2015).

Four stages could be distinguished for European eels that were exposed to chilling in a slurry of ice and water. In the first stage, the eels swim normally. During the second stage, they attempt to escape from the ice water. In the third stage, the eels press their nose to the wall or corner of the tank (Lambooij et al., 2002a) while showing clonic muscle contractions. In the fourth stage, they lay down on their abdominal side on the bottom of the box and do not have their characteristic resting position (S-shape), but continue to breathe. Unconsciousness was induced in most eels at a body temperature of on average 8°C after 12 minutes in the slurry of ice and water. The heart rate decreased and became irregular during live chilling of eel (Lambooij et al.,2002a).

For rainbow trout it takes 9.6 minutes before VERs are lost due to asphyxiation in ice (Robb and Kestin, 2002). In the view of EFSA (2009b) exposure to asphyxia in ice or ice water presents a welfare hazard for conscious trout.

*Method.* Chilling in ice or a slurry of ice and seawater boils down to a transfer of sea bass and sea bream from the seawater in cage or tank to ice flakes or ice water slurry (in ratio ranging from 1:2 to 3:1; (EFSA, 2009f), respectively. Chilling in a slurry of ice and seawater is most commonly used for sea bass and sea bream (EFSA, 2009f). The temperature of the slurry fluctuates from 0 to 2°C. In case of asphyxia in air, sea bass and sea bream are removed from the water and placed in free draining bins or boxes (EFSA, 2009f).

Turbot are transferred from the holding tanks to containers with a mixture of ice and seawater (0 to 4°C) at densities of approximately 300 kg/m$^3$. The fish are kept in these containers for a minimum of 30 minutes (EFSA, 2009g).

In an experimental method, European eels were moved from the holding tank to a slurry of ice and fresh water. The exposure lasted until a body temperature of 5°C was obtained (EFSA, 2009h; Lambooij et al., 2002a). For rainbow trout the method is similar: the fish are transferred into a slurry of ice and fresh water (EFSA, 2009b).

*Effective use.* These methods are very stressful for sea bass, sea bream, turbot, European eel and rainbow trout. Because unconsciousness and insensibility are not induced immediately and all the fish species appear to find asphyxia in air, ice or a slurry of ice and (sea)water aversive, these methods should not be used.

*Monitoring points.* In practice, all these fish species should be monitored for cessation of breathing, loss of body movement and absence of reaction during handling.

### *9.5.2 Carbon dioxide (Atlantic salmon)*

*Neurophysiological and physiological aspects.* Exposure of Atlantic salmon to seawater saturated with carbon dioxide was widely used in Norway to stun

Atlantic salmon prior to killing by exsanguination. EEG recordings show that it took on average 6 minutes before VERs were abolished in Atlantic salmon (Robb et al., 2000). Due to slow induction of unconsciousness and insensibility combined with vigorous escape attempts by fish during the induction period (Robb et al., 2000), the use of carbon dioxide has been prohibited in Norway (Anonymous, 2006).

*Method.* Briefly, in a commercial setting, carbon dioxide is bubbled into a tank filled with seawater until pH levels of about 5.5–6.0 are obtained. This corresponds to $CO_2$ levels of 200–450 $mgl^{-1}$. Fish are transferred to the water and after an exposure of 2-4 minutes the struggling stops. Subsequently, the fish are removed and bled. More information about this method, which is not allowed in Norway, can be found in a report by EFSA (2009c).

*Effective use.* Exposure of Atlantic salmon to seawater saturated with carbon dioxide is highly stressful for this species and, therefore, no information on its effective use is provided.

*Monitoring points.* The use of this method is highly stressful for Atlantic salmon and, therefore, it is not possible to provide monitoring points to control this process with respect to welfare.

### 9.5.3 Salt bath or ammonia to de-slime live European eels

*Neurophysiological and physiological aspects.* Eels make extremely vigorous attempts to escape from salt (Van de Vis et al., 2003) or ammonia (Kuhlmann and Münkner, 1996).

When exposed to a salt bath, it may take longer than 10 minutes, based on the time to abolition of VERs, to induce unconsciousness (Van de Vis et al., 2001). The process of de-sliming causes an osmotic shock in eels that kills them. However, it is possible that the de-slimed eels are eviscerated while still conscious (Verheijen and Flight, 1997).

*Method.* Live eels are placed in a dry tank and sodium chloride (NaCl) or a combination of NaCl and aqueous sodium carbonate ($Na_2CO_3$) is added to the eels. This results in denaturation of the mucus proteins and clotting, damage to the upper layer of skin and opaque eyes.

A 25% ammonia solution, which is an alternative to de-slime live eels, is added at the ratio of 100 kg dry eels/100 ml ammonia solution. The ammonia solution causes denaturation of the mucus proteins, which results in loosening of the mucus layer on the skin and the eyes become opaque or white (Kuhlmann and Münkner, 1996).

Exposing conscious eels to ammonia or a salt bath is considered bad practice in Germany and has been prohibited since 1999 (Anonymous, 1997). It is probable that in 2017 the use of a salt bath will be prohibited in the Netherlands to de-slime eels while they are still conscious.

*Effective use.* A salt bath or the use of aqueous ammonia is highly stressful for eels and, therefore, no information on its effective use is provided.

*Monitoring points.* Given the fact that the exposure of eels to a salt bath or ammonia is highly stressful, it is not possible to provide monitoring points to control this process with respect to welfare.

### 9.5.4 Decapitation (European eel)

*Neurophysiological and physiological aspects.* Decapitation causes death through loss of blood. As judged from recorded EEGs, it may take more than 13 minutes before VERs are abolished after decapitation (Van de Vis et al., 2003).

*Method* The head is separated from the body by a sharp cut resulting in bleeding. Some recoil in the arteries may impede blood flow.

*Effective use.* Decapitation of a conscious eel is highly stressful and, therefore, no information on its effective use is provided. We recommend applying an effective stunning method prior to decapitation.

*Monitoring points.* The method is objectionable on welfare grounds and, therefore, monitoring points are not provided.

### 9.5.5 Exsanguination (turbot)

*Neurophysiological and physiological aspects.* Exsanguination followed by asphyxia in a slurry of ice and water has not been assessed by registration of EEGs. Exsanguinated turbot placed in a slurry of ice and seawater showed escape behaviour and other responses to physical handling. One hour post exsanguination the muscle pH dropped from 7.2 to 6.8–6.9, which is indicative for physical activity (Roth et al., 2007b). Morzel et al. (2002) showed that behavioural responses in turbot were lost within 15–30 minutes after bleeding in an ice slurry. EFSA (2009a) concluded that exsanguination combined with asphyxia in an ice slurry constitutes a considerable welfare risk.

*Method.* In practice, turbot are exsanguinated without stunning them first. The fish are bled straight after netting and transferred into a slurry of ice and water. Exsanguination is performed to remove the blood and to improve the visual quality of the flesh (Roth et al., 2007a) as well as killing the animal prior to further processing such as evisceration and filleting.

*Effective use.* Exsanguination in combination with chilling is highly stressful for turbot and, therefore, no information on its effective use is provided. It is recommended to apply an effective stunning method first.

*Monitoring points.* Monitoring points are not provided, as exsanguination prior to stunning is not recommended.

## 9.6 METHODS USED FOR DEPOPULATION (ALL SPECIES)

Please be aware that information on methods used for depopulation is scarce. To the best of our knowledge, data published in peer-reviewed publications on assessment, descriptions of effective use of these methods, and monitoring points for effective application are not available.

During production of fish, circumstances can occur that require slaughter for legal, health or welfare reasons.

These circumstances include: the detection of a notifiable disease by the competent authority; irreparable failure of a life-supporting production system where welfare compromise might be inevitable; in the event of a serious and untreatable disease or parasite infection (FAWC, 2014). EFSA (2009d) stated that, at present, there is insufficient knowledge of parameters for the sound use of anaesthetics for humane emergency killing. In the view of EFSA (2009e), asphyxia, hypoxia or chilling on ice are not acceptable methods of killing farmed fish. Obviously, studies are needed to establish conditions for methods that can be used for emergency killing.

## 9.7 REFERENCES

Acerete, L., Reig, L., Alvarez, D., Flos, R. and Tort L. (2009). Comparison of two stunning/slaughtering methods on stress response and quality indicators of European sea bass (*Dicentrarchus labrax*). *Aquaculture* 287, 139–144.

Anonymous (2006). Forskrift om slakterier og tilvirkingsanlegg for akvakulturdyr Kapittel 4. In: *kystdepartementet F-o, editor. Nasjonale tilleggsbestemmelser om fiskevelferd.* Oslo, Norway. pp. 13–14.

Anonymous (1997). *Verordnung zum Schutz von Tieren in Zusammenhang mit der Schlachtung oder Tötung – TierSchlV (Tierschutz-Schlachtverordnung),* vom 3. März 1997, Bundesgesetzblatt Jahrgang 1997 Teil I S. 405, zuletzt geändert am 13. April 2008 durch Bundesgesetzblatt Jahrgang 2008 Teil I Nr. 18, S. 855, Art. 19 vom 24. April 2006.

Braithwaite, V. and Ebbesson, L.O. (2014). Pain and stress responses in farmed fish. *Revue scientifique et technique/Office International des Epizooties* 33, 245–253.

Braithwaite, V., Huntingford, F. and Van den Bos, R. (2013). Variation in emotion and cognition among fishes. *Journal of Agricultural Environmental Ethics* 26, 7–23.

Council Regulation (EC) No 1099/2009 (2009). On the protection of animals at the time of killing. *Official Journal of the European Communities* L 303, 1–30.

Damasio, A. and Damasio, H. (2016). Pain and other feelings in animals. *Animal Sentience 2016.059*

Daskalova, A.H., Bracke, M.B.M., Van de Vis, J.W., Roth, B., Reimert, H.G.M., Burggraaf, D. and Lambooij, E. (2016). Effectiveness of tail-first dry electrical stunning, followed by immersion in ice water as a slaughter (killing) procedure for turbot (*Scophthalmus maximus*) and common sole (*Solea solea*). *Aquaculture*, 455, 22-31.

Diana, J.S. (2009). Aquaculture production and biodiversity conservation. *BioScience* 59, 27–38.

EFSA (2004). Welfare aspects of animal stunning and killing methods. EFSA, Parma, Italy, 241 pp.

EFSA (2009a). Species-specific welfare aspects of the main systems of stunning and killing of farmed tuna. *The EFSA Journal* 1072, 1–53.

EFSA (2009b). Species-specific welfare aspects of the main systems of stunning and killing of farmed rainbow trout. *The EFSA Journal* 1013, 1–55.

EFSA (2009c). Species-specific welfare aspects of the main systems of stunning and killing of farmed Atlantic salmon. *The EFSA Journal* 2012, 1–77.

EFSA (2009d). Species-specific welfare aspects of the main systems of stunning and killing of farmed carp. *The EFSA Journal* 1013, 1–37.

EFSA (2009e). General approach to fish welfare and to the concept of sentience in fish. *The EFSA Journal* 954, 1–27.

EFSA (2009f). Species-specific welfare aspects of the main systems of stunning and killing of farmed seabass and seabream *The EFSA Journal* 1010, 1–52.

EFSA (2009g). Species-specific welfare aspects of the main systems of stunning and killing of farmed turbot *The EFSA Journal* 1073, 1–34.

EFSA (2009h). Species-specific welfare aspects of the main systems of stunning and killing of farmed eel. *The EFSA Journal* 1010, 1–52.

Erikson, U. (2011). Assessment of different stunning methods and recovery of farmed Atlantic salmon (*Salmo salar*): isoeugenol, nitrogen and three levels of carbon dioxide. *Animal welfare* 20, 365–375.

Erikson, U., Hultmann, L and Steen, E.j. (2006). Live chilling of Atlantic salmon (*Salmo salar*) combined with mild carbon dioxide anaesthesia – I Establishing a method for large-scale processing of farmed fish. *Aquaculture* 252, 183–198.

Erikson, U., Lambooij, B., Digre, H., Reimert, H.G.M., Bondø and Van de Vis, H. (2012). Conditions for instant electrical stunning of farmed Atlantic cod after de-watering, maintenance of unconsciousness, effects of stress, and fillet quality – A comparison with Aqui-S$^{TM}$. *Aquaculture* 324–325, 135–144.

FAO (2014) The state of world fisheries and aquaculture 2014. http://www.fao.org/3/contents/c235a282-977e-243d239-b231e237-cffee235ccf292f/i3720e3700.htm.

FAWC (2014). Opinion on the welfare of farmed fish at the time of killing. Farm Animal Welfare Committee, London, United Kingdom, 36 pp.

Foss, A., Grimsbo, E., Vikingstad, E., Nortvedt, R., Slinde, E., Roth, B. (2012). Live chilling of Atlantic salmon: physiological response to handling and temperature decrease on welfare. *Fish Physiol Biochem* 38, 565–571.

Jackson, D.C. (2004). Acid-base balance during hypoxic hypometabolism: selected vertebrate strategies. *Respiratory Physiology and Neurobiology* 141, 273–283.

Kestin, S.C., Van de Vis, J.W. and Robb, D.F.H. (2001). A simple protocol for assessing brain function in fish and the effectiveness of stunning and killing methods used on fish. *Veterinary Record* 150, 320–307.

Kuhlmann, H. and Münkner, W. (1996). Gutachterliche Stellungnahme zum tierschutzgerechten Betäuben/Töten von Aalen in größeren Mengen. *Fischer and Teichwirt* 47, 404–410, 445–448, 493–495.

Lambooij, E. (2014). Electrical stunning. *Encyclopedia of Meat Sciences, 2nd Edition* (eds. C. Devine, M. Dikeman). Academic Press, London, UK. pp 407–412.

Lambooij, B., Bracke, M., Reimert, H., Foss, A., Imsland, A. and Van de Vis, H. (2015). Electrophysiological and behavioural responses of turbot (*Scophthalmus maximus*) cooled in ice water. *Physiology and Behavior* 21(149), 23–28. doi: 10.1016/j.physbeh.2015.05.019.

Lambooij, E., Grimsbø, E., Van de Vis, J.W., Reimert, H.G.M., Nortvedt, R. and Roth B. (2010). Percussion and electrical stunning of Atlantic salmon (*Salmo salar*) after dewatering and subsequent effect on brain and heart activities. *Aquaculture* 300, 107–112.

Lambooij, E., Pilarczyk, M., Bialowas, H., Van den Boogaart, J. G. M. and Van de Vis, J. W. (2007). Electrical and percussive stunning of the common carp (*Cyprinus carpio* L.): neurological and behavioural assessment. *Agricultural Engineering* 37, 171–179.

Lambooij, E., van de Vis, J.W., Kloosterboer, R.J. and Pieterse, C. (2002a). Welfare aspects of live chilling and freezing of farmed eel (*Anguilla anguilla,* L.): neurological and behavioural assessment. *Aquaculture* 210, 159–169.

Lambooij, E., Van de Vis, J.W., Kuhlmann, H., Münkner, W., Oehlenschläger, J., Kloosterboer, R.J. and Pieterse, C. (2002b). A feasible method for humane slaughter of eel (*Anguilla anguilla* L.): electrical stunning in fresh water prior to gutting. *Aquaculture Research* 33, 643–652.

Lines, J. and Kestin, S. (2004). Electrical

stunning of fish: the relationship between the electric field strength and water conductivity. *Aquaculture* 241, 219–234.

Lines, J. and Kestin, S. (2005). Electric stunning of trout: power reduction using a two-stage stun. *Agricultural Engineering* 32, 483–491.

Lines, J.A. and Spence, J. (2014). Humane harvesting and slaughter of farmed fish. *Revue scientifique et technique/Office International des Epizooties* 33, 255–264.

Morzel, M., Sohier, S. and van de Vis, J.W. (2002). Evaluation of slaughtering methods of turbots with respect to animal protection and flesh quality. *Journal of the Science of Food and Agriculture* 82, 19–28.

Nilsson, B. and Nordström, C.H. (1977). Rate of cerebral energy consumption in concussive head injury in the rat. *Journal of Neurosurgery* 47, 274–281.

Poli, B.M., F. Scappini, G., Parisi, G., Zampacavallo, M., Mecatti, P., Lupi, G., Mosconi, G., Giorgi and Vigiani, V. (2004). Traditional and innovative stunning slaughtering methods for European seabass compared by the complex of the assessed behavioural, plasmatic and tissue stress and quality indexes at death and during shelf life. In *Proceedings of the 34th WEFTA Conference* 2004, Lubeck, Germany.

Robb, D.F.H., Kestin, S.C. (2002). Methods used to kill fish: field observations and literature reviewed. *Animal Welfare* 11, 269–282.

Robb, D.F.H. and Roth, B. (2003). Brain activity of Atlantic salmon (*Salmo salar*) following electrical stunning using various field strengths and pulse durations. *Aquaculture* 216, 363–369.

Robb, D.H.F., Wotton, S.B., McKinstry, J.L., Sorensen, N.K. and Kestin, S.C. (2000). Commercial slaughter methods used on Atlantic salmon: determination of the onset of brain failure by electroencephalography. *Veterinary Record* 147, 298–303.

Rose, J.D. (2002). The neurobehavioural nature of fishes and the question of awareness and pain. *Fisheries Science* 10(1): 1–38.

Rose, J.D., Arlinghaus, R., Cooke, S.J., Diggles, B.K., Sawynok, W., Stevens, E.D. and Wynne, C.D.L. (2014). Can fish really feel pain? *Fish and Fisheries* 15, 97–133.

Roth, B., Birkeland, S. and Oyarzun, F. (2009). Stunning, pre slaughter and filleting conditions of Atlantic salmon and subsequent effect on flesh quality on fresh and smoked fillets. *Aquaculture* 289, 350–356.

Roth, B., Slinde, E. and Robb, D.F.H. (2007a). Percussive stunning of Atlantic salmon (*Salmo salar*) and the relation between force and stunning. *Aquacultural engineering* 36, 192–197.

Roth, B., Imsland, A., Gunnarsson, S., Foss, A. and Schelvis-Smit, R. (2007b). Slaughter quality and rigor contraction in farmed turbot (*Scophthalmus maximus*); a comparison between different stunning methods. *Aquaculture*, 272, 754–761.

Roth, B., Moeller, D., Slinde, E. (2004). Ability of electric field strength, frequency and current duration to stun farmed Atlantic salmon (*Salmo salar*) and pollock (*Pollachius virens*) and relations to observed injuries using sinusoidal and squarewave AC. *North American Journal of Aquaculture* 65, 208–216.

Roth, B., Veland, J.O., Moeller, D. Imsland, A. and Slinde, E. (2002). The effect of stunning methods on rigor mortis and texture properties of Atlantic salmon (Salmo salar). *Journal of Food Science* 67, 1462–1466.

Shoubridge, E.A. and Hochachka, P.W. (1980). Ethanol: novel end-product in vertebrate anaerobic metabolism. *Science* 209, 308–309.

Sneddon, L. U. (2002). Anatomical and electrophysiological analysis of the trigeminal nerve in a teleost fish, *Oncorhynchus mykiss*. *Neuroscience Letters* 319, 167–171.

Sneddon, L. U. (2003). The evidence for pain in fish: the use of morphine as an analgesic.

*Applied Animal Behaviour Science* 83, 153–162.

Sneddon, L.U., Elwood, R.W., Adamo, S.A. and Leach, M.C. (2014). Defining and assessing animal pain. *Animal Behaviour* 97, 201–212.

Van de Vis, H., Abbink, W., Lambooij, B. and Bracke, M. (2014). Stunning and killing of farmed fish: How to put it into practice? In: *Encyclopedia of meat sciences 2e, Vol. 3.* (C. Devine and M. Dikeman, editors-in-chief). Elsevier, Oxford. pp. 421–426.

Van de Vis, H., Bialowas, H., Pilarczyk, Machiels, M., Reimert, H., Veldman, M and Lambooij, B. (2006). Comparison of commercial and experimental slaughter of farmed carp (*Cyprinus caprio*) with respect to development of rigor mortis and flesh quality. In: *Seafood research from fish to dish* (eds. J.B. Luten, C. Jacobsen, K. Bekaert, A. Sæbø and J. Oehlenschläger). Wageningen Academic Publishers, Wageningen, The Netherlands. pp. 201–210.

Van de Vis, H., Kestin, S.C., Robb, D., Oehlenschläger, J., Lambooij, B., Münkner, W., Kuhlmann, W., Kloosterboer, K., Tejada, M., Huidobro, A., Tejada, M., Otterå, H., Roth, B., Sørensen, N.K., Aske, L., Byrne, H. and Nesvadba, P. (2003). Is humane slaughter of fish possible for industry? *Aquaculture Research* 34, 211–220.

Van de Vis, J.W., Burggraaf, D., Reimert, H., Lambooij, E. (2013). Implementation of electrical stunning of eel (*in Dutch*). IMARES: Yerseke (Rapport/IMARES C089/13). 23 pp.

Van de Vis, J.W., Oehlenschläger, J., Kuhlmann, H., Münkner, W., Robb, D.F.H. and Schelvis- Smit, A.A.M. (2001). Commercial and experimental slaughter of eel (*Anguilla anguilla,* L.): effect on quality and welfare. In: *Farmed fish quality* (eds. S.C. Kestin and P.D. Warriss). Blackwell, Oxford, UK. 234–257.

Verheijen, F.J., Flight, W.F.G. (1997). Decapitation and brining: experimental tests show that after these commercial methods for slaughtering eel *Anguilla anguilla* (L.), death is not instantaneous. *Aquaculture Research* 28, 361–366.

chapter ten

# Slaughter without stunning

*Antoni Dalmau[1] and Haluk Anil[2]*

## Learning objectives

- To understand the physiological aspects of slaughter without stunning.
- To become familiar with the different restraining systems in cattle and small ruminants and their main risk factors.
- To learn about the best practices during restraining and neck cutting.
- To learn the signs of unconsciousness and death.

## CONTENT



## 10.1 INTRODUCTION

Slaughter without stunning for human food is widely practised on religious grounds around the world. These religious rites can be, for instance, the shechita, halal and jhatka methods practised by Jews, Muslims and Sikhs, respectively. The first two methods involve severing all the major blood vessels in the neck, whereas, the jhatka method involves decapitation of animals. Shechita and halal methods are used on a wide scale throughout the world but jhatka is restricted to South Asia only, so it will be not considered here. It

[1] IRTA. Animal Welfare Subprogram, Veïnat de Sies, s/n, 17121 Monells, Spain
[2] School of Veterinary Science, University of Bristol, Langford BS40 5DU, United Kingdom

should be noted that some local religious authorities accept a pre-slaughter stunning method, provided it does not kill the animals based on their interpretations of the religious requirements (Velarde et al., 2014). These cases, including post-cut stunning, are not treated in the present chapter. Bovine, small ruminants and poultry are the main species of farm animals subjected to slaughter without stunning and, in consequence, the ones treated in this chapter.

Slaughter without stunning does not induce immediate loss of consciousness in animals (EFSA, 2013a,b,c). During the time until consciousness is finally lost, animals can be subjected to pain and distress. As stated by Gibson et al. (2015), any delay between the neck-cut and onset of unconsciousness could result in pain and suffering associated with restraint, tissue damage caused by the cut and stimulation of nociceptors in the wound, and distress associated with aspiration of blood in the respiratory tract. The time to loss of consciousness and onset of death varies among species and even between individuals within a species due to factors that can interfere, such as quality of the knife, effectiveness of the cut (depriving the brain of oxygenated blood supply), position of the animal, flow of the blood, and so on.

## 10.2 PHYSIOLOGICAL ASPECTS OF SLAUGHTER WITHOUT STUNNING

Slaughter without stunning does not induce immediate loss of consciousness. Weight of blood is about 8% of live weight of an animal, about 18% of total cardiac output flows through the brain at any one time (EFSA, 2004) and, normally, about 50% of total blood is removed by exsanguination. In mammals, consciousness is lost if 30–40 % of the total blood volume is lost or if blood pressure drops to below 35 to 50 mmHg (Gregory, 2005). Brain responsiveness measured by means of evoked responses is gradually reduced and diminished after slaughter without stunning. In fact, animals are gradually rendered unconscious as brain perfusion becomes insufficient to sustain normal function. Anil et al. (2006) found that 25% of total blood volume in cattle is lost after 17 seconds, although the process is much quicker in sheep (Anil et al., 2004). According to Levinger (1995), in sheep this time could be reduced to 5 to 6 seconds, although normal ranges vary from 2 to 20 seconds (von Holleben et al., 2010). According to Barnett et al., (2007), broiler chickens lose about 40% of their total blood volume within 30 seconds after neck cutting. It is suggested that on average they lost consciousness between 12 and 15 seconds after the cut, although the same author (Barnett et al., 2007) described a case lasting more than 25 seconds to the onset of unconsciousness. There are no specific studies in turkeys, but according to EFSA (2013c), turkeys are more resilient to brain ischaemia and therefore times to onset of unconsciousness and death are expected to be significantly longer when compared with chickens.

Prolonged time to onset of unconsciousness in cattle, in comparison to other mammals or poultry, is explained by a supplementary system of vascularization of the brain. One key element of this system is the presence of vertebral arterial anastomosis in cattle. Vertebral arteries are protected within

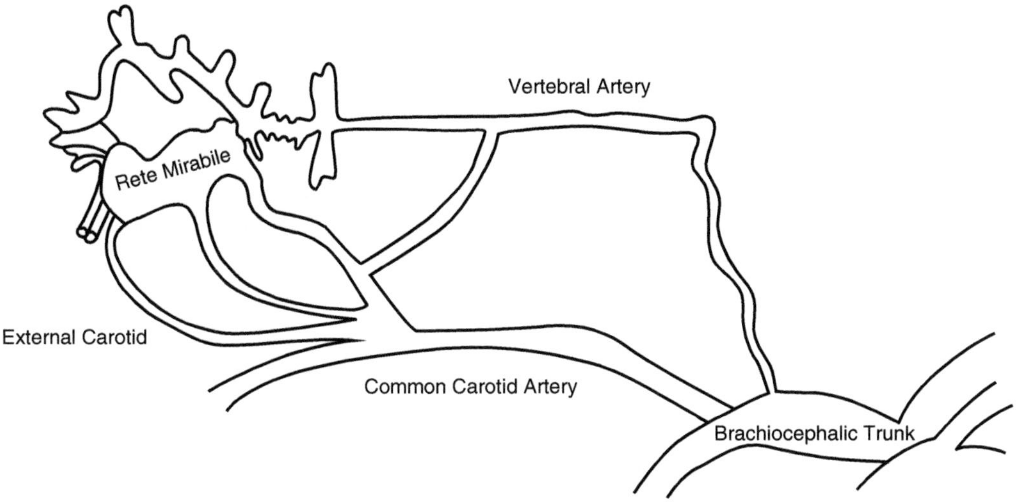


**Figure 10.1:** *Rete mirabile* and circulation vessels to the brain of cattle

the foraminae of the cervical vertebrae, so, even after cutting both carotid arteries by neck cutting, part of the blood supply to the forebrain is maintained via the vertebral–occipital anastomosis and *rete mirabile* (Figure 10.1).

The *rete mirabile*, more extensive in cattle than sheep and goats, is a vascular network that receives branches from the carotid and vertebral arteries. In addition, an additional anastomosis, called the occipito–vertebral anastomosis, may supply blood to *rete mirabile* and the brain. This branch connects occipital and vertebral arteries and contributes, due to continuous blood supply, to the delay in the loss of brain function after neck cutting in cattle. Although poultry have vertebral arteries (Mead, 2004), they do not have the very complex network seen in cattle. On the other hand, sheep and goats do not have the supplementary flow of blood provided by the vertebral arteries (von Holleben et al., 2010). Another point to take into account is based on differences in brain size, arterial cross sectional area and blood volume. The weight of the brain relative to the total body weight in small ruminants is 0.26% and in cattle only between 0.07 and 0.08%. This implies that a lower proportion of total blood volume is necessary to perfuse the brain of cattle compared to small ruminants. In addition, the carotid arteries of adult cattle may be too small relative to total blood volume to allow for sufficiently fast bleed outs and a drastic loss in blood pressure (von Holleben et al., 2010).

In addition, there is a wide variation in the time to loss of brain responsiveness following exsanguination in cattle. The variation may be caused by factors such as individual differences in anatomy, age and susceptibility to develop swellings in cut ends of carotid arteries (e.g. clots causing carotid occlusion/aneurysm). Daly et al. (1988) suggested two explanations. First, the differences between animals in the proportion of total cerebral blood flow which is contributed by the vertebral arteries. Second, the amount of blood reaching the brain via the vertebral arteries after slaughter is very close to the minimum blood flow necessary to sustain electrical activity in the brain cortex, so that slight differences in individuals will result in large variations. According to von Holleben et al. (2010), the estimated range of time to loss of consciousness in most cattle is between 5 and 90 seconds after the cut, but even under laboratory conditions possible resurgence of consciousness has been assumed for more than 5 minutes. Gibson et al. (2015) found a range from 1 to 257 seconds in cattle to final collapse, similar to the ranges described by Gregory et al. (2010) in the same abattoir with the same slaughterman. However, in the study by Gregory et al. (2010) 8% of the animals took more than 60 seconds to final collapse, being 4% in Gibson et al. (2015).

During slaughter without stunning, the purpose of the cut is to induce unconsciousness followed by death. In the conscious animal, the cerebral cortex integrates posture and movement. Collapse, which manifests when a freely standing animal falls to the ground, is the earliest indication of approaching unconsciousness after the neck-cut (EFSA 2013a,b). However, an animal that has collapsed after a dramatic loss of blood pressure may have the capacity to regain consciousness as a result of the body's own counterregulation mechanisms (EFSA 2013a,b). In some cases, animals may exhibit loss of posture as a result of the loss of significant proportions of circulating blood volume but subsequently suffer carotid artery occlusion and, as a consequence, recover consciousness (EFSA 2013a). For instance, in small ruminants, Levinger (1961) found that animals collapsed but were able to regain posture when the carotid arteries were clamped. In any case, the signs used to assess the onset of unconsciousness in animals would be as follows:

- Sustained loss of posture (collapse or loss of balance) can be used only in animals that are free standing or lightly restrained in the upright position. Therefore, loss of posture cannot be determined in animals that are severely restrained and/or rotated. In fact, when possible, only complete and permanent loss of posture without attempt to regain posture can be used as an outcome of unconsciousness (EFSA 2013a,b).
- Loss of muscle tone, resulting in relaxed body, may be used to recognize the onset of unconsciousness. However, involuntary muscle jerks may occur occasionally. Depending upon the type and severity of restraint, loss of tone in neck and leg muscles could be used as an alternative when the whole body cannot be assessed (EFSA 2013a,b).
- Cessation of rhythmic breathing. Rhythmic breathing can be recognized from regular flank movements in ruminants, and abdominal (vent) movements in poultry. Therefore, sustained absence of rhythmic breathing (absence of a respiratory cycle – inspiratory and expiratory movements) can be used as an outcome of unconsciousness (EFSA 2013a,b,c).
- Cessation of breathing can also be used as an outcome of unconsciousness in animals. However, since the trachea is also severed at

the time of neck cutting at slaughter without stunning, absence of breathing cannot be assessed from air movement at the external nostrils or beak of animals (EFSA 2013a,b,c).

- Absence of the pupillary reflex. The constriction of the pupils (miosis) in response to focusing or shining a torch at the pupils will be absent in unconscious animals (EFSA 2013a,b,c).
- Absence of palpebral reflex. The blinking response is elicited by touching or tapping a finger on the inner/outer eye canthus or eyelashes of mammals and will be absent in unconscious animals (EFSA 2013a,b).
- Absence of corneal reflex. The blinking response elicited by touching or tapping the cornea, will inevitably be absent in unconscious ruminants and birds (EFSA 2013a,b,c).
- Absence of response to threatening movements. Conscious animals respond either by blinking or attempting to move away from threatening movements (or clapping) of hands close to the eye. This fear response will be absent in unconscious animals (EFSA 2013a,b,c).

However, according to EFSA, breathing and muscle tone are the preferred indicators, while loss of posture, corneal and palpebral reflexes can be additional indicators that should not be relied upon alone in ruminants (EFSA 2013a,b; Figure 10.2). In poultry, it is recommended to use breathing and corneal or palpebral reflex. Muscle tone might be used as an additional indicator, but it should not be solely relied upon (EFSA 2013c; Figure 10.3).

When the function of the brain stem is sufficiently impaired as a result of blood loss, respiration and cardiac activity will

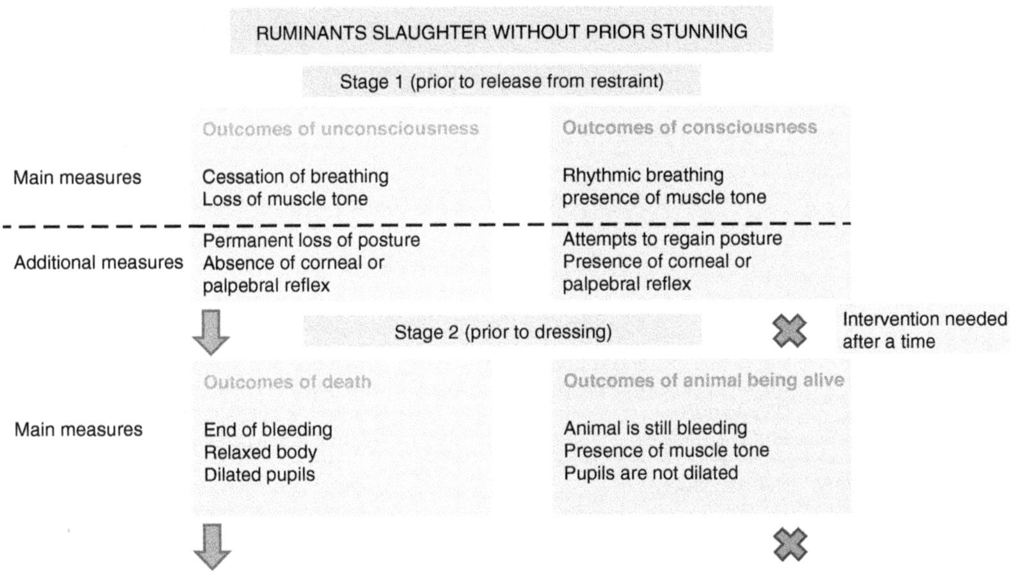


**Figure 10.2:** Measures to assess consciousness/unconsciousness and signs of death in cattle, sheep and goat slaughtered without prior stunning. Based on EFSA 2013a,b

cease over time (EFSA 2013a,b,c). The main clinical signs of death are permanent absence of respiration (and also absence of gagging) and absence of a pulse and cardiac activity (EFSA 2013a,b,c). Other signs are:

- Dilated pupils (midriasis), which require close examination of the eyes (EFSA 2013a,b).
- Relaxed body of the animal. Complete and irreversible loss of muscle tone leads to this state, which can be recognized from the limp carcass (EFSA 2013a,b,c).
- End of bleeding. Slaughter leads eventually to cessation of bleeding, with only minor dripping, from the neck-cut wound. However, formation of aneurysm and occlusion of the carotid artery (known as carotid artery ballooning) may prevent blood flow from the neck-cut wound, and this should not be mistaken for end of bleeding (EFSA 2013a,b,c).

According to EFSA (2013a,b,c), the preferred indicators for the assessment of death in slaughter without stunning are end of bleeding, relaxed body and dilated pupils (Figure 10.2). Live birds can be recognized from the presence of breathing or of the corneal and palpebral reflexes, pupils that are not fully dilated, continued bleeding or the presence of muscle tone and body movements (EFSA 2013c; Figure 10.3).

## 10.3 RESTRAINING

The restraint system allows the operator to perform a sufficient cut to the ventral surfaces of the neck that severs the jugular veins and carotid arteries. The restraint should be designed to allow the wound site to remain open to enable profuse bleeding, while preventing further stimulation of the wound. The ideal restraining method for slaughter depends on the animals (species and size) to be slaughtered, the method of slaughter (with or without stunning) and the capabilities of the staff. There are some basic principles of restraint with regard to animal welfare which have to be fulfilled (von Holleben et al., 2010; Grandin, 2013, 2005; Rosen 2004):

POULTRY SLAUGHTER WITHOUT PRIOR STUNNING

Prior scalding

| | Outcomes of death | Outcomes of animals being alive |
|---|---|---|
| Main measures | Permanent absence of breathing<br>Absence of corneal or palpebral reflex<br>Dilated pupils<br>End of bleeding | Presence of breathing<br>Presence of corneal or palpebral reflex<br>Pupils are not dilated<br>Animal is still bleeding |
| Additional measures | Relaxed carcass | Presence of muscle tone |

**Figure 10.3:** Measures to assess consciousness/unconsciousness and signs of death in poultry slaughtered without prior stunning. Based on EFSA 2013c

- Provision of non-slip flooring surfaces leading up to and in the pen.
- Raceways/entrance ways should be curved and avoid sharp corners to prevent balking and allow easy access into the pen.
- Sufficient lighting to minimize balking.
- Equipment should be engineered to minimize noise that could cause agitation/distress to the animals.
- Restraining itself must cause as little stress/strain as possible and be as short as possible.
- A restraining device or method must suit the size and species and type of animals slaughtered and must not cause injuries. The pressure applied by the device should not be excessive (i.e. struggling is often a sign of excessive pressure).
- The restraining method should not cause defense movements or flight reactions of the animal.
- Restraining for slaughter without stunning needs to make sure that the neck can be stretched, so that an optimum cut is possible.
- Mechanical and chemical stimuli (e.g. bloodborne metabolites) on the wound have to be minimized as long as the animals have not yet lost consciousness.
- Adequate post-cut restraint is vital for correct bleeding. Animals must be unconscious before release from the restraint.

Scared animals secrete stress pheromones that function as alarm signals to other animals. In case of severe stress conditions, the alarm pheromones circulating in blood may cause other animals to baulk. Cleaning and washing the facilities to eliminate stress pheromones (blood, saliva or urine) may help to calm the animals. A negative air pressure in the restraining pen withdraws the blood smell away from the entrance and might facilitate animal movement. Finally, restraining must not cause negative impact on bleed out, carcass or meat quality and should match the intended slaughter speed, achieving good working conditions. Restraint systems for slaughter without stunning often incorporate both body and head/neck holder devices that allow the operation of the specific slaughter method.

Cattle can be restrained either in an upright position, rotated by 90° lying on their side or rotated by 180° lying on their back. Rotating may also occur to other angles than 90 or 180° depending on practical and religious reasons. Restraining cattle by suspending their hind legs causes stress and pain and is not acceptable according to animal welfare standards (von Holleben et al., 2010). Upright restraining is possible either in a box or a pen with the neck being stretched or lifted by means of a halter and lateral straps or chains. More complicated technical equipment uses mechanical systems such as chin lifts as head holders. Many systems have a back-pusher/tailgate and belly plate that further confines the animal. To improve entry it has been recommended that when loading the animal, the belly plate should be recessed into the floor (flush with the floor surface). Another way to restrain cattle in an upright position is by using a double rail (centre track) conveyor restrainer, in which the animals are placed with their legs straddling, not touching the ground, and their bodyweight being supported under the brisket and belly.

When the animal reaches the front of the restrainer the head is stretched by a chin lift and then the cut is performed (Grandin, 1988).

Restraining cattle on their back is practised in rotating pens, in which the head is restrained, the body confined laterally and the animals turned on their backs. In this position the cut is performed, and afterwards the animal is rotated a little backwards to be released from the pen and then hoisted and shackled. Different aspects must be taken into account with these systems, which differ from slaughterhouse to slaughterhouse, such as the head restraint or chin lift, neck yoke, pressure of side walls and head lift, mechanical control and smooth operation of turning, changing direction, angle and speed (Levinger, 1995). In inverted designs, the chin lift should be applied prior to the process of inversion to reduce struggling (Gregory, 2005). Restraining cattle on their side is also possible using the same rotating pens that turn them on their back. The rotating pens are then turned to 45 or 90° or positions in between and the cut is performed while cattle are tilted (von Holleben et al., 2010). Upright restraint of cattle during slaughter without stunning is judged to be a better method even though rotating pens have been improved. However, the cut is made upwards against the ventral aspect of the neck and this makes the cut more awkward (Gregory, 2005; Figure 10.4). Furthermore, due to the action of the cut, it is possible to incompletely severe the carotids on the contralateral side to the operator. Therefore, when slaughtering animals in upright restraints it is important to assess the success of the neck-cut. A greater level of skill may be required to achieve an appropriate cut and manage the post-cut period with slaughter in the upright versus the inverted positions. In fact, the DIALREL project reported that the mean number of cuts performed was higher in cattle restrained in upright (9) compared to 180° (5), 90° (3) and 45° (1) pens (Velarde et al., 2014). Another common complaint of upright systems is that animals might be over restrained. Some upright systems have design flaws, which hinder good restraint, such as excessive pressure on the animal, poorly designed head holder or chin lift or hyperextension of the head (von Holleben et al., 2010), which could all cause discomfort, pain and suffering. The DIALREL project reported that during spot visits, 63% of cattle restrained in upright pens showed struggling compared to 37% in inverted pens (Velarde et al., 2014), potentially due to excessive pressure from the chin lift, back-pusher and belly plate. Grandin and Regenstein (1994) reported reactions of the animals due to irritation of the wound (i.e. if the wound touches the metal parts of the neck frame). This provoked active movements and it may also slow down exsanguination. It is also important to reduce the pressure on the animal's body immediately after the cut to achieve good bleed out and ensure quick loss of consciousness. Finally, another disadvantage of upright pens from the operators perspective is that when performing the cut they are more likely to be covered with blood because of their position relative to the cut (Gregory, 2005).

The advantage of the 180° rotating systems is that they often provide good presentation of the ventral surface of the neck for the neck incision. However, animals inverted on their backs for slaughter in rotating pens can have in some

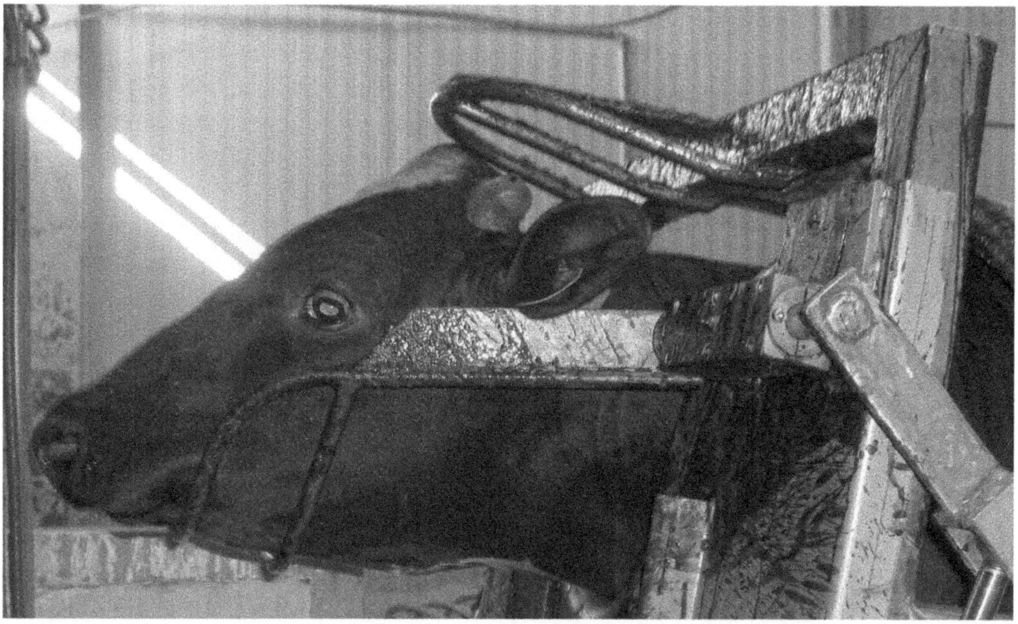

**Figure 10.4:** Head restraining of cattle slaughtered in upright position

cases a longer time interval from entering to full restraint (although it depends a lot on the rotatory system used) and can show more vigorous and longer periods of struggling and increased number of vocalizations (although this depends on the time being restrained) compared to animals restrained in upright position. In fact, Warin-Ramette and Mirabito (2010) reported in a study of rotating restraint systems in France that the frequency of vocalizations was directly linked to the time spent by the animals in inverted position. Some studies described more laboured breathing, increased foaming at the mouth and greater serum cortisol concentrations and haematocrit in cattle restrained in rotatory pens compared to those slaughtered in an upright position (Koorts, 1991; Dunn, 1990), although more modern rotatory systems can reduce these differences by reducing the time animals need to be restrained before slaughtering. In any case, much of the public and scientific debate on the welfare aspects of rotating restraint systems have focused on issues regarding unnatural posture, abdominal pressure on visceral tissues and stress during inversion. The slaughter of cattle in dorsal recumbency is prohibited in some countries (Grandin and Regenstein, 1994), such as in Slovakia, Denmark and the United Kingdom where upright position is mandatory, or the Netherlands where lateral recumbency but not inversion is allowed for slaughter without stunning. It has been suggested in cattle that, when inverted, bleeding might be impaired (Adams and Sheridan, 2008), due to the weight of the abdominal organs pressing the diaphragm and major veins. The added pressure on the heart may decrease stroke volume (compare "cardiac tamponade") and on the veins may impair venous reflux (Adams and Sheridan, 2008). Aspiration of blood and refluxing gut content after the incision is also considered a welfare concern after slaughter without stunning. However, although this problem has been historically associated with the inverted position, it can occur with the upright position for both halal and shechita slaughter as well (Gregory et al., 2009).

In lateral recumbence, cattle can be less stressed compared with lying on their back, as the rumen does not press on the diaphragm and therefore does not cause breathing difficulties. Nevertheless some pressure on the internal organs would still be present even in lateral restraint (Petty et al., 1991). Experiences during the DIALREL project revealed that lateral recumbence can help to avoid some problems like pressure on the aorta, major veins and diaphragm (von Holleben et al., 2010). However, other difficulties may arise as the performance of the cut has to be adapted to this position. In fact, Velarde et al. (2014) reported that during the DIALREL abattoir spot visits the restraint to cut interval for cattle was longer in animals restrained at 45° and 90° compared to those inverted 180° or in the upright position.

Therefore, both rotating and upright restraint systems have strengths and weaknesses. Specific animal welfare concerns of rotating systems are delays in operation between entry and slaughter, and pain/stress/distress from being restrained in an unnatural position, while upright restraints can cause pain and distress to the animal if excessive pressure is applied, and more skill is required to perform a successful neck-cut. In both rotating and upright systems the design of the neck yoke and chin lift has an important impact on the performance of the cut. The chin lift should provide good access to the neck, allow for efficient cutting and bleeding, avoid excessive neck tension (which could be painful) and should not obscure the face and eyes (for assessment of consciousness) of the animals.

Sheep and goats can be restrained either in an upright position, lying on their side or lying on their back (Levinger, 1995). Rotating is also used at angles other than 90 or 180°. As in cattle, restraining animals by suspending their hind legs must be avoided. In fact, according to Blackmore and Delany (1988), in sheep, bleeding is slightly more rapid in a recumbent position than when suspended in a vertical position. Holding or lifting sheep by grasping their wool should not be done (von Holleben et al., 2010).

During shechita slaughter, each chicken is restrained manually by a person holding both its legs in a raised hand and supporting its back, with its wings folded, on the opposite forearm and other hand (Barnett et al., 2007). The shochet is then able to extend the bird's head in his left hand with his thumb against the ventral surface of the bird's upper neck close to the beak and cut all the blood vessels with the knife in his right hand. The bird is then passed to a third person who places it into a bleeding cone. Halal methods are to place the bird in a cone or shackle before performing the cut, placing the birds in lateral recumbency (von Holleben et al., 2010) or even with animals hoisted.

## 10.4 NECK CUT

Slaughter without stunning requires an accurate cut to the throat with a sharp knife by means of an incision below the angle of the jaw that involves severing of major blood vessels in the ventral neck region (skin and vessels cut simultaneously). The two carotid arteries and jugular veins are severed simultaneously with the oesophagus, trachea and vagus nerves (Figure 10.5). Although this practice (i.e. cutting of oesophagus and trachea) has been suggested as not being optimal with regard to hygiene reasons in the EU (Regulation [EC] No 853/2004), it is allowed in ritual slaughter. Neck cutting for ritual slaughter by shechita is carried out by trained and certified people using a very sharp specialized knife (chalaf) in an uninterrupted transverse stroke to sever the neck vessels and surrounding tissues. Halal slaughter is performed by people without any training or certification by using an ordinary slaughter knife of varying length and can be either similar to shechita cut or by retrograde cut (gash stick) applying a stab followed by incision, severing the tissues. In poultry cutting, both carotid arteries, compared to cutting one common carotid artery and/or one jugular vein, induced impaired brain function most rapidly (Gregory and Wotton, 1986), so a complete cut of carotids is always suggested in slaughter without stunning.

As loss of unconsciousness is not immediate after the cut, it must be assumed that the cut and consequences of the cut are painful for animals as noniceptors allocated in the area of the incision are stimulated (Kavaliers, 1989). In fact, researchers in New Zealand showed neck cutting to be noxious (Gibson et al, 2009a,b), and according to EFSA 2004, any cut intended to kill the animal by rapid bleeding will greatly activate the protective nociceptive system for perceiving tissue damage and cause the animal to experience a sensation of pain. Lambooij and Kijlstra (2008) supported that the neck-cut itself causes the sensation of pain since this area of incision has a high density of pain receptors. Gibson et al. (2009a,b) described for the first time by means of electroencephalogram (EEG) that the act of slaughter by ventral-neck incision is associated with noxious stimulation that would be expected to be perceived as painful (Mellor et al., 2009; Figure 10.6). During spot visits (DIALREL, 2009) reactions to the cut observed in cattle were vocalizations or exhalation (as long as the trachea was intact), retracting movements, struggling and shivering. In sheep, only struggling was observed and sometimes, shivering. Reactions of poultry were retraction movements and wing flapping. However, these reactions depend on individual factors, even under optimal conditions to perform the cut. Lambooij and Kijlstra (2009) described for instance that in some animals, a temporary acute shock may block the sensation or expression of pain.

On the other hand, there are factors that can increase the pain perception, such as performing multiple cuts (more noniceptors are affected), changes of direction of the cut, blunt knife (extension of the lesion), insufficient length of the knife, wound manipulation, presentation of the neck in a position disturbing a good cut or flow of blood (i.e. increased flexibility of the skin due to insufficient tension) and any other circumstance that delays the loss of consciousness in the animal and increases inflammatory pain (that can be mitigated with a good throat cut; Brooks and Tracey, 2005; Woolf, 2004). Especially in sheep but also in cattle, an additional aspect may be thick wool or coat, which may have to be

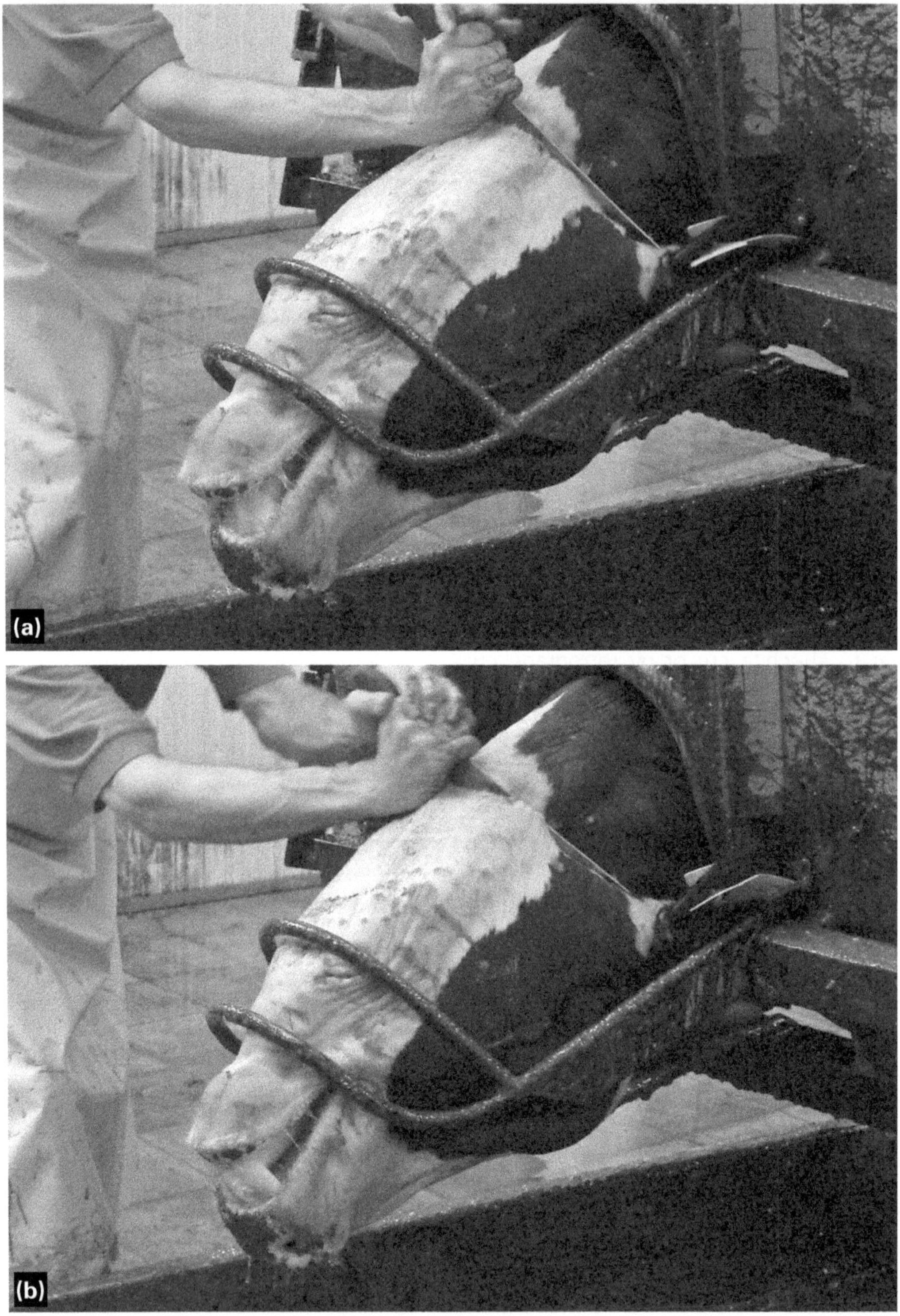

**Figure 10.5: a–f** Sequence of a cut performed in a rotatory pen

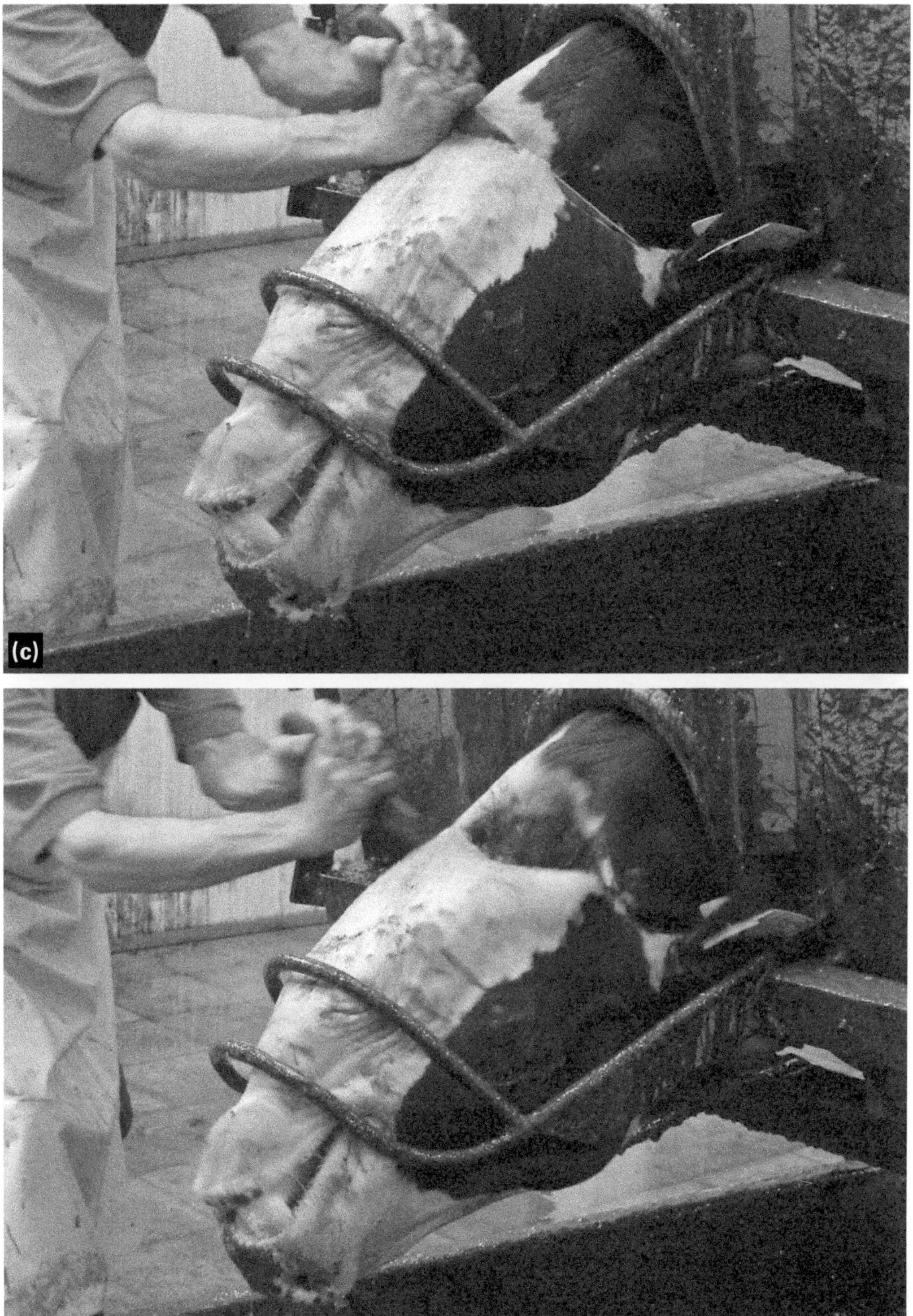
(c)
(d)

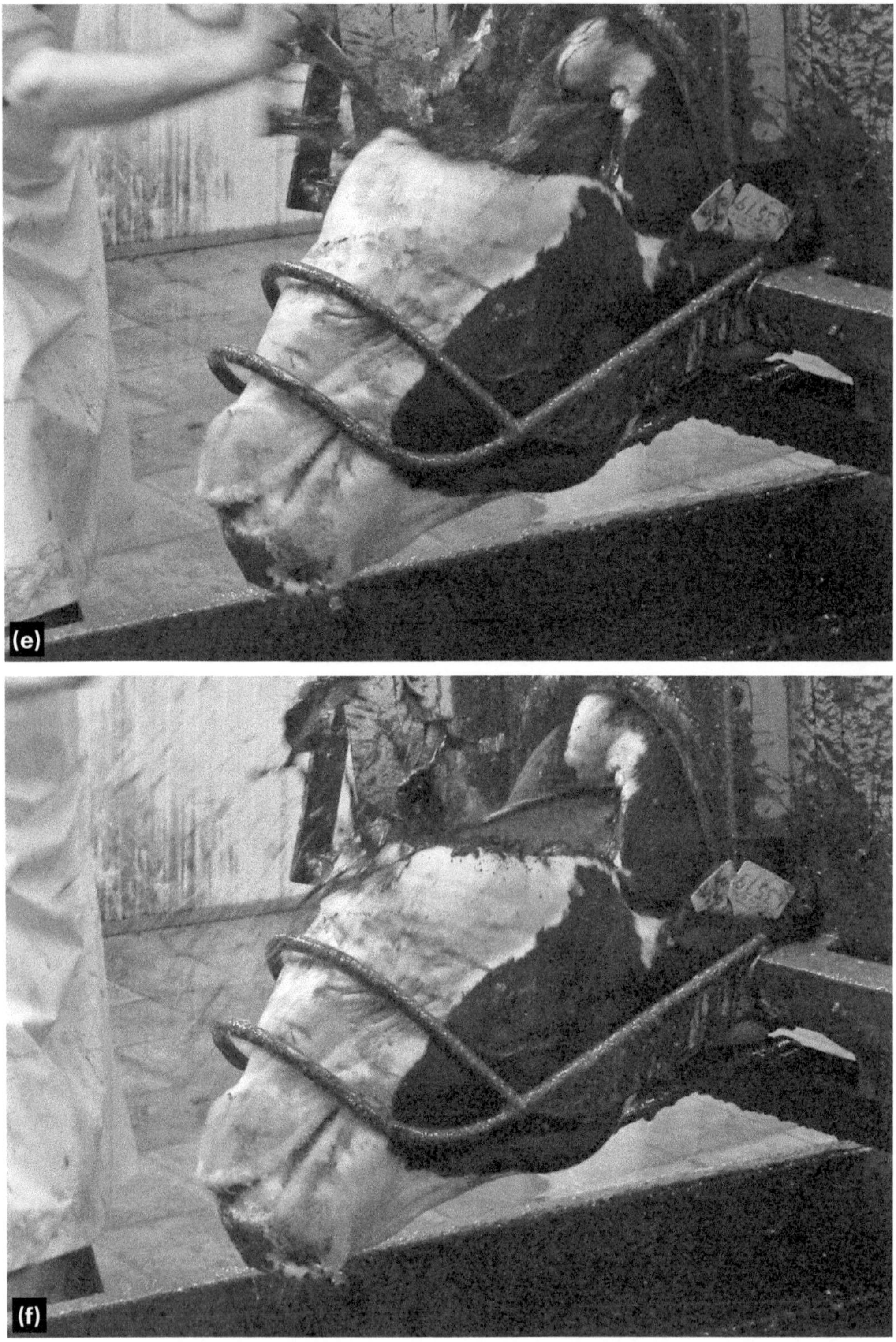
(e)
(f)

separated before the cut. Otherwise, this would constrain the knife during cutting and could cause blunting of the knife. In fact, the quality of the cut, including sharpness of the knife and capability to perform a swift uninterrupted cut within a very short time, is often mentioned, especially in the context of shechita (Rosen, 2004; Lieben, 1925). On the other hand, a bad quality of cut can have as a consequence vasoconstriction, clotting and ballooning or presence of false aneurysms, especially in cattle (Gregory et al., 2006; Anil et al., 1995a,b; Graham and Keatinge, 1974). Gregory et al. (2008) found a prevalence of large (> 3 cm outer diameter) false aneurysms in cattle carotid arteries of 10% for both shechita and halal slaughter.

Several mechanisms have been proposed to impair bleeding in cattle. First, the occlusion of the severed artery by surrounding tissues, and in this respect it should be noted that the artery is elastic and has a tendency to spring back into its connective tissue sheath on being cut. Second, platelets can aggregate at the cut end of the carotid, and this leads to the rapid production of a white clot which can plug the artery. Third, when the cut is made, the artery can go into an annular spasm. These factors will tend to impede blood flow from the caudal cut end, and together might produce a ballooning effect in the severed vessel. False aneurysms develop within a few seconds of cutting carotid arteries; they take the form of encapsulated haematomas and can be distinguished in situ from normal arteries by their swollen size. As a consequence, the animal with a beating heart has a cardiac output sufficient to maintain high blood pressure and consciousness. In addition, according to Gibson et al. (2015), additional mechanisms are associated with a delay in the time to unconsciousness during slaughter without stunning in cattle, such as the retraction of the carotids without occlusion, carotid retraction during respiratory movements (during inhalation bleeding can be reduced momentarily) and physical occlusion of the vessels of the neck, either due to the position of the head relative to the body or to physical pressure on the neck from the metalwork of the pen restricting blood loss.

The prevalence of false aneurysm formation is not influenced by the method of restraint (rotating pen; manual casting into lateral recumbency; upright restraint pen; shackled by one leg and lowered into lateral recumbency; Gregory et al., 2008), but it can be associated with the quality of the cut (Gregory et al., 2012a,b). One of the aspects to take into account in the quality of the cut is the number of cuts performed, as each time the knife touches the surface of the wound or more tissue is damaged, more nociceptors are activated. The average number of cuts in cattle reported by Gregory et al. (2008) was 3.2 during shechita and 5.2 during halal slaughter. The number of cuts reported in the DIALREL project ranged from 1 to 60 sweeps of knife. Velarde et al. (2014), assessing ten halal non-stun abattoirs for cattle, reported significant variation in the number of cuts, between operators and restraint systems, with more cuts performed on cattle restrained in the upright position (nine cuts) compared to inversion at 180° (five cuts) and 90° (three cuts). However, a large variation can be found between slaughterhouses. In sheep, according to the experience of the DIALREL, the number of cuts ranged from one to six.

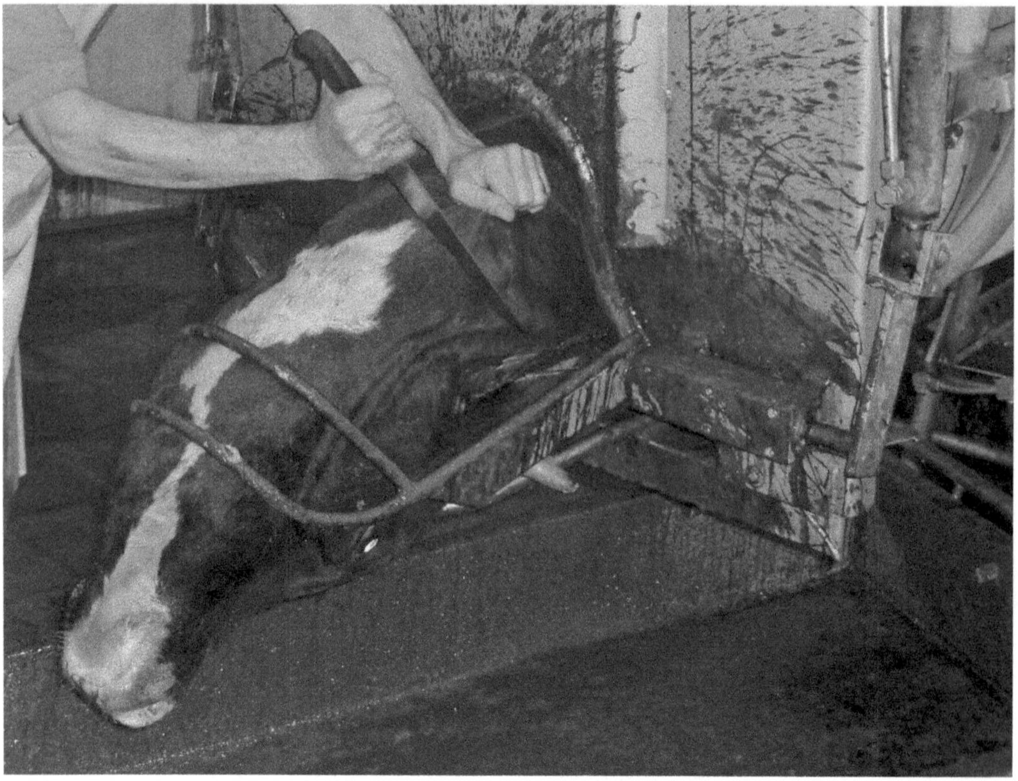

**Figure 10.6:** Neck cutting for halal in cattle restrained in a rotatory box

For poultry, usually only one cut was performed.

Gibson et al. (2015) propose that delays in the onset of unconsciousness in cattle can be reduced by performing the neck-cut at the higher position on the neck, and this would minimize subsequent distress. In fact, cutting the neck at a position corresponding to the first cervical vertebra (C1) compared to the conventional position (between C2 and C4) can almost eliminate false aneurysm development, thereby minimizing the risk of arrested exsanguination (Gregory et al., 2012a,b). Gibson et al. (2015) studied the effect of the cut position in relation to the number of trachea rings situated cranially to the cut. The trachea rings were counted from the arytenoid cartilage onwards, but not including the cricoid cartilage, and the cuts were categorized as either low neck-cut (more than 2.5 trachea rings) or high neck-cut (with less than two trachea rings). Results showed that performing a high neck-cut reduced the time to final collapse in halal slaughter cattle without stunning in comparison to a low neck-cut (13.5 and 18.9 seconds, respectively). In the same study, 9 and 5% of cattle that received a low neck- and high neck-cut regained posture on all four feet, respectively.

In any case, a correct flow of the blood after cutting is a critical point and re-opens an obstructed artery, with a secondary cut when needed, as well. Rosen (2004) mentioned the importance of correct post-cut restraint with regard to correct bleed out and time to loss of consciousness. In the Netherlands, representatives of the Jewish and Muslim communities with the Ministry of Economic Affairs, have agreed on a period where, if the animal is still conscious 40 seconds after the neck-cut, it should be immediately stunned (Tyler, 2012). Similar recommendations are given in the final report of the DIALREL project (DIALREL, 2009).

## 10.5 RECOMMENDATIONS FOR GOOD PRACTICES

The DIALREL project concluded its activities with a set of recommendations for improved practices to be adopted during slaughter without stunning. The document proposed good animal welfare practices during religious slaughter, including restraining, neck cutting and post-cut management. The main recommendations are summarized below.

### *10.5.1 Restraining methods*

- Animals must be restrained only when slaughter can be performed without any delay.
- The restraining device, including both the body and head restrainers or method, must suit the type of animal slaughtered.
- Due care must be taken during loading the animal into the restraining system to minimize stress and injury. The restraint device and surrounding area must have adequate lighting, flooring should be non-slip and the parts in contact with the animal should have smooth, rounded surfaces.
- All restraining devices should use the concept of optimal pressure. The device must hold the animal firmly enough to facilitate slaughter without struggle or undue delay. Optimal pressure might be assessed by the absence of struggling behaviour and vocalization during the restraint, and the absence of any injuries and bruises caused by the restraining method.
- The head restraint must be such that it provides good access to the neck for effective neck cutting and bleeding out and it must be such that it is set with the proper amount of neck tension to optimize slaughter.
- The head restraint must be designed to avoid mechanical stimuli (such as physical contact or scraping) and chemical stimuli (such as contamination with stomach content) on the surface of the wound during the conscious period.
- The design of the head restraint must not obscure the front of the head and should also allow good access to the eyes to check for signs of reflexes and sensibility.
- When rotary pens are used, the head of the animal must be restrained before the start of the turning process. The turning operation should proceed smoothly and quickly without interruption to reduce as much as possible the period of animals being restrained in unnatural positions.
- During neck cutting, the head of sheep and goats (and small calves) may be stretched manually in addition to the mechanical restraining of the body and it is recommended

that the head continues to be supported during the early stages of bleeding.

- When using an upright restraint for cattle the belly plate, if used, must be operated according to the concept of optimal pressure to support the animal without lifting it off the ground.

### *10.5.2 Neck cutting*

- The neck cut must be performed without any delay.

**Figure 10.7:** Animal hoisted with signs of consciousness

- Both carotid arteries and both jugular veins must be cut.
- Each animal should be neck-cut by a single swift or continuous back-and-forward movement of the knife without interruption.
- The knife used must be sufficiently long for each type of animal to minimize the need for multiple cuts.
- The knife must be sharp for each animal. Emphasis on training slaughter persons to improve their knife sharpness is recommended.

### 10.5.3 *Post-cut management*

- There must be no interference with the wound until the animal is unconscious, including mechanical and chemical stimuli.
- The cut should be assessed for complete sectioning of vessels and for the efficiency of bleeding through a strong flow and seeing the pulsating effect of the heartbeat on this flow. When inspecting the wound, unnecessary contact with the severed edge of the skin must be avoided. Thus, visual inspection is preferable.
- The animal must be assessed as unconscious before it can be released from the restraint (Figure 10.7). It is suggested that the signs of unconsciousness are checked at least twice, for cattle between 30 and 40 seconds post-cut, and for sheep between 15 and 25 seconds post-cut.
- In the event of inefficient bleeding or prolonged consciousness being exhibited during repeated checks after neck cutting, animals should be stunned with a suitable method as soon as possible, even if this requires the religious authorities to declare the animal as non-kosher or haram. Optimally, this should be done within 45 seconds post-cut for cattle, or within 30 seconds for small ruminants and poultry.
- When the cut is performed in a 180° inverted position in cattle, it may be preferable to turn the box to a position between 180 and 90° directly after the cut for better access to the head of the animal.

In addition to the mentioned recommendations, specific training of slaughtermen and abattoir staff, including management in key areas (such as animal handling, restraint, knife sharpening, animal physiology, signs of stress and pain, times to unconsciousness and signs of loss of consciousness), is vital to ensure good animal welfare.

## 10.6 REFERENCES

Adams, D.B. and Sheridan, A.D. 2008. Specifying the risk to animal welfare associated with livestock slaughter without induced insensibility. Australian Government Department of Agriculture, Fisheries and Forestry, Canberra, Australia, 81 pp. http://www.daff.gov.au/__data/assets/pdf_file/0019/1370332/animal-welfare-livestock-slaughter.pdf

Anil, M.H., McKinstry, J.L., Wotton, S.B. and Gregory, N.G. 1995a. Welfare of calves. 1. Investigation into some aspects of calf slaughter. *Meat Science* 41: 101–112.

Anil, M.H., McKinstry, J.L., Gregory, N.G., Wotton, S.B. and Symonds, H. 1995b. Welfare of calves. 2. Increase in vertebral artery blood flow following exsanguination by neck sticking and evaluation of chest sticking as an alternative slaughter method. *Meat Science* 41: 113–123.

Anil, M.H., Yesildere, T., Aksu, H., Matur, E., McKinstry, J.L., Erdogan, O., Hughes, S. and Mason C. 2004. Comparison of religious slaughter of sheep with methods that include pre-slaughter stunning, and the lack of differences in exsanguination, packed cell volume and meat quality parameters. *Animal Welfare* 13: 387–392.

Anil, M. H.; Yesildere, T.; Aksu, H.; Matur, E.; McKinstry, J. L.; Weaver, H. R.; Erdogan, O.; Hughes, S.; Mason, C. (2006): Comparison of Halal slaughter with captive bolt stunning and neck cutting in cattle: exsanguination and quality parameters. Animal Welfare 15, 325-330.

Barnett, J.L., Cronin, G.M. and Scott, P.C. 2007. Behavioural responses of poultry during kosher slaughter and their implications for the birds' welfare. *Veterinary Record* 160: 45–49.

Blackmore, D.K. and Delany, M.W. 1988. Slaughter of stock – a practical review and guide. Veterinary Continuing Education, Massey University. Auckland, New Zealand.

Brooks, J. and Tracey, I. 2005. From nociception to pain perception: imaging the spinal and supraspinal pathways. *Journal of Anatomy* 207: 19–33.

Daly, C.C., Kallweit, E. and Ellendorf, F. 1988. Cortical function in cattle during slaughter: Conventional captive bolt stunning followed by exsanguination compared with shechita slaughter. *Veterinary Record* 122: 325–329.

Dialrel 2009. Religious slaughter, improving knowledge and expertise through dialogue and debate on issues of welfare, legislation and socio-economic aspects. http://www.dialrel.eu/dialrel-results

Dunn, C. S. 1990. Stress reactions of cattle undergoing ritual slaughter using two methods of restraint. *Veterinary Record* 126: 522–525.

EFSA. 2004. Welfare aspects of animal stunning and killing methods – Scientific Report of the Scientific Panel for Animal Health and Welfare on a request from the Commission related to welfare aspects of animal stunning and killing methods. http://www.efsa.europa.eu/de/scdocs/doc/45.pdf

EFSA AHAW Panel (EFSA Panel on Animal Health and Welfare). 2013a. Scientific Opinion on monitoring procedures at slaughterhouses for bovines. *EFSA Journal* 11(12): 3460. doi:10.2903/j.efsa.2013.3460

EFSA AHAW Panel (EFSA Panel on Animal Health and Welfare). 2013b. Scientific Opinion on monitoring procedures at slaughterhouses for sheep and goats. *EFSA Journal* 11(12): 3522. doi:10.2903/j.efsa.2013.3522

EFSA AHAW Panel (EFSA Panel on Animal Health and Welfare). 2013c. Scientific Opinion on monitoring procedures at slaughterhouses for poultry. *EFSA Journal* 11(12): 3521. doi:10.2903/j.efsa.2013.3521

Gibson, T.J., Johnson, C.B., Murrel, J.C., Hull, C.M., Mitchinson, S.L., Stafford, K.J., Johnstone, A.C. and Mellor, D.J. 2009a. Electroencephalographic responses of halothane anaesthetized calves to slaughter by ventral-neck incision without prior stunning. *New Zealand Veterinary Journal* 57: 77–83.

Gibson, T.J., Johnson C.B., Murrell, J.C., Chambers, J.P., Stafford, K.J. and Mellor, D.J. 2009b. Components of electroencephalographic responses to slaughter in halothane-anaesthetized calves: Effects of cutting neck tissues compared with major blood vessels. *New Zealand Veterinary Journal* 57: 84–89.

Gibson, T.J., Dadios, N. and Gregory, N.G. 2015. Effect of neck cut position on time to collapse in halal slaughtered cattle without stunning. *Meat Science* 110: 310–314.

Graham, J.M. and Keatinge, W.R. 1974.

Responses of inner and outer muscle of the sheep carotid artery to injury. *Journal of Physiology* 247: 473–482.

Grandin, T. 1988. New concepts in livestock handling. Proceedings of the 3rd International Symposium on Livestock Environment, Toronto 25–27.

Grandin, T. 2005. Restraint methods for holding animals during ritual slaughter. In: Luy J. et al. [ed.] *Animal Welfare at Ritual Slaughter.*

Grandin, T. 2013. Making slaughterhouses more humane for cattle, pigs, and sheep. *Annual Review of Animal Biosciences* 1: 491–512.

Grandin, T. and Regenstein, J.M. 1994. Slaughter: Religious slaughter and animal welfare: A discussion for meat scientists. *Meat Focus International.* March, pp. 115–123.

Gregory, N.G. 2005. Recent concerns about stunning and slaughter. *Meat Science* 70: 481–491.

Gregory, N.G., Fielding, H.R., Von Wenzlawowicz, M. and Von Holleben, K. 2010. Time to collapse following slaughter without stunning in cattle. *Meat Science* 85: 66–69.

Gregory, N.G., Schuster, P., Mirabito, L., Kolesar, R. and McManus, T. 2012a. Arrested blood flow during false aneurysm formation in the carotid arteries of cattle slaughtered with and without stunning. *Meat Science* 90(2): 368–372.

Gregory, N.G., Shaw, F.D., Whitford, J.C. and Patterson-Kane, J.C. 2006. Prevalence of ballooning of the severed carotid arteries at slaughter in cattle, calves and sheep. *Meat Science* 74: 655–657.

Gregory, N.G., Wenzlawowicz, M. v., Alam, R.M., Anil, H., Yesildere, T. and Silva-Fletcher, A. 2008. False aneurysms in carotid arteries of cattle and water buffalo during shechita and halal slaughter. *Meat Science* 79: 285–288.

Gregory, N.G. Wenzlawowicz, M. v., Holleben, K.V. 2009. Blood in the respiratory tract during slaughter with and without stunning in cattle. *Meat Science* 82: 13–16.

Gregory, N.G., von Wenzlawowicz, M., von Holleben, K., Fielding, H.R., Gibson, T.J. Mirabito, L. and Kolesar, R. 2012b. Complications during shechita and halal slaughter without stunning in cattle. *Animal Welfare* 21: 81–86.

Gregory, N.G. and Wotton, S.B. 1986. Effect of slaughter on the spontaneous and evoked activity of the brain. *British Poultry Science* 27: 195–205.

Kavaliers, M. 1989. Evolutionary aspects of the neuro-modulation of nociceptive behaviors. *American Zoology* 29: 1345–1353.

Koorts, R. 1991. The development of a restraining system to accommodate the Jewish method of slaughter (shechita). M.Dip.Tech. Technikon Witwatersrand, Johannesburg.

Lambooij, B. and Kijlstra, A. 2008. Ritueel slachten en het welzijn van herkauwers en pluimvee [Ritual slaughter and animal welfare]. Rapport 161, Animal Sciences Group van Wageningen UR, Lelystad.

Lambooij, B., Kijlstra, A. 2009. Ritueel slachten en het welzijn van herkauwers en pluimvee. *Tijdschrift voor Diergeneeskunde* 134: 984–989.

Levinger, I.M. 1961. Untersuchungen zum Schächtproblem. DVM thesis, University of Zürich, Switzerland.

Levinger, I.M. 1995. Shechita in the light of the year 2000. Critical view of the scientific aspects of methods of slaughter and shechita. Maskil L'David: Jerusalem, Israel.

Lieben, S. 1925. Ueber das Verhalten des Blutdruckes in den Hirngefäßen nach Durchschneidung des Halses. Schächtschnitt der Juden. *Monatsschrift Tierheilkunde* 31: 481–496.

Mead, G.C. 2004. Poultry meat processing and quality. Woodhead Publishing, Cambridge, UK.

Mellor, D.J., Gibson, T.J. and Johnson, C.B. 2009. A re-evaluation of the need to stun calves prior to slaughter by ventral-neck incision: An introductory review. *New Zealand Veterinary Journal* 57: 74–76.

Petty, D.B., Hattingh, J. and Ganhao, M.F. 1991. Concentration of blood variables in cattle after shechita and conventional slaughter. *South African Journal of Science* 65: 397–398.

Regulation (EC) No 853/2004 of the European Parliament and of the Council. 2004. Laying down specific hygiene rules for on the hygiene of foodstuffs. *Official Journal of the European Union* L139/55.

Rosen, S.D. 2004. Physiological insights into Shechita. *The Veterinary Record* 154: 759–765.

Tyler, J. 2012. Dutch compromise on ritual slaughter. Radio Netherlands Worldwide https://www.rnw.org/archive/dutch-compromise-ritual-slaughter

Velarde, A, Rodríguez, P., Dalmau, A., Fuentes, C., Llonch, P., von Holleben, K.V., Anil, M.H., Malbooij, J.B., Pleiter, H., Yesildere, T. and Cenci-Coga, B.T. 2014. Religious slaughter: evaluation of current practices in selected countries. *Meat Science* 96: 278–287.

von Holleben, K., Wenzlawowicz, v. N., Gregory, H., Anil, A., Velarde, P., Rodriguez, B., Cenci Goga, B., Catanese, B. and Lambooij, B. 2010. Report on good and adverse practices – Animal welfare concerns in relation to slaughter practices from the viewpoint of veterinary sciences. Dialrel report. Deliverable 1.3.

Warin-Ramette, A. and Mirabito, L. 2010. Use of rotating box and turned-back position of cattle at the time of slaughter. 61st Annual Meeting of the European Association for Animal Production 56. European Federation of Animal Science: Heraklion, Crete, Greece.

Woolf, C. J. 2004. Pain: Moving from symptom control toward mechanism-specific pharmacologic management. Annals of Internal Medicine 140: 441–451.

chapter eleven

# Animal welfare at depopulation strategies during disease control actions

*Marien Gerritzen[1] and Troy Gibson[2]*

**Learning objectives**

- Guidance for contingency planning with respect for animal welfare.
- Being aware of the key factors and responsibilities in contingency planning and actions.
- Knowledge of the available killing methods.
- Insight into animal welfare issues related to the killing method.

## CONTENT



[1] Wageningen University and Research Centre, Livestock Research Department Animal Welfare, De Elst 1, 6708 WD Wageningen, The Netherlands
[2] Department of Production and Population Health, Royal Veterinary College, Hawkshead Lane, Hertfordshire AL9 7TA, United Kingdom

## 11.1 INTRODUCTION

Animal production is sometimes impaired with outbreaks of diseases that are a major hazard for the industry, economics or even human safety, if the disease is zoonotic. When there is an outbreak of a notifiable disease, it is necessary to kill affected animals at the farm where the disease was identified and in adjacent properties in order to prevent, eradicate or contain the spread of an exotic pathogen that may pose a risk to animal or human health or the economy. While every effort needs to be made to ensure that the disease is contained promptly, the affected animals must be handled and killed humanely. How the competent authorities deal with contagious diseases is regulated in international and national regulations and guidelines (OIE, 2010; Defra, 2011). One of the important measures that are taken for disease eradication purposes is depopulation of infected and contact farms. Depopulation includes all activities from sealing off the farm, the actual killing of all involved animals and the removal of all carcasses and biological materials. The starting point in this is that, according to EU Council Regulation No 1099/1999, the competent authority responsible for a depopulation operation must establish an action plan to ensure compliance with the rules laid down in EU regulations, before the commencement of the operation.

Killing animals for depopulation reasons on farms is a complex process that involves different interests and conflicts between various stakeholders. Conditions differ from regular slaughter for human consumption because killing for disease control purposes is performed on farms with minimum or no handling and restraining facilities. Furthermore, depopulation is usually performed under time pressure and, especially when it concerns highly infectious or zoonotic diseases, the required speed of action and safety measures require good planning, good communication, skilled personnel and accurate monitoring procedures.

## 11.2 CONTINGENCY PLANNING

The key issue in a contingency plan is to prevent the disease from spreading. To do so it is crucial to have an early warning system, which includes disease surveillance, disease reporting and epidemiological analysis. Second, an early response is to establish exclusion and surveillance zones and it is essential to minimize the number of infected farms. Important issues in an action plan are information on roles and responsibilities, communication, farm situation and farm size, available methods and monitoring procedures. Information on contingency planning in case of a notifiable animal disease can be found at http://web.oie.int, https://www.gov.uk. The contingency plan should also include standard operating procedures and list of resources and expertise required to deal with the crisis.

### *11.2.1 Competent authority*

The competent authority is responsible for the depopulation operation and will follow an action plan to ensure that protocols are followed as foreseen. The first practical step is that the farm will be closed which means that all movements

**Figure 11.1:** Example of farm situation

from and to the farm of animals, products, materials and people are limited and registered (Figure 11.1).

Before starting the depopulation action, business operators should be briefed about the situation of the premises, including:

- size of the premises;
- location and access roads;
- location of the farmer's house;
- distance to surrounding farms and houses;
- identifying farmer's family and pets;
- number and type of buildings;
- species, number and age of animals.

Furthermore, it is crucial that a communication strategy is talked through and accepted by all responsible parties.

### *11.2.2 Business operators*

In accordance with standard operating procedures, business operators plan in advance the killing of animals and related operations. This plan is presented to and needs to be approved by the competent authorities. In this action plan a number of different issues need to be taken into account:

- how many animals are on the farm?
- what are the housing conditions?
- which killing methods are available?
- acceptance on behalf of the owner;
- resources needed (people, equipment, etc).

### *11.2.3 Planning and performing the killing procedures*

The standard operating procedures should take into account the equipment manufacturer's recommendations, define the key parameters for the available killing methods and specify the measures to be taken when methods do not work out as planned. Furthermore, killing and related operations should only be carried out by persons with the appropriate level of competence to do so without causing the animals any avoidable pain, distress or suffering.

For animal welfare reasons, the order of killing should be taken into account, and is as follows:

- First, animals that are visibly sick, wounded, not able to walk or in any way suffering should be killed first.
- Second, animals affected by disease, giving priority to those that have not yet been weaned, females that have just given birth or that are in a lactation period, pregnant and restless animals or potentially dangerous animals.
- Finally, animals that have been in direct contact with those affected by disease giving priority to non-weaned animals, females that have just given birth, pregnant or restless animals.

Before starting any killing procedure, it should be checked that the killing equipment is working correctly, that safety precautions are taken, that communication equipment is working and that all staff are ready. Stunning or killing equipment must be regularly serviced and maintained in good working condition according to the manufacturer's instructions. For example, overheating of the barrel of captive bolt guns due to continuous use will lead to failures. This can be overcome by using several captive bolts on a rotational basis, for example.

### *11.2.4 Monitoring procedures*

It is important to be aware of the risks to animal welfare and to monitor what is done. To be in control and to be able to respond, it is important to measure the pre-set conditions of the killing method (e.g. gas concentration, gas distribution, temperature or electrical parameters) during the whole procedure and adjust the settings if required. If procedures are not working as planned, for instance, if animals are not killed with a single procedure, backup killing methods should be available and used. In any event, death should be confirmed in animals by a suitably qualified person.

During the procedure, personnel should be supervised on a regular basis to ensure that they are able to do the job properly and handle animals with care, and operators should be rested at regular intervals to avoid fatigue leading to the risk of compromising animal welfare. With any procedure, it is crucial to confirm that animals are dead before they are moved for destruction to any other place on or away from the premises.

The final, but important, step in a contingency plan is reporting on the depopulation activity. Evaluation of the procedures, the process, the applied killing methods and communication can be used as a learning tool and will help to improve upcoming depopulation actions. The report of a depopulation action should include, in particular:

- the reasons for the depopulation;
- the number and the species of animals killed;

- the stunning and killing methods used;
- the monitoring results;
- a description of the difficulties encountered and, where appropriate, solutions found to alleviate or minimize the suffering of the animals concerned.

## 11.3 DEPOPULATION METHODS FOR POULTRY

Different killing methods are available for the depopulation of poultry farms. Methods differ in working mechanisms, capacity and applicability. Making a decision on the method to be used in a particular situation depends strongly on the farm situation. Depopulation methods for poultry can be divided into methods that require handling of animals (gas-filled containers, electrocution systems) and methods to kill poultry in their sheds by whole house gassing. Since poultry in normal situations are not used to different people entering the buildings, the entrance of people will cause a fear reaction. Therefore, to minimize distress it is important to enter the buildings with a minimal number of people and disturbance. The handling of live animals will introduce fear, distress and a risk of painful injuries. Measures should be taken to minimize stress in birds, for example, dimming of lights, lifting of feeders and water troughs and use of temporary partitions to divide birds into small and manageable group sizes. Furthermore, handling of infected poultry will increase the risk of exposure of workers to the disease. The need for handling or the possibility to prevent handling of live birds should therefore be taken into account.

### *11.3.1 Mobile electrocution systems*

Mobile electrocution systems (Figure 11.2) are based on conventional electrical water bath stunning systems used in poultry slaughterhouses. Birds are caught and taken manually to the killing unit located nearby where they are suspended by their legs on the shackle line and electrocuted in an electrified water bath. The advantage of the system is that it is relatively cheap, easy to adjust the settings and applicable in many situations. The disadvantage of these systems is that live birds will have to be caught which means that large number of people are involved and they have intensive contact with live birds. This limits the capacity, is a risk for reduced animal welfare and increases the possibility of spread of pathogens due to handling of the infected birds outside the growing buildings. Catching, handling and movement of birds frequently cause wing flapping in birds and airborne dispersal of infective agents.

To monitor good practice, it is a requirement that the pre-set parameters (EC 1099/2009), i.e. electrical waveform, frequency and voltage, number of birds in the water bath and line speed are appropriate to killing poultry species. All the birds should be checked to ensure they are killed by the procedure and any surviving ones should be killed using a backup method. The device settings should be checked and adjusted, if required. It is important to adjust the height of the water bath according to the size of birds to ensure even the smallest of birds' (runts) heads are immersed adequately in the bath. It is also important to ensure birds remain in contact with the

**Figure 11.2:** Mobile electrocution line in shipping container

water bath for long enough to receive sufficient electric current required to induce cardiac arrest.

## *11.3.2 Containerized gassing*

Gas-filled containers of different sizes but also of different materials, modified wheelie bins (Figure 11.3), large bags and shipping containers can be used to kill poultry of different ages, from day-old chicks to breeders, and different species (i.e. chickens, ducks, geese) with different gasses like carbon dioxide, nitrogen or argon, or mixtures of these gasses. The system is prefilled with the gas to the required lethal gas concentration. Gas-filled containers or culling bags are placed inside or outside the poultry house depending on access to the building and the possibility of removing the containers full of dead birds after the operation.

The birds are caught by hand and placed into the containers. It is, however, regulated that the systems are filled with birds only in a single layer (EC 1099/2009). A second layer of birds should be placed in the containers when all birds in the system are confirmed dead. Monitoring of birds to determine unconsciousness or death in a gas-filled container is not possible during operation, as operators' health and safety may be compromised. However, a good indication should be based on visual observation (using ideally positioned video cameras, for example) and, in this sense, all movements should have

**Figure 11.3:** Gas-filled containers

ceased before introducing a second layer of birds. The advantage of containerized gassing is that the containers are easy to move, and the method is controllable and adjustable to the situation. The disadvantages are the intensive contact with live birds by a relatively large group of personnel and the relatively small capacity per system; the capacity is limited by the size of the containers and by the number of workers. From a bird welfare point of view, high carbon dioxide concentrations will induce a short but intense period of reduced welfare because they are painful to inhale and induce a strong feeling of breathlessness (McKeegan et al., 2007). High concentrations of nitrogen are effective in killing birds and do not cause aversive or painful reactions during the induction of unconsciousness. A negative aspect of nitrogen is that it requires very low residual oxygen concentrations to be effective (preferably below 2% $O_2$). These low oxygen concentrations cause severe wing flapping and although birds are unconscious at that point, the convulsions can have an effect on stability of the gas concentration.

It is essential that birds stay in the gas concentration until death is confirmed. Birds that are not dead and are taken out of the gas concentration will recover after some time. Monitoring gas concentration in the container and increasing the concentration if required is important to safeguard animal welfare. It is also important to note that when batches of birds are introduced into the container they will displace certain amount of gas and therefore operators should wear appropriate gas monitoring equipment for safety reasons.

### 11.3.3 Whole house gassing

Whole house gassing using carbon dioxide has in recent years been applied on a large scale especially for the control of highly pathogenic avian influenza (HPAI). The advantages of different gasses and gas mixtures, as well as the different gassing methods, have been reviewed previously (Gerritzen et al., 2006; Gerritzen 2006; Raj et al., 2006; Sparks et al., 2010; McKeegan et al., 2011). During whole house gassing, carbon dioxide concentration is gradually increased from 0 to 45% at bird level. The gas is injected from a tanker by using one or more injection points into the house. The location and position of the injection points is strongly dependant on the dimensions of the building and house condition (e.g. deep litter, aviary, cache housing). The way carbon dioxide is administrated to a poultry house strongly influences possible animal welfare risks. Carbon dioxide injected into a poultry house in liquid form will vaporize utilizing the heat from the warm air. When using a single injection point, gas concentration will not increase uniformly throughout the whole building, which implies that not all birds will be affected by the gas at the same time. Birds in the vicinity of the gas entry point will be instantly exposed to high carbon dioxide concentrations, whereas, those remaining away from the gas inlet will be exposed to rising concentrations of carbon dioxide. A negative aspect of this method is that the temperature around the birds can drop below freezing (e.g. −50 °C) and remain at that level for several minutes. Therefore, conscious birds can suffer from extreme cold or freezing. If carbon dioxide is injected under pressure at multiple injection points the temperature drop may not be more than 10–15°C. High-pressure multi injection points lead to a gradual increase of the carbon dioxide in the whole building, which decreases the risk of exposure of conscious birds to high carbon dioxide concentrations.

The advantages of whole house gassing are that the method can be fast, has the potential to kill birds in very large numbers and there is almost no contact between humans and infected live birds and materials, which improves biosecurity. From bird welfare point of view, the main advantage is that birds are killed in their housings without being handled and separated from the flock. The adverse effects of catching, handling and restraining are eliminated with whole house gassing methods. In comparison with the containerized gassing systems using carbon dioxide, the negative effect on birds being exposed to high carbon dioxide concentrations when conscious does not appear because during whole house gassing procedures, carbon dioxide increases gradually which induces unconsciousness before animals are exposed to high levels (Gerritzen et al., 2006, Gerritzen et al., 2007). The disadvantage of whole house gassing is that the procedure is less controllable, will result in low temperature spots and uneven, slow distribution of gas concentrations. The buildings need to be sealed to a large extend but there will always be a leakage of gas. Therefore, it is very important to measure gas concentrations at critical points through the whole building during processing. After ventilating out the gas, effectiveness of the gassing procedure should be assessed by checking for signs of life, such as breathing movements. Birds that have survived the procedure should be killed as soon as possible. How to kill survivors depends on the number of birds: a small number can be killed by lethal injection or neck dislocation or with a non-penetrative captive bolt (CPK, Ted or Zephyr). For larger numbers of birds, containerized gassing or re-gassing the whole building can be considered.

### *11.3.4 Gas-filled foam*

A new method of application of whole house gassing is the use of gas-filled foam (Figure 11.4). In the US, air-filled high density water-based foam has been tested and conditionally approved by the United States Department of Agriculture (USDA) Animal and Plant Health Inspection Service (APHIS) for use with floor-reared poultry. A blanket of high density foam is created and spread to cover all the birds. The dense foam blocks the airways resulting in death by suffocation (Benson et al., 2007). In general, suffocation is not accepted as a humane method for killing animals, including poultry. In order to investigate and solve the problem of suffocation, different studies have been carried out (Gerritzen and Sparrey, 2008; Raj et al., 2008, McKeegan et al., 2013). In these studies, the feasibility and bird welfare implications of using different expansion ratios, bubble sizes and gasses have been investigated. The expansion ratio determines the amount of water contained within the foam. A low expansion ration will lead to wet foam with a small bubble size. A large expansion ratio will create a "dry" foam with a larger bubble size (Figure 11.5). Foam with an expansion ratio between 1:250 and 1:350 seems to be the optimum compromise between foam stability, water content, bubble size and wetness.

Large bubble sizes (e.g. > 10 mm diameter) help to deliver more gas at the level of birds in the house. When using an inert gas such as 100% nitrogen in the foam to create very low oxygen concentrations, the birds die very quickly (within 1.5 minutes; McKeegan et al., 2013). However, low oxygen levels induce strong convulsions by which the foam can be broken down. The nitrogen released from the bubbles can then

**Figure 11.4:** Foam production in poultry house

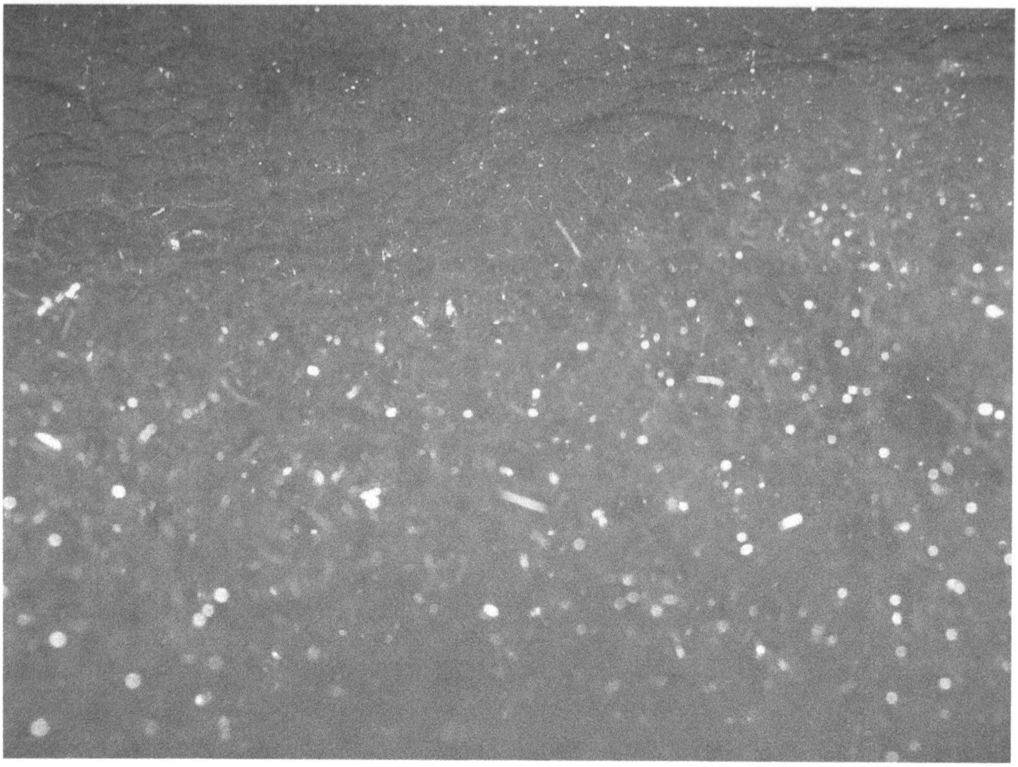

**Figure 11.5:** Gas-filled high expansion foam

mix with air resulting in increases in the oxygen levels and birds surviving the treatment. Therefore, the foam production capacity is required to be much larger than the rates of breakdown of bubbles and dilution of the gas, which may not be possible in large poultry houses.

The advantage of gas-filled foam is that it is applicable as a whole house treatment in open or difficult to seal buildings (Figure 11.6) to kill large groups of poultry without the need for handling of live birds. The method is very effective and will lead to a very quick induction of unconsciousness and death.

The disadvantages include, first, that the amount of equipment needed is substantial, but still under development. Second, the carcasses, following the breaking down of the foam after a few hours, will be very wet and thus more difficult to handle.

## 11.4 DEPOPULATION METHOD FOR CATTLE, SHEEP AND PIGS

The methods used for depopulation of livestock (cattle, sheep and pigs) are dependent on the situation, legislation, species, age and skills of the operators involved, and include: a blow to the head in neonates; shooting (rifle and shotgun); penetrative and non-penetrative captive bolt stunning); lethal injection; and bleeding (ventral neck incision or thoracic sticking; Galvin et al., 2005).

**Figure 11.6:** Foam in a wooden poultry building

### *11.4.1 Captive bolt*

Penetrating captive bolt guns (Figure 11.7) are widely used for rendering livestock unconscious and insensible on farms prior to achieving death by bleeding or destruction of the brain and upper spinal cord by pithing (see Chapter 6 for more details). Non-penetrating captive bolt guns should not be used for stunning adult ruminants or pigs in disease control programmes, as they are less effective in reliably inducing unconsciousness and insensibility. Penetrative captive bolt guns are designed to fire a retractable steel bolt into the head. The bolt penetrates the cranium and enters the brain. Unconsciousness and insensibility is achieved due to brain concussion caused by the impact of the bolt on the cranium and by structural damage to the brain (Gibson et al., 2015a). For effective use in a disease control programme, it is important that:

1. an operating procedure for the control programme is in place and agreed by all operators;
2. the operation is performed in isolation from the general public;
3. operators wear the appropriate protective clothing for the disease control programme;
4. steps are taken to avoid environmental contamination;
5. the correct captive bolt gun/ cartridge combination is used for the species or animal class to be killed;
6. backup captive bolt guns or other methods are readily available in case of gun failure;
7. spare cartridges are readily available;

8. animals are sufficiently restrained to prevent escape or excessive movement that could complicate the shot;
9. animals are shot in the appropriate position for the species;
10. after shooting, animals are assessed for signs of consciousness/unconsciousness and if present are immediately reshot;
11. after, the shot animals are killed by an appropriate secondary procedure and monitoring is continued;
12. procedures are in place for the disposal of carcasses without compromising biosecurity and animal or human health;
13. the captive bolt guns are cleaned and maintained after use;
14. after use, the clean captive bolt guns and cartridges are stored in a secure, dry environment.

The specific shooting positions vary depending on the species (see Chapter 6 for further details):

- cattle are shot on the front of the head at the crossover of lines drawn between the opposing eyes and horns;
- sheep can be shot on the highest point of the head (crown position) with the gun aiming towards the back of the jaw;
- pigs are shot on the front of the head, approximately 20 mm above the level of the eyes, aiming towards the tail.

Care must be taken when shooting pigs with captive bolt guns, as pigs have thick frontal bones and sinuses underneath. In older animals, there can also be a bony ridge running down the centre of the

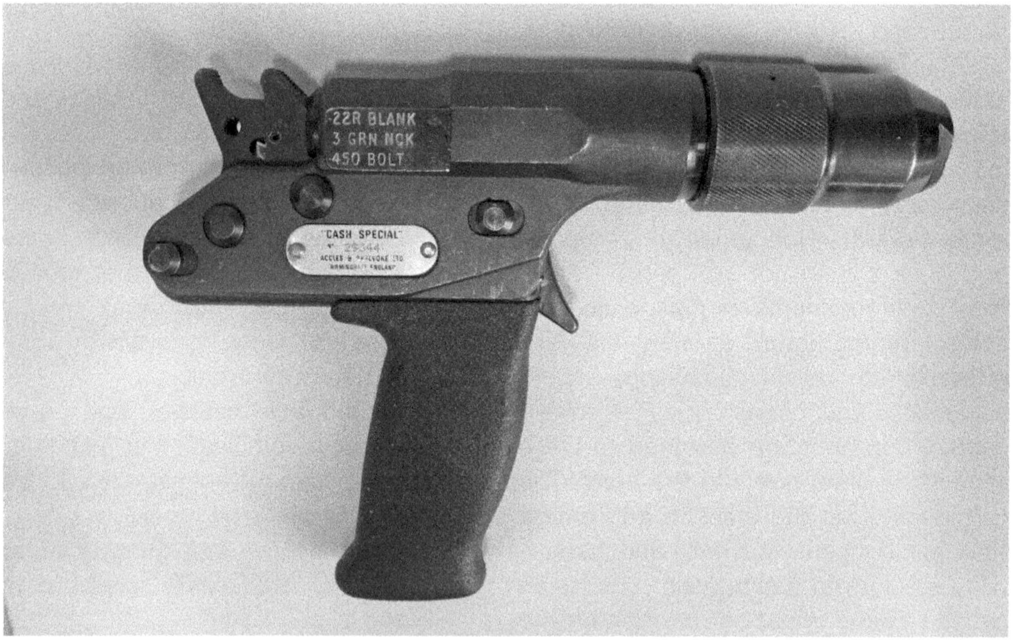


**Figure 11.7:** The .22 calibre Cash Special is a model of captive bolt widely used on farms and by veterinarians for livestock

head, which can reduce bolt penetration. When shooting pigs, it is recommended to use the heaviest gun/cartridge combination available and the maximum bolt length to achieving effective stunning. This should be determined according to the age and weight of pigs to be killed. Many operators refuse to shoot pigs greater that 80 kg with a captive bolt, preferring instead to use electrical stunning/killing or free bullet (including shotgun).

When killing livestock with captive bolt, it is recommended that the shot is immediately followed by a secondary procedure of either pithing (insertion of a metal or plastic rod into the bolthole to cause physical destruction of brain tissue) or exsanguination (sticking). Captive bolt stunning with a secondary procedure (pithing and bleeding) has been previously suggested as a practical and humane method for the killing of livestock on farm. Previously, pithing was used during commercial slaughter for human consumption, however its practice was discontinued following identification of the link between bovine spongiform encephalopathy (BSE) and human Creutzfeldt-Jakob disease (CJD; Bruce et al., 1997; Hill et al., 1997), as pithing can result in the dispersal of central nervous system (CNS) tissue fragments into distal organs and tissues (Daly et al., 2002). However, in depopulation operations, the carcass will be disposed of and will not enter the human food supply chain, so limiting this risk. However, it should be noted that there is a risk of aerial dispersal and leakage of CNS tissues from the bolthole, which could contain pathogens. This could result in contamination of equipment, personnel and the environment. Disposable pithing rods have been developed that remain within the head after use and plug the bolthole to reduce leakage of blood and CNS tissue. In all disease control operations, it is highly advisable, if not mandatory for biosecurity and human health, for all operators to wear suitable disposable protective clothing. Furthermore, equipment should be appropriately cleaned and disinfected between sites and steps taken to minimize environmental contamination. This might involve disinfection of surfaces, protection of surfaces with disposable coverings or even removal of contaminated topsoil.

During depopulation operations, the advantages of cartridge-operated penetrating captive bolt are that: it can be performed in the field with basic restraint and fencing; it is safer for the operators than free bullet; it is relatively simple to perform; it is considered more economical than killing by chemical agents; and finally, if performed correctly it is an effective method for inducing unconsciousness and insensibility in livestock. However, if the shot is inaccurate or insufficient kinetic energy is transferred to the head or brain it can result in ineffective stunning or recovery of consciousness prior to the application of a killing method. Recent work has shown that poor marksmanship can be a major factor leading to poor welfare outcomes in sheep and alpacas shot with penetrating captive bolt guns (Gibson et al., 2012, 2015a). In addition, the performance of captive bolt stunning in rendering animals insensible can be affected by the following factors: (a) selection of the appropriate captive bolt/cartridge combination for the species and/or animal class; (b) experience of the stunner operator; (c) storage of the stunner and cartridges in damp conditions; and (d) regular maintenance of the stunner, including removing excessive carbon build-up, regular rotation of the rubber buffers (recuperator sleeves) and replacement of damaged firing pins and bolts (Grandin, 1980, 1994, 2002; Gibson et al., 2014). Furthermore, the restraint of animals prior to shooting has a direct impact on the success of marksmanship and captive bolt performance in rendering sheep insensible leading to death. It has been suggested that when shooting and pithing it is best to closely pack small groups of sheep of no more than 15 animals (Anon, 2003). For cattle, it is important that the animals are adequately restrained; this can be in mobile or field-built stunning pens, mobile crushes or head bales, or with chemical sedation (e.g. xylazine).

When performed correctly with the appropriate gun/cartridge combinations for the species, captive bolt should function as a single-step killing method. However, if the shot is misplaced or bolt velocity is degraded then there is the potential that the animal will be incompletely concussed and could regain sensibility. During the 2001 foot-and-mouth disease outbreak in the United Kingdom (UK), there were a small number of unconfirmed reports of secondary procedures not being correctly performed, resulting in some animals showing signs of incomplete concussion. For this reason, many government agencies recommend or mandate that a secondary killing procedure is performed when captive bolt is used for disease control.

When animals are shot with a captive bolt, depending on the species they often develop severe tonic–clonic convulsions. Although potentially unsightly for the general public, it is known that the onset of these convulsions is associated with the onset of brain concussion (Blackmore and Newhook, 1982; Blackmore, 1984). When pigs are shot with captive bolt they convulse violently. Care should be taken as the convulsions in unrestrained animals can result in injury to the operators. After captive bolt shooting, some animals can be completely unconscious while still having a functioning heart, which can continue to beat for a prolonged period. This can occur when an animal is insufficiently pithed or when the carotid arteries are incompletely severed or when a secondary procedure is not used. This can be further protracted for up to 15 minutes with the development of agonal gasping, which is due to the residual medullary activity, resulting in gasps periodically oxygenating the blood, which in turn strengthens the pulse. Only for unconscious sheep after captive bolt shooting, thoracic or cardiac compression can be used when the animal is in lateral recumbency to hasten death. If a heartbeat can still be detected in the unconscious animal, compression can be performed with a clenched fist for younger animals, or for larger animals by directly kneeling on the left lateral surface of the chest in the region of the heart. It is important that the pressure is continued and of sufficient force to suppress cardiac function. This must be maintained until all cardiac function has ceased in addition to the continued absence of signs of consciousness for a period of at least 1 minute. Compression should only be attempted when the animal is deemed to be irreversibly unconscious.

When shooting large numbers of animals with captive bolt in a relatively short period of time, the gun barrels can become extremely hot from repeat firing. Gibson et al. (2014) reported that when guns were fired at four shots per minute, the guns reached a temperature of 88.8°C over 500 shots. Associated with this temperature increase was a decrease in bolt velocity and accumulation of carbon and silica around the washers and bolt, which leads to reduced efficacy. It is recommended that if captive bolt guns are repeatedly fired within a single session, they should be allowed to cool and be cleaned after every 500 shots to ensure their efficacy. Operators must examine the gun between firings to check for full bolt retraction. When the bolt has not fully retracted, the expansion chamber can be enlarged, which results in low velocity shots. This can be caused by damage to the washers or the build-up of carbon and silica in the expansion chamber, over the washers and around the bolt. When the washers are damaged, they must be replaced immediately to allow correct resetting of the captive bolt. If the bolt is not retracting due to carbon and silica build-up, the gun can be reset by pressing the bolt and muzzle into a block of wood. This should only be performed as a short-term measure until the captive bolt gun can be stripped and fully cleaned.

Neonates and young stock, during depopulation operations have in the past been killed with manual blunt force trauma to the forehead. When performed correctly this can be an effective method, however there is significant risk of causing incomplete concussion and suffering. The issues are: 1) to be effective, it must involve a single blow to the correct position on the cranium for the species/type, of sufficient force to produce immediate depression and destruction of brain tissue; 2) if insufficient kinetic energy is delivered to the cranium there is the potential for incomplete concussion, which could lead to pain and distress; 3) compared to other methods it is less reproducible between animals; 4) as a method it is unsightly and can be unpleasant for operators and observers; and 5) it requires a level of skill that most stockpersons and veterinarians would be unlikely to possess if they infrequently perform the procedure (Sharp et al., 2015).

Penetrative and non-penetrative captive bolt can be used to dispatch neonate pigs, sheep and cattle. Care must be taken when using penetrative captive bolts, as in young stock the bolt can penetrate through the head of the animal and can cause injury to the operator. Several non-penetrative captive bolt guns have been developed for poultry and rabbits that can be used for the dispatch of neonates (Raj and O'Callaghan 2001; Erasmus et al., 2010a,b; Widowski et al., 2008). As with killing adult livestock, it is essential that the appropriate captive bolt gun type, cartridge/air-pressure/canister and shooting position is used for the animal type to be shot. Captive bolt killing of neonates can be a single-step method, however, animals should be monitored after shooting and if there are any signs of incomplete concussion they should receive a second shot in the correct position and/or a secondary procedure to ensure or hasten death (e.g. pithing or bleeding). There have been concerns about the use of captive bolt for killing neonates (Svendsen et al., 2008). The skulls of neonates have not fully ossified and sutures are not fused; it has been theorized that this results in the kinetic energy from the moving bolt being dissipated throughout the skull before being transferred to the brain (Sharp et al., 2015). Because of this it is essential that captive bolt guns and power sources deliver sufficient energy to reduce the likelihood of incomplete concussion and that neonates after shooting are routinely checked for signs of recovery.

There are several spring- or band-powered captive bolt guns that have been developed for the dispatch of rabbits and small birds. These should not be used for the dispatch of neonate mammalian livestock as they are very low velocity weapons which impart insufficient kinetic energy to the head to reliably induce unconsciousness leading to death.

### *11.4.2 Free bullet*

Free bullet is considered to be a killing method requiring no secondary procedure(s) provided the bullet is fired into the cranium to destroy the brain. However, it is important that a backup procedure (e.g. lethal injection, pithing, bleeding, etc) is available in case of animals showing signs of life. Animals are generally shot at close range, however, for extensively reared, stressed or dangerous livestock it can be more appropriate to shoot from a distance. This can be done with a high velocity rifle with a telescopic sight. It is also advisable to use a silencer to minimize noise and hence disturbance to other animals. In any event, the gun and ammunition should be appropriate for the distance of shooting and species of animals. It is important to ensure that the marksmen are trained and licensed, and the ammunition is designed and constructed such that they do not exit the head of the animals.

During depopulation operations, rifles, pistols and shotguns have previously been used. With these firearms it is important that the individual operating the device is experienced in its operation and holds the required firearms licence. Safety is the most important consideration; when animals are killed by free bullet there is the potential for ricochets either off bone or from solid objects. As such, animals should not be shot in enclosed spaces or in areas with substantial solid objects (e.g. metal gates). Sandpits, earth floor and walls or rubber tyre enclosures can be used to minimize the risk of ricochets and prevent further travel of the projectile if it leaves the head of the animal. All operators must be behind the muzzle of the firearm; the animal should either be restrained or held in a calm environment to prevent excessive movement that could complicate the shot or result in injury to the operator(s) and finally, the appropriate firearm and cartridge should be used for the animal type to be killed (Gibson et al., 2015b).

To ensure effective killing, it is recommended that animals are shot in the head, aiming to maximize damage to the structures of the brainstem (midbrain, pons and medulla). Millar and Mills (2000) reported that to maximize damage to the pons and medulla in horses shot with a 0.32 calibre pistol, the shot needs to be positioned correctly and not deviate in angle from the midline path. In addition to the position and angulation of the shot, the projectile must have sufficient kinetic energy to ensure penetration and sufficient damage to the brain to produce instantaneous death. As with captive bolt, pigs can be difficult to kill with free bullet. This is due to the anatomy of the skull and frontal sinus.

Pigs should be shot with free bullet in the same position as for captive bolt. To ensure accurate shot placement it may be necessary to restrain the animal (e.g. nose snare), however this can cause further stress and agitation and put the operators at risk of ricochets. The calibre of firearm used depends on the age of the animal. For adult sows and boars a 12-gauge shotgun at short range is often used.

### *11.4.3 Chemical agents*

Lethal injection with a chemical agent, can produce rapid onset of unconsciousness and insensibility and death in livestock and is considered less unsightly for the public than captive bolt or free bullet. The compounds used are classed as prescription-only veterinary medicines (POM-V), which can only be prescribed by and, depending on the compound, administered by a veterinary surgeon. Chemical killing agents are often given after pre-medication with sedatives and/or anaesthetics (e.g. xylazine, ketamine), which physically restrain the animal to prevent injury to operators and improve the animal's welfare. Barbiturate-based chemical agents, for example, euthatal® (Figure 11.8), are widely used in the veterinary profession for the euthanasia of animals. When given intravenously as a rapid overdose, barbiturates are effective in producing a smooth and quick, irrecoverable death. Barbiturates are administered at doses of between 100 and 200 mg/kg bodyweight, however it is generally advised to give 200 mg/kg to livestock. Other chemical agents such as T-61, a combination of embutramide, mebezonium iodide and tetracaine hydrochloride, can be used for the euthanasia of livestock. However, the humaneness of this has been questioned. Concerns with T-61 include the potential for pain and irritation during rapid injection, and paralysis which can result in the suppression of respiration prior to the onset on unconsciousness (EFSA, 2004). Because of these concerns, T-61 is no longer manufactured in the United States (AVMA, 2013).

Intravenous injection of the chemical agent is the most widely used delivery route. The jugular veins are the most commonly used vessels for injection of barbiturates for cattle and sheep, while for pigs, an ear vein can often be more easily accessed. In sheep, it may be necessary to clip the wool over the jugulars to help administration of the compound. When killing livestock with chemical agents, it is essential that the animal is sufficiently restrained to allow safe access to the required vein(s). This can either be with chemical restraints (sedatives), physical restraint or a combination of both. During the 2001 UK foot-and-mouth disease outbreak, cattle were often restrained in mobile crushes or head bales to allow direct venepuncture and rapid infusion with pentobarbital agents. An intracardiac route may be used in previously anaesthetized animals (e.g. with potassium chloride) and intraperitoneal (small animals) routes can also be used provided the contents of the injection do not cause irritation or pain during the induction of unconsciousness in animals.

After delivery of the agent, it is essential that the animals are monitored to ensure death. The major advantages of barbiturates compared to captive bolt and free bullet is that it is considered less unsightly to the general public;

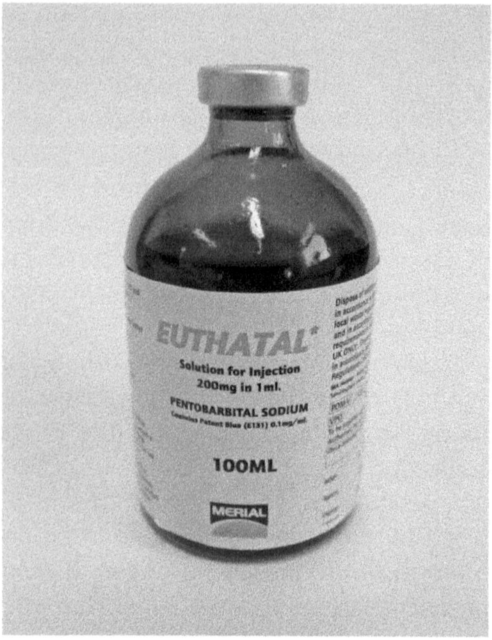


**Figure 11.8:** Licensed commercially available solution of pentobarbital sodium, marketed as "Euthatal®" in the UK

if performed correctly, it results in a smooth induction of insensibility leading to death and has fewer risks for the operator than free bullet. However, the disadvantages are that it is a relatively expensive method and often requires a veterinarian to administer the agent.

## 11.5 REFERENCES

AVMA, 2013. AVMA Guidelines for the Euthanasia of Animals, 2013 Edition, pp. 29–30.

Benson, E., Malone, G.W., Alphin, R.L., Dawsion, M.D., Pope, C.R. and Van Wicken, G.L. 2007. Foam-based mass emergency depopulation of floor-reared meat-type poultry operations. *Poultry Science* 86, 219–224.

Blackmore, D.K. 1984. Differences in behavior between sheep and cattle during slaughter. *Research in Veterinary Science* 37, 223–226.

Blackmore, D.K. and Newhook, J.C. 1982. Electroencephalographic studies of stunning and slaughter of sheep and calves 3. The duration of insensibility induced by electrical stunning in sheep and calves. *Meat Science* 7, 19–28.

Bruce, M.E., Will, R.G., Ironside, J.W., McConnell, I., Drummond, D., Suttie, A., McCardle, L., Chree, A., Hope, J., Birkett, C., Cousens, S, Fraser, H. and Bostock, C.J. 1997. Transmission to mice indicates that 'new variant' CJD is caused by the BSE agent. *Nature* 389, 498–501.

Council Regulation (EC) No 1099/2009 2009. Council Regulation No 1099/2009 on the protection of animals at the time of killing. *Official Journal of the European Union* L303, 1–30.

Daly, D.J., Prendergast, D.M., Sheridan, J.J., Blair, I.S. and McDowell, A. 2002. Use of a marker organism to model the spread of central nervous system tissues in cattle and the abattoir environment during commercial stunning and carcass dressing. *Applied and Environmental Microbiology* 68, 791–798.

Defra.gov.uk 2011. Contingency plan for exotic notifiable diseases of animals in England. https://www.gov.uk/government/publications/contingency-plan-for-exotic-notifiable-diseases-of-animals

EFSA 2004. Welfare aspects of animal stunning and killing methods. Scientific Report of the Scientific Panel for Animal Health and Welfare on request from the Commission related to welfare aspects of animal stunning and killing methods (Question No. EFSA-Q-2003-093). European Food Safety Authority (AHAW 04-027).

Erasmus, M.A., Lawlis, P., Duncan, I.J. and Widowski, T.M. (2010a). Using time to insensibility and estimated time of death to evaluate a nonpenetrating captive bolt, cervical dislocation, and blunt trauma for on-farm killing of turkeys. *Poultry Science* 89(7), 1345–1354.

Erasmus, M.A., Turner, P.V., Nykamp, S.G. and Widowski, T.M. (2010b). Brain and skull lesions resulting from use of percussive bolt, cervical dislocation by stretching, cervical dislocation by crushing and blunt trauma in turkeys. *Veterinary Record* 167(22), 850–858.

Galvin, J.W., Blokhuis, H., Chimbombi, M.C., Jong, D. and Wotton, S. 2005. Killing of animals for disease control purposes. *Revue Scientifique et Technique–Office International des Epizooties* 24: 711–722.

Gerritzen, M.A., 2006. Acceptable methods for large scale on-farm killing of poultry for disease control. PhD Thesis, Utrecht University, Faculty of Veterinary Medicine.

Gerritzen, M.A., Lambooij, E., Stegeman, J.A. and Spruijt, B.M. 2006. Killing poultry during the 2003 avian influenza epidemic in the Netherlands. *Veterinary Record* 159, 39–42.

Gerritzen, M.A., Lambooij, B., Reimert, H., Stegeman, A. and Spruijt, B. 2007. A note on behaviour of poultry exposed to increasing carbon dioxide concentrations. *Applied Animal Behaviour Science* 108, 179–185.

Gerritzen, M.A. and Sparrey, J. 2008. A pilot study to assess whether high expansion CO2-enriched foam is acceptable for on-farm emergency killing of poultry. *Animal Welfare* 17, 285–288.

Gibson, T.J., Ridler, A.L., Lamb, C.R., Williams, A., Giles, S. and Gregory, N.G. 2012. Preliminary evaluation of the effectiveness of captive-bolt guns as a killing method without exsanguination for horned and unhorned sheep. *Animal Welfare* 21, 35–42.

Gibson, T.J., Mason, C.W., Spence, J.Y., Barker, H. and Gregory, N.G. 2014. Factors affecting penetrating captive bolt gun performance. *Journal of Applied Animal Welfare Science* 1–17.

Gibson, T.J., Whitehead, C., Taylor, R., Sykes, O., Chancellor, N.M. and Limon, G. 2015a. Pathophysiology of penetrating captive bolt stunning in Alpacas (*Vicugna pacos*). *Meat Science* 100, 227–231.

Gibson, T.J., Bedford, E.M., Chancellor, N.M. and Limon, G. 2015b. Pathophysiology of free-bullet slaughter of horses and ponies. *Meat Science* 108, 120–124.

Grandin, T. 1980. Mechanical, electrical and anesthetic stunning methods for livestock. *International Journal for the Study of Animal Problems* 1, 242–263.

Grandin, T. 1994. Farm animal-welfare during handling, transport, and slaughter. *Journal of the American Veterinary Medical Association* 204, 372–377.

Grandin, T. 2002. Return-to-sensibility problems after penetrating captive bolt stunning of cattle in commercial beef slaughter plants. *Journal of the American Veterinary Medical Association* 221, 1258–1261.

Hill, A.N., Desbruslais, M., Joiner, S., Sidle, K.C.L., Gowland, I., Collinge, J., Doey, L.J. and Lantos, P. 1997. The same prion strain causes vCJD and BSE. *Nature* 389, 448–450.

McKeegan, D.E.F., McIntyre, J., Demmers, T.G.M., Lowe, J.C., Wathes, C.M., van den Broek, P.L.C., Coenen, A.M. and Gentle, M.J. 2007. Physiological and behavioral responses of broilers to controlled atmosphere stunning—Implications for welfare. *Animal Welfare* 16, 409–426.

McKeegan, D.E.F., Sparks, N.H.C., Sandilands, V., Demmers, T.G.M., Boulcott, P. and Wathes, C.M. 2011. Physiological responses of laying hens during whole house killing with carbon dioxide. British Poultry Science 52, 645–657.

McKeegan, D.E.F., Reimert, H.G.M., Hindle, V.A., Boulcott, P., Sparrey, J.M., Wathes, C.M., Demmers, T.G.M., Gerritzen, M.A. 2013. Physiological and behavioral responses of poultry exposed to gas-filled high expansion foam. *Poultry Science* 92(5), 1145–1154.

Millar, G.I. and Mills, D.S. 2000. Observations

on the trajectory of the bullet in 15 horses euthanased by free bullet. *Veterinary Record* 146(26), 754–757.

OIE 2010. Terrestrial Animal Health Code. Chapter 7.6. Killing of animals for disease control purposes. http://web.oie.int/eng/normes/mcode/en_chapitre_1.7.6.pdf

Raj, A.B.M. and O'Callaghan, M. (2001). Evaluation of a pneumatically operated captive bolt for stunning/killing broiler chickens. *British Poultry Science* 42, 295–299.

Raj, A.B.M., Sandilands, V. and Sparks N.H.C. 2006. Review of gaseous methods of killing poultry on farm for disease control purposes. *Veterinary Record* 159, 229–235.

Raj, A.B.M., Smith, C. and Hickman, G. 2008. Novel method of killing poultry with dry foam created using nitrogen. *Veterinary Record* 162, 722–723.

Sharp, T.M., McLeod, S.R., Leggett, K.E.A. and Gibson, T.J. (2015). Evaluation of a spring-powered captive bolt gun for killing kangaroo pouch young. *Wildlife Research* 41(7), 623–632.

Sparks, N.H., Sandilands, V., Raj, A.B., Turney, E., Pennycott, T. and Voas, A. 2010. Use of liquid carbon dioxide for whole-house-gassing of poultry and implications for the welfare of the birds. *Veterinary Record* 167, 403–407.

Svendsen, O., Jensen, S.K., Karlsen, L.V., Svalastoga, E. and Jensen, H.E. (2008). Observations on newborn calves rendered unconscious with a captive bolt gun. *Veterinary Record* 162(3), 90–92.

Widowski, T.M., Elgie, R.H. and Lawlis, P. (2008). Assessing the effectiveness of a non–penetrating captive bolt for euthansia of newborn piglets. Paper presented at the Allen D. Leman Swine Conference, St. Paul, Minnesota, pp. 107–111.

chapter twelve

# European Council Regulation (EC) 1099/2009 on the protection of animals at the time of killing

*Rebeca Garcia Pinillos*[1]

**Learning objectives**

- Understand the basics of European legislation for the protection of animals at the time of killing 1099/2009, key concepts, definitions and key actors.
- Cover the legal requirements for necessary training and competency for those engaged in handling, stunning and slaughter of animals as well as the supervision and monitoring requirements.
- Monitoring and inspection of European legislation for the protection of animals at the time of killing, including standard operating procedures and key relevant official controls.

**CONTENT**



[1] Veterinary Surgeon, London, United Kingdom

## 12.1 LEGAL REQUIREMENTS TO PROTECT WELFARE AT SLAUGHTER IN THE EU. KEY CONCEPTS AND DEFINITIONS

Protection of the welfare of animals, as sentient beings, is a matter of public concern affecting consumers (EU, 2006). The main legislation relevant for welfare at slaughter in Europe is Regulation 1099/2009 on the protection of animals at the time of killing. This regulation was published in 2009 and replaced Council Directive 93/119/EC on the protection of animals at the time of slaughter and killing. The objective of the regulation was to ensure a harmonized approach in relation to animal welfare at the time of killing (EU, 2009). Regulation 1099/2009 includes the underpinning basic five freedom principles to protect animal welfare under the general requirements for slaughter (Table 12.1).

In 2004 and 2006, the European Food Safety Authority (EFSA) published two opinions on the welfare aspects of the main systems of stunning and killing the main commercial species of animals (EFSA 2004, 2006). This also triggered a need to review existing legislation on welfare at slaughter to take into account those scientific opinions. In 2007, the World Organisation for Animal Health (OIE) adopted a chapter on welfare at slaughter within the Terrestrial Animal Health Code. The EU is one of the Members of this organization and it has committed to adopt the chapters into EU legislation (EU, 2009). Regulation 1099/2009 fulfilled this by incorporating the OIE chapter on welfare at slaughter in the European Law.

This regulation also includes a requirement for countries exporting to the EU, under the framework of Regulation (EC) No 854/2004, to take into account the

**Table 12.1** Five freedom principles applied to animal welfare at the time of slaughter and killing

| Five freedoms (FAWC, 2009) | 1099/2009 Article 3 (EU, 2009) |
|---|---|
| Freedom from hunger and thirst – by ready access to fresh water and a diet to maintain full health and vigour. | Do not suffer from prolonged withdrawal of feed or water. |
| Freedom from discomfort – by providing an appropriate environment including shelter and a comfortable resting area. | Are provided with physical comfort and protection, in particular by being kept clean in adequate thermal conditions and prevented from falling or slipping. |
| Freedom from pain, injury or disease – by prevention or rapid diagnosis and treatment. | Are protected from injury. (...) do not show signs of avoidable pain or fear. |
| Freedom to express normal behaviour – by providing sufficient space, proper facilities and company of the animal's own kind. | Are handled and housed taking into consideration their normal behaviour. (...) do not exhibit abnormal behaviour. |
| Freedom from fear and distress – by ensuring conditions and treatment which avoid mental suffering. | Are prevented from avoidable interaction with other animals that could harm their welfare. |

general and additional requirements on welfare at slaughter. To verify this they have to include verification of compliance within the health certificate accompanying meat imported into Europe (EU, 2009). This helps to promote welfare standards globally and support the implementation of the OIE standards on animal welfare at slaughter. Regulation 882/2004 acknowledges that animal welfare, alongside animal health, is a factor that contributes to the quality and safety of food (EU, 2004).

### *12.1.1 Key concepts and definitions*

Legislation is there to ensure that those responsible for the killing of animals do so in a way that avoids pain and minimizes distress and suffering of the animals. Prevention of pain, suffering and distress are the three key pillars to monitor and preserve good welfare at the time of killing during every operation on live animals (Figure 12.1). It is also important to ensure that animals are rendered immediately unconscious and insensible prior to slaughter. Regulation 1099/2009 describes consciousness of an animal as *"its ability to feel emotions and control its voluntary mobility"*. Sensibility on the other hand is described as *"its ability to feel pain"*.

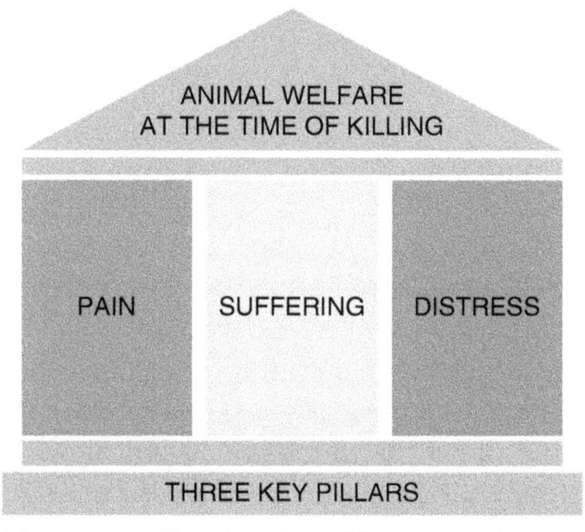


**Figure 12.1:** The three key pillars to monitor and preserve good welfare at the time of killing.

*Stunning* is defined in Regulation 1099/2009 as "any intentionally induced process, which causes loss of consciousness and sensibility without causing avoidable pain, including any process resulting in instantaneous death". This means that some stunning methods, such as free bullet, will lead to immediate death whilst others will only cause reversible unconsciousness. The latter are known as "simple stunning" methods.

Stunning methods are classified into four categories: mechanical methods, electrical methods, gas methods and other methods. Regulation 1099/2009 introduces *key parameters* for stunning methods. This is to ascertain that monitoring procedures include relevant information on the necessary stunning parameters to ensure effective stunning (Figure 12.2). The regulation also establishes *conditions of use* for each stunning method. These explain the species and situations for which each stunning method is legally permitted for slaughter of animals.

Regulation 1099/2009 stipulates that the food business operator must perform regular checks to ensure effective stunning and slaughter of animals. To assess stunning effectiveness, it is important to use *monitoring indicators* to detect signs of unconsciousness and death. Therefore, EFSA published in 2013 four scientific opinions on the welfare of cattle, pigs, sheep and goats, and poultry during the slaughter process. The opinions propose toolboxes for

| SUMMARY CHECKLIST FOR STUNNING | | |
|---|---|---|
| | mechanical stunning | |
| ✔ position and direction of shot | | |
| ✔ appropriate velocity | | |
| ✔ maximum stun to stick/kill interval(s) | | |
| + Plus | | |
| ✔ penetrative captive bolt devices | exit length and diameter of the bolt | |
| | | |
| | electrical stunning | |
| ✔ minimum current (A or mA) | | |
| ✔ minimum voltage (V) | | |
| ✔ maximum frequency (Hz) | | |
| ✔ minimum time of exposure | | |
| ✔ maximum stun-to-stick/kill interval(s) | | |
| ✔ frequency of calibration of the equipment | | |
| ✔ optimisation of the current flow | | |
| ✔ prevention of electrical shocks before stunning | | |

**Figure 12.2:** Summary checklist for stunning key parameters (EU, 2012a)

monitoring welfare at slaughterhouses using animal-based indicators, with their corresponding outcomes of consciousness, unconsciousness or death. The toolboxes provide monitoring indicators for three different key stages of the slaughter operation. They also propose a sample size calculator for monitoring stunning in slaughterhouses (EFSA, 2013a–e).

Work is in progress to define scientifically measurable *animal-based indicators* for monitoring welfare of farm animals (EFSA, 2012). Regulation 1099/2009 identifies key operations at slaughterhouses: handling and care before they are restrained; restraint for the purpose of stunning or killing; stunning; assessment of effective stunning; shackling and killing. All of these operations have to be carried out without causing avoidable pain, suffering or distress. For this it is important to monitor animal welfare through these operations.

Regulation 1099/2009 also requires food business operators to monitor the effect of stunning equipment by checking unconsciousness. However, it is important to know that these indicators have to be considered alongside technical parameters in some of the stunning methods, for example in water bath electrical stunners. In summary, animal-based indicators may be useful to assess animal welfare at slaughterhouses on operations involving live animals other than stunning (see Table 12.2). The animal-based indicators will vary depending on the check undertaken, for example, immediate collapse will be an indicator for unconsciousness following stunning and permanent absence of breathing will be the indicator to ascertain death in animals.

### 12.1.2 Derogations

European law accounts for a number of areas where derogations from the main welfare at slaughter provisions are enabled to account for cultural and sporting events, small businesses, emergencies, private domestic consumption and religious rites. The regulation applies to all vertebrate animals, however it does not include reptiles and amphibians.

Special provisions are in place for small-scale slaughterhouses, mobile slaughterhouses and private consumption; slaughter for the purpose of religious rites, emergency killing (including dangerous animals) and situations where the competent authority considers that compliance may affect human health or slow down the process of eradication of an animal disease. However some basic requirements still apply; for example, where slaughter without stunning takes

**Table 12.2** Slaughter operations established by Regulation 1099/2009 Article 7, 2

| Slaughter operation | Animal state[1] | | | What to monitor[2] | | |
|---|---|---|---|---|---|---|
| | LC | LU | D | AW | U | D |
| Handling and care of animals before they are restrained | X | | | X | | |
| Restraint of animals for the purpose of stunning or killing | X | | | X | | |
| Stunning of animals | X | X | X | X | X | X |
| Assessment of effective stunning | X | X | X | X | X | X |
| Shackling or hoisting of live animals | | X | X | X | X | X |
| Bleeding of live animals | | X | X | X | X | X |
| Slaughtering without stunning[3] | X | X | X | X | X | X |

[1]Animal state: indicates the possible state each animal may undergo during a particular slaughter operation. It may be possible that all four states manifest for each indicator and monitoring must account for this during the process. LC: live conscious; LU: live unconscious; D: death.
[2]What to monitor: AW: animal welfare; U: signs of unconsciousness; D: signs of absence of life.
[3]The regulation refers to slaughter without stunning as "slaughtering in accordance with Article 4(4)".

place for the purpose of religious rites, the regulation still requires animals to be slaughtered at a slaughterhouse and that the animals do not present any signs of consciousness or sensibility before being released from restraint and do not present any sign of life before undergoing dressing or scalding (EU, 2009). There are also transitional provisions for requirements requiring long-term planning and investment. Member states were provided with 10 years to comply with such requirements.

Key areas for improvement identified at the time of drafting Regulation 1099/2009 were the use of carbon dioxide for pigs and the use of water bath stunners for poultry. Both methods are extensively used across the world and the preparation work to develop the 2009 regulation indicated that implementing changes would not be economically viable at the time. Equally, recommendations for fish were not included as further scientific evidence and economic evaluation were required (EU, 2009). European law also identifies farmed fish welfare at slaughter and restraint by inversion of cattle for the purpose of religious rites, areas that require further scientific, social and economic information (EU, 2009).

## 12.2 SUPERVISION AND MONITORING: THE ROLE OF THE BUSINESS OPERATOR

Under Regulation 1099/2009 a "business operator" is defined as "*any natural or legal person having under its control an undertaking carrying out the killing of animals or any related operations falling within the scope of this Regulation*". Regulation 1099/2009, in line with the hygiene package, places the responsibility for welfare of animals in the care of the business operator. For this, the legislation requires business operators and any person involved in the killing of animals to comply with legal requirements and to do so without causing avoidable pain, distress or suffering. The business operators shall take all necessary measures to avoid pain and minimize distress and suffering to animals during the slaughtering or killing process. The regulation considers that, for example, the use of practices which do not reflect the state of the art might induce by negligence or intention pain, distress or suffering to the animals which could be avoidable (EU, 2009).

In 2004, the European Union adopted the "food hygiene package", which amended the controls of food safety. A relevant change introduced by this package was to ensure that food operators through the food chain would bear primary responsibility for food safety. The new regulations merged, harmonized and simplified a number of previous directives. The "hygiene package" comprises three basic Acts, Regulations (EC) No 852/2004 and 853/2004, addressed to food business operators (FBO) and Regulation (EC) No 854/2004, along with Regulation (EC) No 882/2004 on official controls, to competent authorities (CA). Slaughterhouses are subject to approvals under these regulations and Regulation 1099/2009 establishes a link to ensure that animal welfare concerns are integrated within this process, accounting for the construction, layout and equipment used (EU, 2009). This acknowledges that animal welfare is not a standalone topic, but an integral part of the process.

Under Regulation 1099/2009, *business operators* have a number of key legal responsibilities including:

1. Establish a representative sample to check the efficiency of their stunning practices, taking into account the homogeneity of the group of animals, equipment used, personnel involved, and so on.
2. Develop standard operating procedures for all stages of the production cycle, including unloading, lairaging, handling of live animals, restraining, stunning and slaughter (see section 12.4).
3. Maintain all equipment used with live animals, such as unloading equipment (i.e. ramps), lairage equipment (i.e. drinkers), restraining or stunning equipment. This may require cleaning, checking and replacement of worn-out parts, calibration, and so on, as well as keeping maintenance records for at least one year.
4. Ensure that there is backup stunning equipment in place in case routine stunning fails.
5. Ascertain the maximum capacity in lairage pens and the maximum throughput of animals achievable whilst preserving good animal welfare.
6. Ensure that animals are unloaded as soon as possible after arrival.
7. During the time of lairaging, provide physical comfort, clean and adequate thermal conditions, protection from injury, pain and suffering. They shall also avoid suffering from prolonged withdrawal of feed or water; handle and house animals taking into consideration their normal behaviour and prevent avoidable interaction with other animals that could harm their welfare.
8. Appoint an animal welfare officer; this should be a member of staff with the authority and technical

competence to monitor operations and advise personnel as necessary.

9. Establish and monitor key parameters relevant to the stunning methods in use and each of the slaughter lines.
10. Ensure that the layout and construction of their slaughterhouse and equipment used is in compliance with the law.
11. Accept only animals for which the slaughterhouse is approved.
12. Ensure that his/her personnel are competent and appropriately trained.
13. The business operator should submit to the competent authority (a) the maximum number of animals per hour for each slaughter line; (b) the categories of animals and weights for which the restraining or stunning equipment available may be used; (c) the maximum capacity for each lairage area.

*Manufacturers* of restraining and stunning equipment are required to provide detailed instructions for users of the equipment. This should include both the operation, including key parameters, and maintenance (EU, 2009).

The *animal welfare officer* (AWO) plays a key role for animal welfare at slaughter. This requirement appeared as a result of the experience gained in some Member states, such as the United Kingdom, where it was common practice for business operators to appoint a specifically qualified person to coordinate and follow up the implementation of animal welfare operating procedures in slaughterhouses. This was seen as a positive influence leading to animal welfare benefits.

The regulation requires business operators of high throughput premises to designate as an AWO a person under their direct authority and in a position to require personnel to undertake remedial actions to ensure compliance with welfare rules (EU, 2009). The duties of the AWO must be written out by the business operator. However, whilst animal welfare-related activities are performed by the AWO, the responsibility for following EC rules lies with the business operator. The AWO should have sufficient authority and technical competence to provide relevant guidance to slaughter line personnel. There are two key tasks that the AWO is legally required to fulfil:

1. Hold a certificate of competence for each of the operations for which he/she is responsible.
2. Keep a record of actions taken to improve AW at the slaughterhouse for at least one year. For this the AWO must be involved in the development and implementation of standard operating procedures.

In addition, the AWO or a person under their supervision has to systematically assess the welfare conditions of each consignment of animals upon arrival, in order to identify the priorities and welfare needs of the different animals as well as the measures to be taken. To fulfil the roles and responsibilities of the animal welfare officer, it is key that this member of staff has sufficient authority, technical competence, knowledge of the legal requirements and skills to cascade relevant information to slaughter line personnel.

Whilst the role of the AWO is mainly operational, this member of staff is also best placed to advise the business operator in relation to slaughterhouse improvements and how they may affect animal welfare.

## 12.3 SUPERVISION AND MONITORING: THE ROLE OF THE COMPETENT AUTHORITY

Welfare at slaughter is inevitably linked to food hygiene controls at slaughterhouses. The latter were updated in 2004 with the "hygiene package", a series of regulations to lay down specific hygiene rules for food of animal origin. These sets of regulations establish the main legal responsibilities on the business operator to ensure food safety. Regulation 1099/2009 intends to better integrate welfare at slaughter control within the rest of official controls. The "hygiene package" was adopted in April 2004 and comprises a series of four regulations applicable since 1 January, 2006. They are provided for in the following key Acts:

a. Regulation (EC) 852/2004 on the hygiene of foodstuffs, 29 April 2004.
b. Regulation (EC) 853/2004 laying down specific hygiene rules for food of animal origin, 29 April 2004.
c. Regulation (EC) 854/2004 laying down specific rules for the organization of official controls on products of animal origin intended for human consumption, 29 April 2004.
d. Directive 2004/41/EC repealing certain directives concerning food hygiene and health conditions for the production and placing on the market of certain products of animal origin intended for human consumption and amending Council Directives 89/662/EEC and 92/118/EEC and Council Decision 95/408/EC, 21 April 2004 (EU, 2016).

The hygiene rules take particular account of, amongst others, the following principles:

e. primary responsibility by the food business operator;
f. controls across the food chain, starting with primary production;
g. general implementation of management procedures based on the HACCP principles;
h. registration or approval for certain food establishments.

The role of the competent authority is to perform *official controls* on the application of the rules provided by EC legislative framework, including those relating to animal welfare (EC Regulation 854/2004). Official control means any form of control that the competent authority performs for the verification of compliance with food law, including animal health and animal welfare rules (EC Regulation 854/2004, Art. 2[a])

Regulation 882/2004 establishes the basis for "*official controls* (i.e. *any form of control that the competent authority (...) performs for the verification of compliance with (...) animal welfare rules*)" (EU, 2004). This regulation establishes the basis for seven key functions (i.e. inspection, monitoring, surveillance, sampling and analysis, verification and audit) within the competent authority, which might be used to verify compliance with animal welfare rules (Table 12.3).

Regulation 1099/2009 allows Member states to maintain and in some areas,

**Table 12.3** Official control functions to verify compliance with animal welfare rules

| | |
|---|---|
| Inspection | The examination of any aspect of (...) animal welfare in order to verify that such aspect(s) comply with the legal requirements of (...) animal welfare rules. |
| Monitoring | Conducting a planned sequence of observations or measurements with a view to obtaining an overview of the state of compliance with feed or food law, animal health and animal welfare rules. |
| Sampling/ analysis[1] | Taking feed or food or any other substance (including from the environment) relevant to the production, processing and distribution of feed or food or to the health of animals, in order to verify through analysis compliance with (...) rules. |
| Surveillance | A careful observation of one or more feed or food businesses, feed or food business operators or their activities. |
| Audit | A systematic and independent examination to determine whether activities and related results comply with planned arrangements and whether these arrangements are implemented effectively and are suitable to achieve objectives. |
| Verification | Checking, by examination and the consideration of objective evidence, whether specified requirements have been fulfilled. |

adopt more stringent national rules without prejudice to the internal market. It also establishes a number of requirements on the competent authorities of Member states to ensure:

1. The delivery of certificate of competence (CoC) and training courses is consistent. For this purpose Member states competent authorities may grant, suspend or withdraw CoCs.
2. The set-up and implementation of a penalties system applicable to breaches of legal requirements on welfare at slaughter. For this, Member states need to create national legislation stating the penalties, which should be effective, proportionate and dissuasive for the different types of breaches.

The evaluation of animal welfare requires scientific knowledge, practical experience and cooperation between the interested parties. Regulation 1099/2009 establishes a means of cooperation to ensure sufficient independent scientific support to assist the *competent authority via Article 20.* This Article lays down a number of functions where an independent scientific support might assist officials such as: the approval of slaughterhouses, the development of stunning methods, the development of guides to good practice and opinions on user instructions developed by manufacturers of stunning and restraining equipment. They may also provide recommendations

[1] Regulation 882/2004 definition of sampling for analysis does not cite animal welfare rules, however this is still a key function, which might be used by Member states to verify animal welfare rules. Some may have specific provisions allowing officials to take samples in national legislation.

related to inspections and audits and opinions on the capacity and suitability of separate bodies and entities to provide final examination for the purpose of certificates of competence (EU, 2009).

The regulation makes flexible provisions for the scientific support to be provided via a network, which may spread across a number of Member states. To facilitate this, the regulation requires Member states to identify a single contact point and make it publicly available via the Internet. The role of the contact point is to share technical and scientific information and best practices regarding the implementation of this regulation with its counterparts and the Commission (EU, 2009).

## 12.4 STANDARD OPERATING PROCEDURES

The daily slaughterhouse operations and practices largely influence animal welfare. To ensure that it is managed correctly, slaughterhouse business operators are required to develop standard operating procedures for all stages of the production cycle. This includes all the slaughter operations listed in Table 12.2. Regulation 1099/2009 defines standard operating procedures (SOPs) as a set of written instructions aimed at achieving uniformity of the performance of a specific function or standard (EU, 2009).

SOPs should be tailor-made for specific establishments, however food business operators are legally required to include a number of areas for specific operations:

1. Stunning operation: Checks on stunning and corrective action if stunning fails – SOPs must specify the necessary checks to ensure that animals do not present any signs of consciousness or sensibility in the period between the end of stunning and death.
2. Killing and related procedures (see list in Table 12.1): The SOP has to take into account the manufacturer's recommendations for stunning and restraining equipment. SOPs must also define for each stunning method the key parameters (see Figure 12.2) necessary to induce effective stun.
3. All operations: The SOP must describe the responsibilities of the animal welfare officer.

Standard operating procedures should also include the following information:

- clear objectives, describing what the aim for the operation is;
- responsible persons, specifying the name of the members of staff responsible for this operation and its supervision;
- modus operandi, describing the procedure step-by-step;
- measurable criteria, including the number of animals in each sample and the stunning method key parameters;
- monitoring indicators (i.e. to detect absence of signs of consciousness or life), describing how the checks will be undertaken, by whom and on what frequency;
- corrective action, with detail of the steps, equipment or any other measures necessary to address any problems identified during the checks;

| Standard Operating Procedure<br><br>Activity or task | | Ref: |
|---|---|---|
| | | Issue No. |
| | | Issue Date. |
| Prepared by: | Authorised by: | Page |

**Objective**

Describe what the operation aim is
(e.g. to stun sheep humanely)

**Responsibility**

Named member of staff
(e.g. Peter K. – sheep licensed slaughtermen)

**Control measures**

Describe any controls which might be adhere to when following this procedure
(e.g. refer to the manufacturer's instructions sheet for the correct loading procedure and correct cartridge)

**Procedure**

Describe the procedure step by step

**Monitoring procedure**

Describe how monitoring will take place, by whom, numbers and frequency

**Corrective action**

Describe action to take in case of procedure failure

**Records**

Insert a check list to be completed by responsible plant staff with a verification area for the AWO.

**Figure 12.3:** Blank example SOP

- records, designed in a way that data collection can be consistent and auditable.

Whilst there are many examples of SOPs it is important that these should be tailored to the individual slaughterhouse. The reason for this is that SOPs should be practical documents (explaining how to do a task), adapted to the specific working ways of each individual establishment to ensure that they are fully integrated within the operation and are part of the day-to-day work rather than an additional burden. There is guidance available for business operators, such as a sample SOP for the use of water bath electrical stunners developed under the framework of the EU Welnet Project (EU, 2012b). Guides of good practice may also include SOPs (BMPA, 2014). Figure 12.3 shows an example of a blank SOP.

## 12.5 TRAINING AND COMPETENCY OF SLAUGHTERHOUSE PERSONNEL

Slaughterhouse personnel must be trained and skilled. Regulation 1099/2009 describes *competence* as the *"knowledge of the basic behavioral patterns and the needs of the species concerned as well as signs of consciousness and sensibility. It also includes technical expertise with regard of the stunning equipment used"*. The expectation is that knowledge of those undertaking slaughterhouse operations with live animals will go beyond the operation of the machinery and extend to the species' own behaviour, needs and physiology (Figure 12.4).

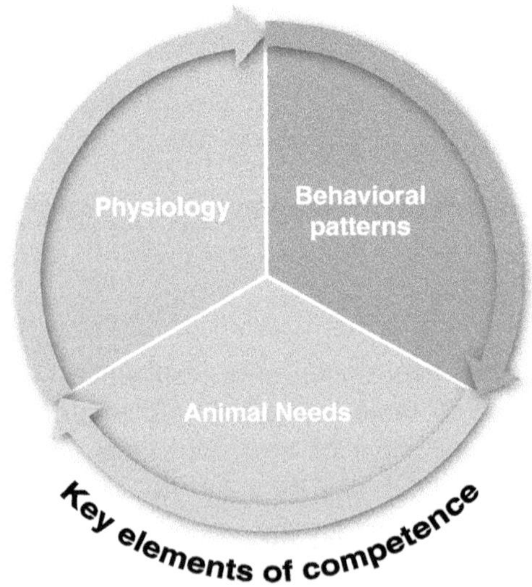


**Figure 12.4:** Key elements of competence with regards to animal welfare

At slaughterhouses, personnel carrying out operations related to live animals from unloading until death (i.e. handling and care, restraining, stunning, assessment of stunning, shackling or hoisting and bleeding of conscious or unconscious animals) are required to have a *certificate of competence* (CoC) relevant to the operations and species they work with. Regulation 1099/2009 requires that the bodies issuing CoCs are accredited to ensure consistent standards across the Member states. There are a number of tools that may be available to underpin personnel knowledge.

Regulation 1099/2009 requires Member states to encourage the development and dissemination of *guides to good practice (GGPs)*, drawn up by organizations of business operators (see example in Figure 12.5) in consultation with other parties, including the competent authority (EU, 2009). The purpose of the GGPs is to help business operators comply with some of the legal requirements, such as, for instance, the development and implementation of standard operating procedures.

The guides to good practice should cover operation and monitoring procedures at the time of slaughter. They should also take into account scientific opinions of the Member states' independent *scientific support* for welfare at killing, be validated by the competent authority and submitted to the Commission to ensure there is a registration system. Where there are no industry representatives drafting GGPs the competent authority may take the initiative and produce the guidance at official level, however the

bmpa
British Meat Processors Association

GUIDE TO GOOD PRACTICE:
WELFARE AT SLAUGHTER

THE WELFARE OF ANIMALS AT THE TIME OF KILLING

| CONTENTS | Paragraph Nos. |
|---|---|
| Preface | |
| Introduction | 1.1-1.6 |
| Definitions | 2.1- 2.3 |
| Staff and Training | |
| Animal Welfare Officer | 3.1- 3.8 |
| Certificates of Competence (Licensing) | 3.9 – 3.24 |
| Supervision and training | 3.25- 3.26 |
| Standard Operating Procedures | 4.1 – 4.5 |
| Management Checks of Restraining and Stunning | 5.1 – 5.7 |
| Procedures | |
| Basic Management Checks | 6 |
| Operations | |
| Arrival & Unloading | 7.1 – 7.8 |
| Lairages | 7.9 |
| Lairage Facilities | 7.10 – 7.41 |
| Field Lairages | 7.42 – 7.45 |
| Handling and moving animals | 7.46 – 4.52 |
| Passageways and races | 7.53 – 7.59 |
| Slaughterhouse waste | 7.60 |
| Restrainers, stunning and stun kill methods | 7.61 – 7.76 |
| Captive bolt stunning | 7.77 – 7.85 |
| Percussion stunners | 7.86 – 7.89 |
| Checking for proper stunning | 7.90-7.91 |
| Recognising a proper stun for captive bolt or percussion stunner | 7.92 |
| Electrical stunning and killing | 7.93 – 7.100 |
| Head only stunning | 7.101 – 7.106 |
| Low voltage systems | 7.107 – 7.118 |
| Head to body stunning and killing | 7.119 – 7.111 |
| Electrical stunning and killing of adult cattle | 7.112 |
| How to recognise a proper stun or stun kill using electricity | 7.113 – 7.115 |
| Stun-killing of pigs using CO2 | 7.116 – 7.118 |

GSP: Welfare at Slaughter – Final 1.0 Page 2 of 113 March 2014

**Figure 12.5:** Example of guides to good practice for red meat animals from the United Kingdom (BMPA, 2014). (http://www.bmpa.uk.com/_attachments/Resources/3602_S4.pdf)

preference is for those in the sector to take responsibility for developing it.

Section 12.3 above included detail on the EC regulation requirement for Member states to have independent scientific support to provide advice in relation to slaughterhouse approvals, the development of new stunning methods, manufacturer instructions, guides to good practice developed within the Member state, recommendations in relation to inspections, audits and opinions on the suitability of training assessment bodies. The opinions and recommendations produced by independent scientific support might also be useful tools for training and development of personnel.

For AWOs, the European Union published a brochure (EU, 2012b) with a summary of the role and functions, which might also serve as training material for those learning about the job. There is a legal requirement for AWOs to have adequate training and competence. The regulation requires them to have a certificate of competence for all the operations taking place in the slaughterhouse for which they are responsible (EU, 2009).

## 12.6 REFERENCES


BMPA 2014. British Meat Processors Association Guide To Good Practice: Welfare at slaughter. http://www.bmpa.uk.com/_attachments/Resources/3602_S4.pdf (Accessed 24 January 2016)

EFSA 2004. Opinion of the Scientific Panel on Animal Health and Welfare (AHAW) on a request from the Commission related to welfare aspects of the main systems of stunning and killing the main commercial species of animals. *EFSA Journal* 45, 1–29.

EFSA 2006. Opinion of the Scientific Panel on Animal Health and Welfare on a request from the Commission related to "The welfare aspects of the main systems of stunning and killing applied to commercially farmed deer, goats, rabbits, ostriches, ducks, geese and quail". *EFSA Journal* 326, 1–18.

EFSA 2012. Scientific Opinion. Statement on the use of animal-based measures to assess the welfare of animals. EFSA Panel on Animal Health and Welfare (AHAW). *EFSA Journal* 10(6), 2767.

EFSA 2013a. Scientific Opinion on monitoring procedures at slaughterhouses for bovines. Panel on Animal Health and Welfare (AHAW). *EFSA Journal* 11(12), 3460.

EFSA 2013b. Scientific Opinion on monitoring procedures at slaughterhouses for poultry. EFSA Panel on Animal Health and Welfare (AHAW). *EFSA Journal* 11(12), 3521.

EFSA 2013c. Scientific Opinion on monitoring procedures at slaughterhouses for sheep and goats. EFSA Panel on Animal Health and Welfare (AHAW). *EFSA Journal* 11(12), 3522.

EFSA 2013d. Scientific Opinion on monitoring procedures at slaughterhouses for pigs. EFSA Panel on Animal Health and Welfare (AHAW). *EFSA Journal* 11(12), 3523.

EFSA 2013. Sample size calculation tool for monitoring stunning at slaughter. European Food Safety Authority; EFSA supporting publication EN-541. http://www.efsa.europa.eu/sites/default/files/scientific_output/files/main_documents/541e.pdf (Accessed 24 January 2016)

EU 2004. Regulation (EC) No. 882/2004 of the European Parliament and of the Council of 29th April 2004 on official controls performed to ensure the verification of compliance with feed and food law, animal health and animal welfare rules. *Official Journal of the European Union* L165/1–L 165/141.

EU 2006. Protocol (No 33) on protection and welfare of animals (1997). *Official Journal of the European Union* C321 E, 29/12/2006, pp. 0314–0314.

EU 2009. Council Regulation (EC) No 1099/2009 of 24 September 2009 on the protection of animals at the time of killing. *Official Journal of the European Union* L303/1–L303/30.

EU 2012a The Animal Welfare Officer in the European Union. Available at: http://ec.europa.eu/dgs/health_food-safety/information_sources/docs/ahw/brochure_24102012_en.pdf (Accessed 12 January 2016)

EU 2012b. EU WelNet, Coordinated European Animal Welfare Network. Appendix 29. http://www.euwelnet.eu/euwelnet/53430/7/0/80 (Accessed 24 January 2016)

EU 2016. http://ec.europa.eu/food/food/biosafety/hygienelegislation/comm_rules_en.htm (Accessed 26 February 2016)

FAWC 2009. The Five Freedoms. http://webarchive.nationalarchives.gov.uk/20121007104210/http:/www.fawc.org.uk/freedoms.htm (Accessed March, 2016) chapter thirteen

# Future trends to improve animal welfare at slaughter

*Andrea Gervelmeyer[1], Denise Candiani[1] and Frank Berthe[1]*

**Learning objectives**

- Understand crucial elements of stunning methods related to animal welfare.
- Understand the concept of using animal-based measures for assessing good or poor welfare outcomes.
- Understand how stunning effectiveness should be monitored at the slaughterhouse.

**CONTENT**



## 13.1 INTRODUCTION

### *13.1.1 EFSA's role in animal welfare*

The European Food Safety Authority (EFSA) is a decentralized European agency tasked with delivering independent scientific advice and transparent communication on existing and emerging risks in the area of food and feed safety. EFSA can be tasked by the European Commission, by EU Member States (MS) or through self-tasks developed by EFSA. The EFSA Scientific Panel on Animal Health and Welfare (AHAW Panel) provides scientific advice on all aspects of animal health and welfare, including those that have implications

[1] European Food Safety Authority (EFSA), Via Carlo Magno 1/A, 43126 Parma, Italy

for human health, in order to support the science-based development of animal health and welfare standards within the European Union. Assessing risks to the health and welfare of animal populations serves to protect public health, our environment and the economic benefit we derive from them. The scientific opinions of the AHAW Panel are used as a scientific basis for many of the EU legislative measures on animal welfare. For example, Council Regulation (EC) No 1/2005 on the protection of animals during transport is essentially based on conclusions and recommendations of the 2004 EFSA opinion on the welfare of animals during transport (EFSA, 2004a). This opinion has recently been updated (EFSA, 2011) and has contributed to a comprehensive report to the European Parliament proposing additional management options. Similarly, EFSA opinions on welfare aspects of the main systems of stunning and killing (EFSA, 2004b, 2006) have led to Council Regulation (EC) No 1099/2009 on the protection of animals at the time of killing.

### *13.1.2 Use of animal-based measures*

With its work in the area of animal-based measures (ABMs) of welfare, EFSA has contributed to the objectives of the European Union Strategy for the Protection and Welfare of Animals 2012–2015 (EC, 2012). EFSA has carried out a series of assessments on the use of outcome-based animal welfare indicators, rather than input- or resource-based welfare indicators (EFSA, 2012a,b). One major outcome of the EFSA mandates on ABMs is a set of welfare indicators for different livestock species that can be used in future animal welfare surveillance systems (EFSA, 2012c). Future databases containing measurements of animal welfare measures could give a new dimension to EFSA's work, allowing for quantitative welfare risk assessments.

### *13.1.3 Welfare at slaughter*

The central legal document in the EU on welfare at slaughter is Council Regulation (EC) No 1099/2009 on the protection of animals at the time of killing. It defines "stunning" in Article 2 (f) as "any intentionally induced process which causes loss of consciousness and sensibility without pain including any process resulting in instantaneous death". Annex I of the Regulation lists the stunning interventions and related specifications. Article 4 on stunning interventions regulates that "animals shall only be killed after stunning in accordance with the methods and specific requirements related to the application of those methods set out in Annex I of the Regulation" and "that the loss of consciousness and sensibility shall be maintained until the death of the animal".

Article 4 (2) of the Regulation allows the Commission to amend Annex I to this Regulation to take account of scientific and technical progress on the basis of an opinion of EFSA. In response to these articles, EFSA has developed a document that provides guidance to the AHAW Panel on how studies on new or modified stunning methods will be assessed, which also provides useful guidance to researchers for planning and carrying out such studies (EFSA, 2013a).

Article 16 of Council Regulation (EC) No 1099/2009 requires slaughterhouse operators to put in place and implement monitoring procedures in order to check that their stunning processes deliver the expected results in a reliable way. It refers to Article 5, which requires operators to carry out regular checks to ensure that animals do not present any signs of consciousness or sensibility in the period between the end of the stunning process and death. In support of these requirements, EFSA has published several scientific documents that provide toolboxes containing animal-based indicators and their corresponding outcomes of consciousness, unconsciousness or death. The documents also highlight when to apply corrective interventions, and provide guidance on establishing adequate monitoring procedures, including setting a suitable sampling frequency (EFSA, 2013b–e). The following sections provide more details on the scientific advice EFSA has produced to improve animal welfare at slaughter.

## 13.2 GUIDANCE ON ASSESSING NEW SLAUGHTER METHODS

Several studies assessing the efficacy of modified protocols of stunning interventions listed in Annex I or new stunning interventions, have been submitted to the Commission who has requested EFSA to review the studies. When assessing these studies, a lack of harmonization of designing and reporting studies that investigate the effectiveness of the proposed stunning interventions has been observed. This has been identified as a drawback in several EFSA opinions. Therefore, EFSA formulated clear guidance to the AHAW Panel on how studies on modified or new stunning methods will be assessed by EFSA (EFSA, 2013a).

The guidance first highlights general aspects applicable to studies on stunning interventions that should be considered when studying their effectiveness. Well-controlled studies conducted under laboratory conditions are needed as a first step in order to characterize the animal's responses (here: unconsciousness and absence of pain). Such studies under laboratory conditions should use the most sensitive and specific methods available to assess if the stunning intervention leads to unconsciousness and absence of pain until death, using established scientific tools such as electroencephalogram (EEG) or electrocorticogram (ECoG) to measure the onset and duration of unconsciousness, or blood samples to measure the concentrations of hormones indicative of pain, distress and suffering, such as corticosteroids or adrenaline. As most of the sensitive and specific measurements that can be carried out under laboratory conditions, such as EEG, cannot be carried out in slaughterhouses, the laboratory studies should also establish the correlations between these measurements and non-invasive animal-based measures that can be applied in slaughterhouses. The second step of the research of new stunning methods, studies under slaughterhouse conditions, is intended to assess whether the results obtained in the laboratory studies can also be achieved in the slaughterhouse context. The set of eligibility criteria defined in the EFSA guidance document will be applied to both steps of the research on the stunning intervention. Figure 13.1 shows the two steps and the research elements EFSA recommends to cover in each of them.

| Type | Conditions | Elements of research recommended |
|---|---|---|
| I. Proof of concept | Study under controlled laboratory conditions | A. Comprehensive record of stunning intervention and key parameters<br>B. Assessment of onset and duration of unconsciousness by EEG or ECoG<br>C. Assessment of absence of pain, distress and suffering using behavioural and either physiological or neurological animal-based measures<br>D. Comprehensive record of outcome assessment<br>E. Stunning without sticking to establish duration of unconsciousness achievable with simple stunning intervention |
| II. Ground truthing | Study under slaughterhouse conditions | F. Comprehensive record of stunning intervention and key parameters<br>G. Assessment of onset and duration of unconsciousness using animal-based measures<br>H. Assessment of absence of pain, distress and suffering using behavioural and either physiological or neurological animal-based measures<br>I. Comprehensive record of outcome assessment<br>J. Assessment of absence of pain, distress and suffering during restraint/pre-stunning if it deviates from conventional methods and/or is potentially painful |

**Figure 13.1:** Recommended approach for research on stunning interventions (EFSA, 2013a)

The duration of unconsciousness achieved by the stunning method is an important outcome that needs to be measured and reported for studies researching a new or modified legal simple stunning (i.e. reversible) intervention. Therefore, animals should be stunned without exsanguination (i.e. bleeding out by neck cutting [severing the carotid arteries] or sticking [severing the brachiocephalic trunk]) in the proof-of-concept studies under controlled laboratory conditions (Figure 13.1, E). It is important to note that the experimental protocols must apply humane endpoints as specified in various international (e.g. http://www.animalethics.org.au/legislation/international) or European guidelines on the ethical use of animals in research (e.g. Directive 2010/63/EU). Therefore, it needs to be considered if the stun experienced leads to long-term adverse effects. In that case, the animals should be re-stunned and euthanized as soon as they regain consciousness, in accordance with these guidelines.

Studies on stunning interventions should also explain in detail, how and when the onset of unconsciousness and insensibility is measured (Figure 13.1, B, C, G, H). It is required that the methodology used for the determination of the onset and the duration of unconsciousness has previously been accepted in appropriate internationally recognized and stringently peer-reviewed journals. It is also necessary that all data are provided at the individual animal level. In the case of EEG (or ECoG) measurements, all parameters that are crucial to the assessment of the EEG data should be specified (e.g. the EEG recording electrode position on the skull or on the brain itself, the configuration of the electrode [transhemispheric or from the same hemisphere of the brain], the background noise filtration method employed in the data acquisition and analysis). In order to estimate quantitative changes apparent on the EEG (or ECoG), the method used to derive the transformations of EEG data must be described (Figure 13.1, B). In addition, the indicators used to assess onset and duration of unconsciousness should be relevant to that stunning intervention, based on the available scientific knowledge of each indicator's sensitivity and specificity. A clear explanation of how and when the animal-based measures were recorded and analysed should be provided in the methods section of the studies (Figure 13.1, G, H, I). Detailed experimental protocols should be provided to allow assessment of potential limitations of the selected animal-based measures. For example, animals connected to measuring equipment may behave differently, the effect of the sampling procedure or the delay of a physiological response could influence the results obtained with physiological parameters. In addition, exposure of an animal to a new environment can change its behavioural, physiological or autonomic responses. Therefore, selecting the combination of indicators to be used depends upon the design of the study and the animal species on which the stunning intervention is tested. The scoring system applied to categorize/classify the animal-based measures should be clearly defined (Figure 13.1, G, H, I). It is essential that the observers making the measurements are adequately trained and that scoring systems are adapted to the species and the stunning conditions. If applicable, the observers assessing the outcomes should be blinded to the experimental groups (e.g. control and treatment).

For any intervention that does not lead to an immediate onset of unconsciousness and insensibility, the time to loss of consciousness from the beginning of the application of the stunning intervention, and signs of pain, distress and suffering until the onset of unconsciousness should be recorded in all animals and reported as individual animal-level data or mean or median and range and standard deviation or interquartile range (Figure 13.1, B, C, G, H).

It is recommended that the animal-based measures for pain, distress and suffering are first examined under controlled laboratory conditions – for each animal undergoing the stunning procedure – during the exposure of the animal to the procedure/apparatus without the actual stunning. This control or "sham" operation provides a baseline result of the pain, distress and suffering that is due to the handling process. The animal-based measures for pain, distress and suffering should be examined again during exposure of the animal to the full procedure/apparatus including the stunning (Figure 13.1, C, H). Comparison of the two observations differentiates between pain, distress and suffering due to the handling process versus pain, distress and suffering due to the stunning intervention itself. In the absence of avoidable pain, distress and suffering caused by the application of a stunning intervention, the response of animals exposed to the procedure/apparatus without the application of stunning (control or sham operation) should not be significantly different from the response of the animals exposed to the procedure/apparatus with stunning. It is, however, essential that the control/sham operation itself does not lead to peak pain, distress and suffering response levels in the animals, because in that case no further increases in response could be expected within the physiological limits of the species/animal under investigation.

If the pre-stunning handling of animals during the proposed new stunning intervention is very different from that associated with the conventional process – and/or if it is possibly a source of pain, distress or suffering – then the researchers should provide scientific evidence that allows for an assessment of animal welfare (Figure 13.1, J).

The guidance explains the minimum eligibility criteria, reporting quality criteria and further study quality criteria. Only if these criteria are fulfilled, can a given study be considered for assessment as a potential alternative to the stunning methods and related specifications listed in Annex 1 of Council Regulation (EC) No 1099/2009. Information on all the preceding should be provided and will be assessed by the EFSA Panel on Animal Health and Welfare (AHAW), based upon scientific knowledge available at that time.

The eligibility criteria that must be fulfilled by submitted studies are based upon Council Regulation (EC) No 1099/2009 and focus on the stunning intervention and its outcome regarding loss of consciousness and sensibility and absence of pain, distress and suffering. For the intervention, EFSA requires that the key parameters described in the legislation as well as additional parameters needed to fully define and characterize the stunning intervention, that have been identified from previously published scientific data, are reported. An example of the parameters that should be reported for head-only and head-to-body electrical stunning is provided in Table 13.1. For the outcome of the stunning, the study must provide evidence for an immediate onset of unconsciousness and insensibility. An example for how to assess this for electrical stunning is given in Figures 13.2 and 13.3. In case the stunning method does not lead to an immediate loss of consciousness, evidence must be provided for the absence of avoidable pain, distress and suffering until the loss of consciousness and sensibility, and for duration of the unconsciousness and insensibility until the death of the animal. An overview of response types and animal-based measures that can be used to assess pain, distress and suffering during the induction of unconsciousness and insensibility is given in Table 13.2. Two criteria/rules have to be fulfilled before a stunning intervention is considered not to induce pain, distress and suffering before the onset of unconsciousness and insensibility. First, animal-based measures from the behaviour response type and animal-based measures from at least one of the two additional response types presented in Table 13.2 (i.e. physiological or neurological response), that are relevant to the intervention/species, must be indicative of the absence of pain, distress and suffering before the onset of unconsciousness and insensibility. This means that these animal-based measures should not be significantly different between the appropriate control and treatment groups. It also means that the response of animals exposed to the procedure/ apparatus without the application of stunning (control or sham operation) should not be significantly different from the response of the animals exposed to the procedure/apparatus with stunning. The possibility that the control/ sham operation itself has not resulted in a maximum response to pain, distress and suffering in the animals should be demonstrated. Second, these animal-based measures should be consistent at the level of the individual animal, depending upon the species and the coping strategies (that is, consistent with respect to their interpretation).

**Table 13.1** Parameters to be provided when applying a stunning intervention based on head-only and head-to-body electrical stunning (EFSA, 2013a)

| Parameter | Component | Description (all specifications should be in internationally recognized units) |
|---|---|---|
| Minimum current (A or mA) | Current type | The electrical current used to stun animals can be sine or square wave alternating current (bipolar or biphasic) or pulsed direct current (monopolar or monophasic). Define the current type used. |
| | Waveform | The waveform of the current used for stunning animals varies widely and includes clipped or rectified sine or square waves. The proportion of clipping also varies widely. Define the waveform used, including the proportion of clippings, and report the mark–space ratio when pulsed direct current is used. If multiple frequencies and waveforms are used, describe them. |
| | Minimum current [(a)] | Specify the minimum current (A or mA) to which animals are exposed. Explain how this value was obtained. Normally, when using sine wave alternating current the minimum current will be expressed as the root mean square current. When a pulsed direct current is used, the minimum will be expressed as the average current. Describe how the minimum current was calculated. In a multiple-cycle method of a head-to-body stunning system, details should be provided for each cycle. |
| | Latency [(a)] | Specify how soon the minimum current was reached after the intervention was applied to the animal. In a multiple-cycle method of a head-to-body stunning system, details should be provided for each cycle. |

| Parameter | Component | Description (all specifications should be in internationally recognized units) |
|---|---|---|
| Minimum voltage (V) | Exposed minimum voltage (V) [(a)] | Specify the minimum voltage (V) to which animals are exposed. Explain how this value was measured (e.g. peak voltage, peak–peak voltage, root mean square voltage or average voltage). Root mean square voltage is the recommended description of the exposed minimum voltage. In a multiple-cycle method of a head-to-body stunning system, details should be provided for each cycle. |
| | Delivered minimum voltage (V) [(a)] | According to Ohm's law, the amount of voltage required to deliver 1 A will depend upon the electrical resistance in the pathways, which in turn is determined by several factors. Describe how the stunning equipment was set up to deliver the minimum current level to the animal. In a multiple-cycle method of a head-to-body stunning system, details should be provided for each cycle. Describe how the pre-set constant current was applied (e.g. variable voltage/ constant current stunner). |
| Maximum frequency (Hz) | Maximum frequency (Hz) | If applicable, define the maximum frequency (Hz) applied to the animal. In a multiple-cycle method of head-to-body stunning system, details should be provided for each cycle. |
| | Minimum frequency (Hz) | If applicable, define the minimum frequency (Hz) applied to the animal. In a multiple-cycle method of a head-to-body stunning system, details should be provided for each cycle. |
| Minimum time exposure [(a)] | | Define the minimum duration of electrical exposure applied to the animals. In a multiple-cycle method of a head-to-body stunning system, details should be provided for each cycle. |
| Maximum stun-to-stick/kill interval(s) [(a),(b)] | | Describe the maximum stun-to-stick/kill interval and the exsanguination method (blood vessels cut) that have been applied to guarantee unconsciousness and insensibility of the stunned animal until the moment of death (except for proof-of-concept studies where the duration of unconsciousness must be determined without sticking). |
| Frequency of calibration of the equipment | | Provide information on the method used for, and the time intervals between, consecutive calibrations of the equipment. |

| Parameter | Component | Description (all specifications should be in internationally recognized units) |
|---|---|---|
| Optimization of the current flow | Electrode characteristics | The form of the stunning tongs or electrodes and the material are important to overcome the resistance in the pathway. Provide a description of the electrode (form/shape, presence and description of spikes (depth of penetration, wetting). |
| | Electrode appearance | The condition (e.g. corroded) and cleanliness (fat and wool cover, carbonization of dirt) of stunning electrodes contribute to the electrical resistance. Electrodes should be cleaned regularly using a wire brush to prevent build-up of materials. Describe the appearance of the electrodes as well as the method used to clean them between use on individual animals. |
| | Animal restraining | Describe how animals are restrained. |
| Prevention of electrical shocks before stunning | | Explain how the animals are protected from inadvertent, unintentional electrical shocks immediately before the stunning intervention is initiated. For instance, the stunning electrodes could be placed firmly without slipping and held with uniform pressure throughout the duration of stunning to ensure that the current flows uninterruptedly. |
| Position and contact surface area of electrodes | Position of the electrodes | Specify the topographical anatomical position where the electrodes are attached to the animal and the method to hold electrodes in place during the intervention. Placement and application of electrodes should be described and validated. |
| | Type of electrode | Provide information on the type of electrodes used (e.g. tong, wand, etc.). |
| | Animal skin condition | The amount of wool/hair/feathers covering the head at the site of stunning electrode position is critical as the electrical resistance increases with the increasing amount of wool, etc. Provide a description of the study population in relation to the wool/hair/feather cover, and cleanliness of the coat (e.g. clipped or not, breed, wet/dry head). |

(a) Provide information on mean or median and range and standard deviation or interquartile range.
(b) In case of simple stunning.

- Induction of a generalised epileptiform activity in the brain, which can be recognised from the predominance of 8–13 Hz high-amplitude EEG activity, followed by a quiescent EEG.

OR

- An immediate onset of a quiescent EEG.

OR

- No somatosensory, visual or auditory evoked responses or potentials in the brain immediately after the stunning.

**Figure 13.2:** EEG patterns that can be used to ascertain unconsciousness and insensibility in laboratory condition studies (EFSA, 2013a)

1. Presence of tonic seizures after removal of the current;

AND

2. Apnoea during tonic and clonic seizures.

**Figure 13.3:** Sequence of indicators to assess the effectiveness of electrical stunning to be sure that the animal is unconscious and insensible (EFSA, 2013a)

In addition to the eligibility criteria described above, a number of criteria indicative of good reporting (reporting quality criteria) and a number of criteria indicative of presence of biases (methodological quality criteria) need to be met. If they are met, the study on the new or modified legal intervention provides sufficient detail regarding the intervention and the outcome to allow for a conclusion to be reached about the suitability (or lack thereof) of the intervention. In that case, a full assessment of the animal welfare implications of the proposed alternative stunning intervention, including both pre-stunning and stunning phases, and an evaluation of the quality, strength and external validity of the evidence presented would be carried out by the EFSA AHAW Panel.

## 13.3 MONITORING WELFARE AT SLAUGHTER

Article 16 of Council Regulation (EC) No 1099/2009 on the protection of animals at the time of killing requires slaughterhouse operators to put in place and implement monitoring procedures in order to check that their stunning processes deliver the expected results in a reliable way. Article 16 refers to Article 5 which requires operators to carry out regular checks to ensure that animals do not present any signs of consciousness or sensibility in the period between the end of the stunning process and death. Those checks shall be carried out on a sufficiently representative sample of animals and their frequency shall be established taking into account the outcomes of previous checks and any factors which may affect the efficiency of the stunning process. Article 5 also requires operators, when animals are slaughtered without stunning, to carry out systematic checks to ensure that the animals do not present any signs of consciousness or sensibility before being released from restraint and do not present any sign of life before undergoing dressing or scalding.

In support of these requirements, EFSA has provided animal-based indicators that can be used for assessing signs of (a) consciousness, in the case of slaughter with stunning, and (b) unconsciousness and (c) death of the animals, in the case of slaughter without stunning. These indicators have been selected based on their performance (i.e. their sensitivity and specificity, and

**Table 13.2** Overview of response types and animal-based measures associated with pain, distress and suffering during the induction of unconsciousness and insensibility (EFSA, 2013a)

| Response type | Groups of animal-based measures | Example | References |
|---|---|---|---|
| Behaviour | Vocalizations | e.g. number and duration, intensity, spectral components | EFSA, 2005; Le Neindre et al., 2009; Atkinson et al., 2012; Landa, 2012; Llonch et al., 2012a,b, 2013 |
| | Postures and movements | e.g. kicking, tail flicking, avoidance | Jongman et al., 2000; EFSA, 2005; McKeegan et al., 2006; Gerritzen et al., 2007; Velarde et al., 2007; Kirkden et al., 2008; Svendsen et al., 2008; Dalmau et al., 2010; Atkinson et al., 2012; Landa, 2012; Llonch et al., 2012a,b, 2013 |
| | General behaviour | e.g. agitation, freezing, retreat attempts, escape attempts | EFSA, 2005; Velarde et al., 2007; Dalmau et al., 2010; Landa, 2012 |
| Physiological response | Hormone concentrations | e.g. Hypothalamic–pituitary–adrenal axis: corticosteroids, Adrenocorticotrophic hormone; Sympathetic system: adrenaline, noradrenaline | Mellor et al., 2000; EFSA, 2005; Le Neindre et al., 2009; Coetzee et al., 2010; Landa, 2012 |
| | Blood metabolites | e.g. glucose, lactate, lactate dehydrogenase | EFSA, 2005; Vogel et al., 2011; Landa 2012; Mota-Rojas et al., 2012 |
| | Autonomic responses | e.g. heart rate and heart rate variability, blood pressure, respiratory rate, body temperature | Martoft et al., 2001; EFSA, 2005; Gerritzen et al., 2007; von Borell et al. 2007; Rodriguez et al., 2008; Svendsen et al., 2008; Le Neindre et al., 2009; Dalmau et al., 2010; McKeegan et al., 2011; Atkinson et al., 2012; Landa, 2012; Llonch et al., 2012a,b, 2013 |
| Neurological response | Brain activity | e.g. EEG, EcoG | Gibson et al., 2009 |

their feasibility). In addition, EFSA has indicated the most common risk factors and their welfare consequences to determine the circumstances of the monitoring procedures and provided examples of sampling protocols, based on different possible scenarios. This work has been done for a number of different species and slaughter methods, namely for the assessment of consciousness in bovines stunned with penetrative captive bolt, in pigs stunned with the head-only electrical method or carbon dioxide at high concentration, in poultry after water bath stunning or with gas mixtures and in small ruminants after stunning with the head-only electrical method. It has also been done for the assessment of unconsciousness and death for bovines, poultry and small ruminants slaughtered without stunning (EFSA, 2013b–e).

The focus of the welfare monitoring was placed on detecting consciousness, that is, ineffective stunning or recovery of consciousness, as risks of poor welfare can be detected better in this way, instead of looking for outcomes of unconsciousness in animals following stunning. Therefore the indicators were phrased neutrally (e.g. corneal reflex) and the outcomes were phrased either suggesting unconsciousness (e.g. absence of corneal reflex) or suggesting consciousness (e.g. presence of corneal reflex). This is an approach that is commonly used in animal health studies (e.g. when testing for the presence of a disease) but new to animal welfare monitoring in slaughterhouses. Sets of indicators have been proposed to check for signs of consciousness in a given species stunned with a given method, for example for bovines after penetrative captive bolt stunning. A different set of indicators has been proposed for confirming unconsciousness as well as death of animals following slaughter without stunning.

For each indicator, the sensitivity, specificity and feasibility were determined using the scientific knowledge available in published literature and from experts involved with monitoring welfare at slaughter. Based on this information, the most appropriate indicators were selected and a toolbox of indicators to be used in monitoring procedures was proposed. The feasibility of an indicator is determined in relation to the physical aspects of its assessment. These include the position of the animal relative to the assessor, the assessor's access to the animal and the line speed. It is very likely that the feasibility of assessing an indicator is influenced by the stage of the slaughter process, that is, immediately after stunning, at sticking/neck cutting and during bleeding, the animals usually are in different positions and proximity relative to the assessor, which may affect how easily the indicator can be used.

For slaughter with prior stunning, the sensitivity of an indicator is the percentage of truly conscious animals that the indicator identifies as conscious, while the specificity of an indicator is the percentage of truly unconscious animals that the indicator does not identify as conscious. In the case of slaughter without stunning, sensitivity is the percentage of animals truly still conscious or alive that the indicator tests conscious or alive, while specificity is the percentage of truly unconscious or dead animals that the indicator does test as unconscious or dead.

As mentioned previously, the use of

animal-based indicators for the monitoring of welfare at slaughterhouse is similar to the use of a diagnostic or statistical "test" that can either have a positive or negative outcome. In the case of stunning of animals, the major interest is to detect the undesired outcome, namely the presence of consciousness in animals. The toolbox proposes respective indicators and their outcomes. In the case of slaughter without stunning, the interest is to detect whether the animals become unconscious and to detect when the animal dies. However, the indicators applied for this task also have to correctly detect animals as conscious or alive. Therefore, in the case of slaughter without stunning, the toolbox proposes respective indicators detecting the presence of conscious or alive animals and their outcomes.

Each of the toolboxes provides a set of recommended indicators which are considered to have a good level of sensitivity and feasibility. In addition, a set of additional indicators, that are currently considered to have a lower sensitivity or feasibility, has been included. In the future, the use of these indicators might acquire a higher level of sensitivity and feasibility through education, training of personnel responsible for monitoring welfare at slaughter or through changes in the layout of the slaughter lines or changes to existing slaughter practices. The most appropriate set of indicators (at least two indicators should be used) must be chosen from these toolboxes based on a) the expertise of the persons carrying out the monitoring and b) the infrastructure present in the slaughterhouse.

For the monitoring of slaughter with prior stunning, three key stages at which monitoring should be carried out are proposed for bovines, pigs and small ruminants: after stunning (between the end of stunning and shackling), during neck cutting or sticking and during bleeding. For poultry, two key stages at which monitoring should be carried out are proposed: first between exit from the water bath and neck cutting (for stunning with water bath) or during shackling (for stunning with gas mixtures), and then during bleeding. Monitoring should be carried out by repeatedly checking the indicators to detect signs of consciousness at these stages. For the monitoring of slaughter without stunning, all animals should be checked until they become unconscious, before being released from the restraint; death should be confirmed before starting carcass dressing.

The indicators were selected on the basis that consciousness or life in checked animals can be correctly identified by their use. The recommended indicators for monitoring unconsciousness after stunning are shown in Table 13.3. The recommended indicators to check for death are shown in Table 13.4. It is important to note that all personnel performing pre-slaughter handling, stunning, shackling, hoisting and/or bleeding should check all animals to rule out the presence of consciousness following stunning or to confirm unconsciousness and death during slaughter without stunning.

The person in charge of monitoring the overall animal welfare at slaughter (i.e. the animal welfare officer) has to check a certain sample of slaughtered animals for approval. This sample can be calculated based on the throughput rate (the total number of animals slaughtered in the slaughter plant) and

**Table 13.3** Indicators to monitor welfare during slaughter with stunning

| | | After stunning | | During neck cutting or sticking | | During bleeding | |
|---|---|---|---|---|---|---|---|
| Species | Stunning method | Recommended indicators | Additional indicators | Recommended indicators | Additional indicators | Recommended indicators | Additional indicators |
| Bovines | Captive bolt | Posture, breathing, tonic seizure, corneal or palpebral reflex | Muscle tone, eye movements and vocalization | Body movements, muscle tone and breathing | Eye movements, corneal or palpebral reflex and spontaneous blinking | Muscle tone, breathing and spontaneous blinking | No additional indicators |
| Pigs | Head-only electrical stunning | Tonic/clonic seizures, breathing and corneal or palpebral reflex | Spontaneous blinking, posture and vocalizations | Breathing, tonic/clonic seizures and muscle tone | Corneal or palpebral reflex, spontaneous blinking and vocalizations | Breathing and muscle tone | Vocalizations, the corneal or palpebral reflex and spontaneous blinking |
| | Carbon dioxide stunning | Muscle tone, breathing and corneal or palpebral reflex | Response to nose prick or ear pinch and vocalizations | Muscle tone, breathing and vocalizations | Corneal or palpebral reflex and response to nose prick or ear pinch | Muscle tone and breathing | Corneal or palpebral reflex and vocalizations |
| Sheep and goats | Head-only electrical stunning | Tonic/clonic seizures, breathing and corneal or palpebral reflex | Spontaneous blinking, posture and vocalizations | Breathing, tonic/clonic seizures and muscle tone | Corneal or palpebral reflex, spontaneous blinking and vocalizations | Breathing and muscle tone | Corneal or palpebral reflex, spontaneous blinking and vocalizations |
| Poultry | Electrical water bath | Tonic seizures, breathing and spontaneous blinking | Corneal or palpebral reflex and vocalizations | Not applicable | Not applicable | Wing flapping and breathing | Corneal or palpebral reflex, spontaneous swallowing and head shaking |
| | Stunning with gas mixtures | breathing, muscle tone, wing flapping and spontaneous blinking | Corneal or palpebral reflex and vocalizations | Not applicable | Not applicable | Wing flapping, muscle tone and breathing | Corneal or palpebral reflex |

**Table 13.4** Indicators to monitor welfare during slaughter without stunning

| | Prior to release from restraint | | Prior to dressing (or, in the case of poultry, prior to scalding) | |
|---|---|---|---|---|
| Species | Recommended indicators | Additional indicators | Recommended indicators | Additional indicators |
| Bovines | Breathing and muscle tone | Posture and corneal or palpebral reflex | End of bleeding, relaxed body, dilated pupils | No additional indicators |
| Sheep and goats | Breathing and muscle tone | Posture and corneal or palpebral reflex | Bleeding, muscle tone and pupil size | No additional indicators |
| Poultry | Not applicable | Not applicable | Breathing, the corneal or palpebral reflex, pupil size and bleeding | Muscle tone |

the tolerance level (number of potential failures, meaning the animals that are conscious after stunning). A tool with a user-friendly interface and user manual that can be used to calculate the sample size for a specific slaughterhouse situation has been developed by EFSA and is available online (EFSA, 2013f).

In addition, risk factors to animal welfare associated with the different methods and species have been identified. They have been differentiated by their effect on the sampling protocol: risk factors which reduce the quality of the stun and risk factors which reduce the sensitivity of the indicators used. Depending on which of these risk factors are present in a given slaughterhouse context, the sampling protocol applied in that slaughterhouse may require changes. Three scenarios of sampling levels have been identified: standard, reinforced and light sampling, corresponding to normal, tightened and reduced inspections.

## 13.4 OUTLOOK

The EU legal framework provides for numerous useful mechanisms to ensure and, where necessary, improve animal welfare at slaughter. Food business operators, animal welfare officers and national and European authorities need to work hand in hand to achieve the high animal welfare standards expected by European citizens. Information on the performance of the monitoring systems established at slaughterhouses, specifically on the animal-based measures used as indicators therein, should be exchanged with national contact points that are providing scientific support for the implementation of Regulation EC 1099/2009. This will enable a gradual improvement of the monitoring systems and be instrumental in further characterizing the sensitivity and specificity of the indicators in specific slaughterhouse contexts. Ultimately, the analysis of the monitoring results will yield crucial information for improving slaughter processes, but also for identifying if and how existing slaughter methods need to be modified. The European Food Safety Authority is committed to continuous improvement to animal welfare at the time of killing, on the basis of advancements in science and technology, and is actively supporting Member States, food business operators and the Commission along this way.

## 13.5 REFERENCES


Atkinson S, Velarde A, Llonch P and Algers B, 2012. Assessing pig welfare at stunning in Swedish commercial abattoirs using CO2 group-stun methods. *Animal Welfare*, 21, 487–495.

Coetzee JF, Gehring R, Tarus-Sang J and Anderson DE, 2010. Effect of sub-anesthetic xylazine and ketamine ('ketamine stun') administered to calves immediately prior to castration. *Veterinary Anaesthesia and Analgesia*, 37, 566–578.

Council Regulation (EC) No 1099/2009 of 24 September 2009 on the protection of animals at the time of killing. Available from: http://eur-lex.europa.eu/legal-content/EN/TXT/PDF/?uri=CELEX:32009R1099&from=EN

Dalmau A, Rodriguez P, Llonch P and Velarde A, 2010. Stunning pigs with different gas mixtures: aversion in pigs. *Animal Welfare*, 19, 325–333.

Directive 2010/63/EU of the European Parliament and of the Council of 22 September 2010 on the protection of animals used for scientific purposes. Available from: http://eur-lex.europa.eu/LexUriServ/LexUriServ.do?uri=OJ:L:2010:276:0033:0079:en:PDF

EC (European Commission), 2012. European Union Strategy for the Protection and Welfare of Animals 2012–2015. COM(2012) 6 final, Brussels. 12 pp. Available from: http://ec.europa.eu/food/animal/welfare/actionplan/docs/aw_strategy_19012012_en.pdf

EFSA AHAW Panel (EFSA Panel on Animal Health and Welfare), 2004a. Opinion of the Scientific Panel on Animal Health and Welfare (AHAW) on a request from the Commission related to the welfare of animals during transport. *EFSA Journal* 44, 1–36.

EFSA AHAW Panel (EFSA Panel on Animal Health and Welfare), 2004b. Opinion of the Scientific Panel on Animal Health and Welfare (AHAW) on a request from the Commission related to welfare aspects of the main systems of stunning and killing the main commercial species of animals. *EFSA Journal* 45, 1–29.

EFSA (European Food Safety Authority), 2005. Aspects of the biology and welfare of animals used for experimental and other scientific purposes. *EFSA Journal* 292, 1–46.

EFSA AHAW Panel (EFSA Panel on Animal Health and Welfare), 2006. Opinion of the Scientific Panel on Animal Health and Welfare (AHAW) on a request from the Commission related to the welfare aspects of the main systems of stunning and killing applied to commercially farmed deer, goats, rabbits, ostriches, ducks, geese. *EFSA Journal* 326, 1–18.

EFSA AHAW Panel (EFSA Panel on Animal Health and Welfare), 2011. Scientific Opinion concerning the welfare of animals during transport. *EFSA Journal* 9(1), 1966.

EFSA Panel on Animal Health and Welfare (AHAW), 2012a. Scientific Opinion on the use of animal based measures to assess welfare of dairy cows. *EFSA Journal*, 10(1), 2554, 81 pp.

EFSA Panel on Animal Health and Welfare (AHAW), 2012b. Scientific Opinion on the use of animal based measures to assess welfare in pigs. *EFSA Journal*, 10(1), 2512, 85 pp.

EFSA Panel on Animal Health and Welfare (AHAW), 2012c. Guidance on risk assessment for animal welfare. *EFSA Journal*, 10(1), 2513, 30 pp.

EFSA AHAW Panel (EFSA Panel on Animal Health and Welfare), 2013a. Guidance on the assessment criteria for studies evaluating the effectiveness of stunning methods regarding animal protection at the time of killing. *EFSA Journal*, 11(12), 3486, 41 pp. doi:10.2903/j.efsa.2013.3486

EFSA AHAW Panel (EFSA Panel on Animal Health and Welfare), 2013b. Scientific Opinion on monitoring procedures at slaughterhouses for bovines. *EFSA Journal* 11(12), 3460, 65 pp. doi:10.2903/j.efsa.2013.3460

EFSA AHAW Panel (EFSA Panel on Animal Health and Welfare), 2013c. Scientific Opinion on monitoring procedures at slaughterhouses for poultry. *EFSA Journal*, 11(12), 3521, 65 pp. doi:10.2903/j.efsa.2013.3521

EFSA AHAW Panel (EFSA Panel on Animal Health and Welfare), 2013d. Scientific Opinion on monitoring procedures at slaughterhouses for sheep and goats. *EFSA Journal* 11(12), 3522, 65 pp. doi:10.2903/j.efsa.2013.3522

EFSA AHAW Panel (EFSA Panel on Animal Health and Welfare), 2013e. Scientific Opinion on monitoring procedures at slaughterhouses for pigs. *EFSA Journal* 11(12), 3523, 62 pp. doi:10.2903/j.efsa.2013.3523

EFSA, 2013f. Sample size calculation tool for monitoring stunning at slaughter. http://www.efsa.europa.eu/it/supporting/pub/541e

Gerritzen M, Lambooij B, Reimert H, Stegeman A and Spruijt B, 2007. A note on behaviour of poultry exposed to increasing carbon dioxide concentrations. *Applied Animal Behaviour Science*, 108, 179–185.

Gibson TJ, Johnson CB, Murrell JC, Hulls CM, Mitchinson SL, Stafford KJ, Johnstone AC and Mellor DJ, 2009. Electroencephalographic responses of halothane-anaesthetised calves to slaughter by ventral-neck incision without prior stunning. *New Zealand Veterinary Journal*, 57, 77–83.

Jongman EC, Barnett JL and Hemsworth PH, 2000. The aversiveness of carbon dioxide stunning in pigs and a comparison of the CO2 stunner crate vs. the V-restrainer. *Applied Animal Behaviour Science*, 67, 67–76.

Kirkden RD, Niel L, Lee G, Makowska IJ, Pfaffinger MJ and Weary DM, 2008. The validity of using an approach-avoidance test to measure the strength of aversion to carbon dioxide in rats. *Applied Animal Behaviour Science*, 114, 216–234.

Landa L, 2012. Pain in domestic animals and how to assess it: a review. *Veterinarni Medicina*, 57, 185–192.

Le Neindre PGR, Guémené D, Guichet J-L, Latouche K, Leterrier C, Levionnois OMP, Prunier A, Serrie A and Servière J, 2009. Animal pain: identifying, understanding and minimising pain in farm animals. Multidisciplinary scientific assessment, Summary of the expert report. INRA, Paris, 98 pp.

Llonch P, Dalmau A, Rodriguez P, Manteca X and Velarde A, 2012a. Aversion to nitrogen and carbon dioxide mixtures for stunning pigs. *Animal Welfare*, 21, 33–39.

Llonch P, Rodriguez P, Velarde A, de Lima VA and Dalmau A, 2012b. Aversion to the inhalation of nitrogen and carbon dioxide mixtures compared to high concentrations of carbon dioxide for stunning rabbits. *Animal Welfare*, 21, 123–129.

Llonch P, Rodriguez P, Jospin M, Dalmau A, Manteca X and Velarde A, 2013. Assessment of unconsciousness in pigs during exposure to nitrogen and carbon dioxide mixtures. *Animal*, 7, 492–498.

Martoft L, Jensen EW, Rodriguez BE, Jorgensen PF, Forslid A and Pedersen HD, 2001. Middle latency auditory evoked potentials during induction of thiopentone anaesthesia in pigs. *Laboratory Animals*, 35, 353–363.

McKeegan DEF, McIntyre J, Demmers TGM, Wathes CM and Jones RB, 2006. Behavioural responses of broiler chickens during acute exposure to gaseous stimulation. *Applied Animal Behaviour Science*, 99, 271–286.

Mellor DJ, Cook CJ and Stafford KJ, 2000. Quantifying some responses to pain as a stressor. In: GP Moberg and JA Mench (eds.), *The biology of animal stress*, CAB International, pp. 171–198.

Mota-Rojas D, Bolanos-Lopez D, Concepcion-Mendez M, Ramirez-Telles J, Roldan-Santiago P, Flores-Peinado S and Mora-Medina P, 2012. Stunning swine with CO2 gas: controversies related to animal welfare. *International Journal of Pharmacology*, 8, 141–151.

Rodriguez P, Dalmau A, Ruiz-de-la-Torre JL, Manteca X, Jensen EW, Rodriguez B, Litvan H and Velarde A, 2008. Assessment of unconsciousness during carbon dioxide stunning in pigs. *Animal Welfare*, 17, 341–349.

Svendsen O, Jensen SK, Karlsen LV, Svalastoga E and Jensen HE, 2008. Observations on newborn calves rendered unconscious with a captive bolt gun. *Veterinary Record*, 162, 90–92.

Velarde A, Cruz J, Gispert M, Carrion D, de la Torre JLR, Diestre A and Manteca X, 2007. Aversion to carbon dioxide stunning in pigs: effect of carbon dioxide concentration and halothane genotype. *Animal Welfare*, 16, 513–522.

Vogel KD, Badtram G, Claus JR, Grandin T, Turpin S, Weyker RE and Voogd E, 2011. Head-only followed by cardiac arrest electrical stunning is an effective alternative to head-only electrical stunning in pigs. *Journal of Animal Science*, 89, 1412–1418.

von Borell E, Langbein J, Després G, Hansen S, Leterrier C, Marchant-Forde J, Marchant-Forde R, Minero M, Mohr E, Prunier A, Valance D and Veissier I, 2007. Heart rate variability as a measure of autonomic regulation of cardiac activity for assessing stress and welfare in farm animals – A review. *Physiology & Behavior*, 92, 293-316.

chapter fourteen

# Effect of pre-slaughter handling and stunning on meat quality

*Carmen Gallo[1], Luigi Faucitano[2], Marien A. Gerritzen[3]*

## Learning objectives

- Describe the relationship between animal welfare and meat quality within a slaughterhouse.
- Be aware of some specific factors that can affect animal welfare and meat quality of ruminants, pigs and poultry during lairage and slaughter.
- Understand how improvements in animal handling at the slaughterhouse can also reduce meat quality problems.

## CONTENT



[1] Instituto de Ciencia Animal, Facultad de Ciencias Veterinarias, Universidad Austral de Chile, Casilla 567, Valdivia, Chile

[2] Agriculture and Agri-Food Canada, Sherbrooke Research and Development Centre, 2000, College Street, Sherbrooke (QC), J1M 0C8 Canada

[3] Wageningen University and Research Centre, Livestock Research Department Animal Welfare, De Elst 1, 6708 WD Wageningen, The Netherlands

## 14.1 INTRODUCTION

The close relationship existing between handling animals at pre-slaughter and the yield and quality of the meat products can be useful as part of a strategy for improving animal welfare in meat producing species. Poor animal welfare during the pre-slaughter operations can lead to economical losses due to increased mortality, reduced carcass and meat yield and quality and more carcass trimmings due to bruising. As these carcass defects can be the result of poor animal welfare before slaughter, their occurrence can negatively affect consumer acceptance in countries with high welfare standards and, in others, reduce the chances of exporting meat and meat products to countries that strongly enforce animal welfare legislation. In many countries, particularly in South America, meat exportation has provided a good opportunity/economic incentive to make improvements in quality assurance schemes and good animal handling practices that consider their welfare as a component in the production chain on farm, during transport and at slaughter (Gallo, 2009; Huertas et al., 2014; Paranhos da Costa et al., 2012).

Pre-slaughter handling begins on the farm and includes preparation of animals for transport, herding and penning, withdrawal of feed and water, loading on to transport vehicles, journey distance, duration and driving/road conditions, temperature and humidity in the vehicle and finally, waiting time until the animals are unloaded. These factors affect carcass and meat quality individually and cumulatively. Animals that are fractious, raised on farming systems with minimum or no contact with humans and those stressed during these processes are difficult to handle and restrain at slaughter and, as a consequence, become a risk factor causing poor animal welfare outcomes.

The purpose of lairage at the slaughterhouse is to provide a reservoir of animals aimed at maintaining the constant speed of the slaughter line and to allow an opportunity for stressed and (or) fatigued animals to recover from loading and transport stress (Warriss, 1987). Lairage and stunning are of extreme importance in meat producing animals as there are a number of serious hazards for animal welfare as well as for meat quality. Mistakes made at these points can have irreversible effects on meat quality and may offset all efforts made by the production sector to improve performance and animal welfare. Recommended lairage times depend on the animal species and previous conditions on farm and during transport. Precautions must be taken to ensure an adequate handling and environmental control to safeguard the benefit of lairage as resting area, enabling animals to recover from the stress of transport and limit the effects on meat quality variation.

Properly designed and maintained lairage pens and stunning boxes, correct use of stunning methods by trained operators and a short interval between stunning and sticking will improve animal welfare and reduce carcass and meat quality problems.

Training animal handlers throughout the meat processing chain can improve animal welfare and meat quality. Both from an animal welfare and economical point of view, it is important to handle animals with care and to pay attention to lairage conditions, and to the right stunning methods and settings. The objective of this chapter is to discuss the main meat quality problems that can be produced in ruminants, pigs and poultry during the pre-slaughter operations after arrival at the slaughterhouse.

## 14.2 RUMINANTS

The different events involved during pre-slaughter handling of ruminants at a slaughterhouse, such as crowding and mixing in lairage pens, feed and water deprivation, and the use of inadequate aids for driving from unloading through lairage to stunning are all stressful events that affect animal welfare and can also affect meat quality (Ferguson and Warner, 2008; Gallo, 2009; Gregory, 1998; Warriss, 2000).

### *14.2.1 Lairage duration and conditions*

The time animals spend without feed and/or water is important from a welfare point of view as they could be subjected to prolonged thirst and hunger, and also from an economic point of view due to reduction in carcass yield and meat quality (Gallo and Gatica, 1995; Gallo et al., 2003a). Exposing ruminants to various adverse conditions prior to slaughter, like hunger, thirst, adverse climatic conditions, mixing of unfamiliar animals, noise, muscular fatigue and space restrictions during lairage will lead to stress (Grandin, 2007) and hence to high final pH in meat (Ferguson and Warner, 2008; Hood and Tarrant, 1980). Stress will reduce muscle glycogen reserves and minimize lactic acid formation after death, all resulting in a reduced pH decline and production of meat characterized by a pH > 5.8 and dark colour (known as dark, firm and dry [DFD] or dark cutting beef; Hood and Tarrant, 1980). The dark colour of DFD meat negatively affects consumer acceptability of the product; moreover, the high pH and the higher water holding capacity favour the development of bacteria and reduce the shelf-life of meat (Hoffman, 1988; Wirth, 1987).

In order to reduce meat quality problems, lairage time should be kept to a minimum (< 12 h), unless animals have access to feed and water (OIE, 2014). The longer the time ruminants spend in lairage, the higher the risk of carcass weight loss, bruising and dark cutters (Amtmann et al., 2006; Ferguson and Warner, 2008; Gallo et al., 2003a; Strappini et al., 2013; Warner et al., 1998), especially if they have been transported for over 12 hours before (Gallo et al., 2003a). In these conditions, although the provision of water may help ruminants rehydrate reducing carcass yield losses, the negative effects on final pH are usually irreversible (Ferguson and Warner, 2008; Gallo et al., 2003a). Steers transported for 3 to 24 hours, and then submitted to 24-hour lairage showed 9.4 times a higher probability of having a muscle pH > 5.8 (threshold pH value used for classification of DFD meat) than those with 3-hour lairage (Amtmann et al., 2006). Significant losses in lamb and goats carcass yield can result from long lairage times (Carter and Gallo, 2008; Díaz et al., 2014; Warriss et al., 1987). Part of these losses can be explained by a significant dehydration rate and mobilization of body reserves, as evidenced by the increased concentration of betahydroxibutyrate in blood (Kannan et al., 2002; Tadich et al., 2009).

Many events that lead to physical injury can happen during the pre-slaughter handling process ending in lairage. Strappini et al. (2013) registered 1,792 potential bruising events in 52 cows during loading, transport, unloading, lairage and stunning. They found that 91% of these occurred during the 19-hour lairage they were submitted to and were mostly due to social interactions between animals. Besides bruising, mixing animals from different origin results in reduced glycogen reserves due to greater physical activity, such as mounting and butting (Warriss et al., 1984). It appears that in steers, agonistic behaviours in lairage are more frequent after short transport journeys than long ones (3 vs. 24 hours; Gallo, 2009).

In sheep, a decrease in muscle glycogen concentration has been observed when lairage time is extended, although results on ultimate pH of meat are less clear (Carter and Gallo, 2008; Díaz et al., 2014; Jacob et al., 2005; Warriss et al., 1987). Younger age at slaughter and shipment to slaughter immediately after weaning can have an additive effect on stress response in lambs (Tadich et al., 2009). Carter and Gallo (2008) reported almost no glycogen reserves in the muscles of lambs raised in Patagonia, weaned immediately before transport to slaughter (46-hour journey) and slaughtered after 6 to 12 hours in lairage with access to water only. This metabolic muscle condition may result in pH higher than 6.0 and DFD meat in lambs (De la Fuente et al., 2006).

The reported effects of lairage duration on meat quality in ruminants vary between studies and countries in the world because of differences between production and transport conditions, species and category of ruminant (Del Campo et al., 2014).

### *14.2.2 Moving ruminants to stunning*

In order to achieve good livestock handling at the slaughterhouse and move animals efficiently from the lairage pens to the stunning point it is necessary to consider that interactions occur between the three components of this process: the animals, the handlers and the plant installations. Stress and lesions can be minimized by ensuring that the animal handling facilities are designed, constructed and maintained to favour a smooth and efficient movement of livestock and stockpersons should be trained in animal handling according to species-specific behaviour (Grandin, 2007; Grandin and Shivley, 2015; OIE, 2014). Quantifiable behaviour indicators, such as the percentage of cattle slipping, falling and vocalizing while being driven through the race from unloading to stunning, reflect difficulties during handling and are a useful tool to assess the effects of improvements resulting from handler training or facilities design (Gallo et al., 2003b).

The OIE (2014) has provided some principles for the handling of animals

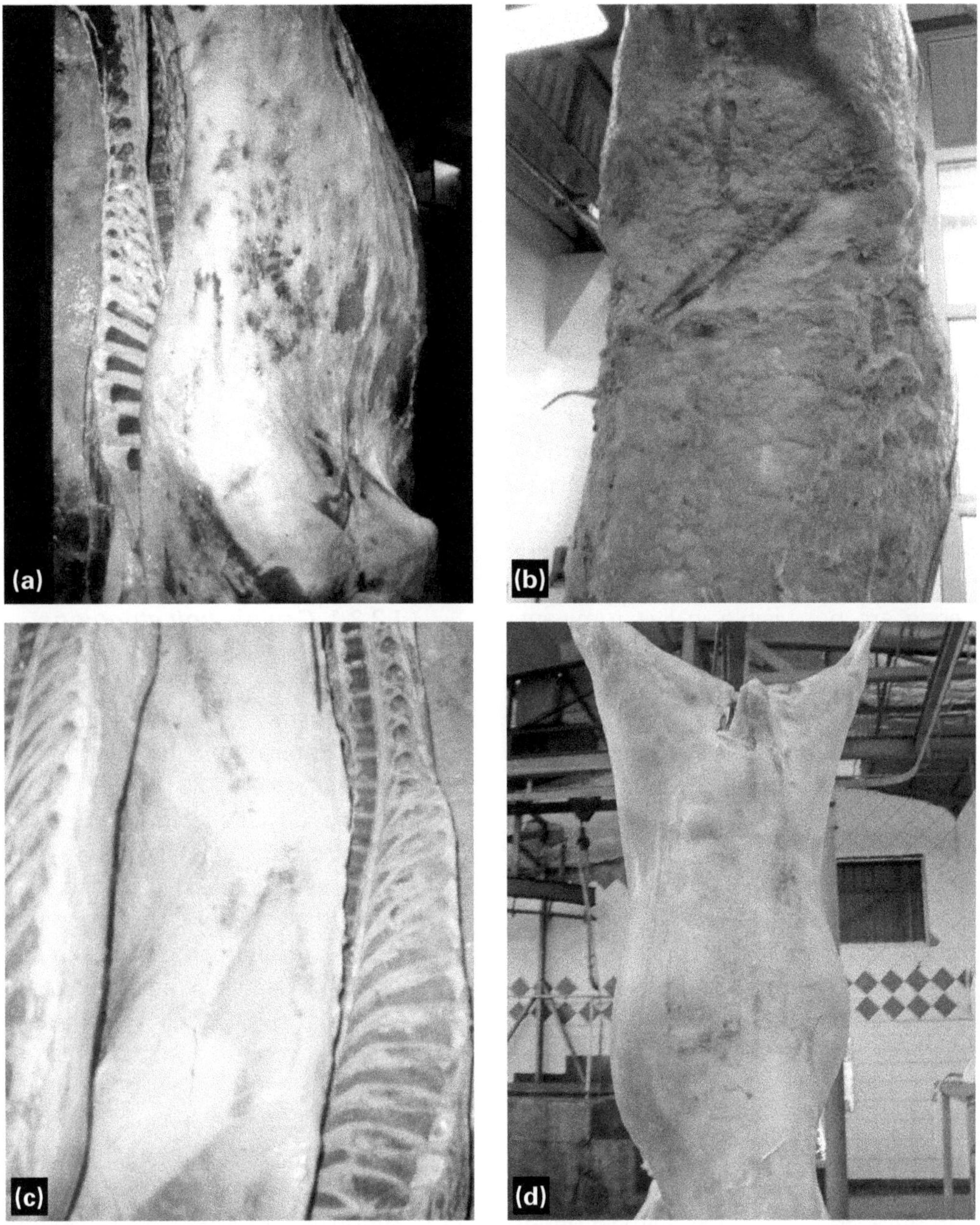


**Figure 14.1:** Carcass bruising produced by: (a) pricking cattle with pointy driving tools; (b) hitting cattle with sticks; (c) prodding cattle with electric goads; (d) lifting lambs by the wool

at slaughterhouses on arrival, in lairage pens and through the raceways leading to the stunning pens. In all cases, the use of procedures that cause pain to the animals, such as pricking or hitting cattle with sticks, is forbidden. The use of sticks can also leave typical marks on the carcasses (Figures14.1a and b). Although the use of electric goads is permitted for cattle, an inappropriate application of this tool (too frequent, too long application or too high voltage) can also leave marks on the carcass (Figure 14.1c). Due to their smaller size, ruminants like sheep and goats are commonly mishandled by being lifted by the wool (Fig.14.1d), horns and other parts of the body. These poor handling practices cause pain and lead to carcass bruising in sheep (Tarumán and Gallo, 2008).

Bruising, besides indicating poor handling conditions, results in economic losses due to trimming, reduced yield and carcass downgrading (Warriss, 2000). In contrast to pigs and poultry, bruises inflicted during the pre-slaughter period are not visible in live ruminants due to the presence of hair and thick skin. Therefore, bruises and marks due to inadequate use of handling aids can only be evidenced on the carcasses of cattle (Strappini et al., 2013) and sheep (Carter and Gallo, 2008; Tarumán and Gallo, 2008) after slaughter. Although attempts have been made to use the colour of bruising, it is often difficult to trace back the time of bruise infliction along the meat production chain, hence efforts should be made to minimize or eliminate the risks throughout the pre-slaughter handling procedures.

### *14.2.3 Stunning of ruminants*

The purpose of stunning animals before slaughter is to avoid causing unnecessary suffering and pain when they are bled and because of this it is a mandatory practice to stun animals prior to slaughter according to the OIE (2014) animal welfare standards. Based on the significant impact of trained handlers at this stage on animal welfare and meat quality, training personnel on stunning practices is considered of particular importance (Gallo et al., 2003b; Huertas et al., 2014).

In order to stun ruminants effectively, they need to be restrained (OIE, 2014) and a fundamental principle to bear in mind is that operators should not restrain the animal if they are not ready to stun. Moreover, stunning operators should not stun an animal if they are not ready to shackle, hoist and bleed immediately after, as this can lead to negative consequences in terms of animal welfare as well as meat quality. Negative effects at this point will also depend on the stunning methods used.

#### *14.2.3.1 Captive bolt stunning*

Captive bolt stunning is the most common method for stunning ruminants. Animals have to be restrained for as short time as possible before stunning (Ewbank et al., 1992; Muñoz et al., 2012; OIE, 2014). If restraining devices are not carefully designed to hold the animal appropriately and/or produce an excessive pressure on the animal, this can lead to discomfort and carcass bruising can be observed if the design includes pinch points or sharp edges (Gallo, 2009; Grandin and Shivley, 2015). Another common problem in stunning boxes for cattle is guillotine doors being dropped on the back of animals at the entrance into the box, leading to bruises with a typical pattern on the back (Strappini et al., 2013). To avoid this bruise type, the guillotine door bottom edges should be cushioned.

#### *14.2.3.2 Electrical stunning*

Electrical stunning methods, head-only or head-to-body (or back/cardiac arrest type), are common methods used for small ruminants. To ensure stunning efficiency, it is important to train operators on the positioning of electrodes, the application time and the voltage/amperage (HSA, 2006b). A frequent problem observed in sheep stunning is the placement of electrodes at wrong anatomical sites, such as on the neck or behind ears, which reduces stunning efficiency (Gallo, 2009; Velarde et al., 2000a,b). Poor or intermittent electrical contact during stunning can produce high voltage spikes that are detrimental to meat quality.

#### *14.2.3.3 Stun-to-stick interval*

A short interval between stunning and bleeding is desirable from an animal welfare and a meat quality point of view (HSA, 2006a,b; OIE, 2014). With head-only electrical stunning, ideally an animal should be bled before 15–20 seconds (HSA, 2006b; OIE, 2014) in order to avoid a return to sensibility as it is a reversible stunning method. In the case of head-to-back electrical stunning or captive bolt stunning, it is recommended to bleed before 60 seconds after stunning application (OIE, 2014). A longer interval can lead to haemorrhages of pin-head to about 1 cm in size (blood splash) in any muscle (Warriss, 2000). Blood splashing is more common in lambs, but has been also recorded in valuable beef cuts, such as *Psoas major* and *Longissimus dorsi* (Gallo, 2009). There is evidence that one of the main reasons for blood splashing is the increase in blood pressure that occurs after the stunning procedure. Exsanguinating the animal as soon as possible after stunning reduces this problem (Warriss, 2000). Results on meat quality and the incidence of haemorrhages in lambs due to electrical stunning are diverse. Paulick et al. (1989) and Vergara and Gallego (2000) reported a faster muscle pH decrease in head-only electrically stunned lambs, whereas other authors failed to find any effect on meat quality, but observed haemorrhages in the heart, neck and rump (Velarde et al., 2003; Vergara et al., 2005).

## 14.3 PIGS

### *14.3.1 Lairage duration and conditions*

The recovery of pigs from stress and the related economic losses due to poor meat quality in lairage depend on lairage time and on the quality of the handling systems, facility design and environment, and whether pigs are mixed or not (Faucitano and Geverink, 2008).

#### *14.3.1.1 Lairage duration*

A number of studies (Pérez et al., 2002; Warriss et al., 1992) concluded that after two to three hours in lairage, physiological basal levels are reached again indicating pigs' full recovery from transport and handling stress and ensuring the production of good quality pork. No or short lairage (15–60 min) is not recommended as it results in higher muscle temperature immediately before slaughter and higher level of lactic acid in the muscle resulting in increased incidence of pale, soft and exudative (PSE) pork (Fraqueza et al., 1998; Shen et al., 2006). On the other hand, longer lairage time proved to reduce the risk to produce PSE pork (2% at 10 h; Guàrdia et al., 2005), but to increase the percentage of dark, firm and dry (DFD) pork (Gispert et al., 2000; Warriss et al., 1998). Guàrdia et al. (2005) reported a 12% risk to produce DFD pork after 3-hour lairage reaching a 19% risk after overnight lairage. The increase in DFD pork in proportion with lairage time is the result of reduced muscle glycogen content at slaughter caused by the combined effect of fasting and fighting (Guàrdia et al., 2009; Nanni Costa et al., 2002).

The fasting condition of pigs on arrival at the slaughter plant may affect their activity in lairage and meat quality variation. Dalla Costa et al. (2016) compared the application of two 24-hour fasting strategies pre-slaughter and reported the production of darker and drier pork in pigs that spent most of their fasting time at the slaughter plant (22 h) compared with those that were mostly fasted at the farm prior to transport (18 h) due to more extended fighting in the lairage pen.

#### *14.3.1.2 Lairage conditions*

Mixing unfamiliar pigs, which is a common practice at high throughput commercial slaughterhouses, inevitably causes some fighting which results in greater risk of DFD pork (Warriss and Brown, 1985). The effects on pork quality can be explained by the glycogen exhaustion caused by the greater physical activity and the significant rise in body temperature that may last for 8 hours after mixing (de Jong et al., 1999; Jones et al., 1994). To limit fighting in lairage it is either recommended to keep pigs in smaller groups (15–40 pigs; Geverink et al., 1996; Rabaste et al., 2007) or to mix very large groups (up to 200 pigs) in the pen (Grandin, 1990). Space allowance also has a big impact on social behaviour of pigs in the lairage pen, regardless of the group size (Moss, 1978; Figure 14.2a,b). Based on these observations and on the evidence that most fighting occurs within the first 30–60 minutes of mixing in lairage (Geverink et al., 1996; Moss, 1978), to restrict fighting between mixed pigs stocking densities of 0.42 $m^2$/pig for short lairage (< 3 h) and 0.66 $m^2$/pig for long lairage (> 3 h) are recommended (SCAHAW, 2002).

Air temperature during lairage has an impact on pork quality variation (Lammens et al., 2007). Lairage temperatures of 15–18°C and relative humidities (RH) of 59–65% are recommended to limit pork quality variation (Honkavaara, 1989). When these environmental conditions are not respected, pigs can either suffer from cold stress (shown by shivering) which may result in DFD pork due to muscle energy depletion to maintain a constant body temperature (Knowles et al., 1998) or heat stress as shown by increased panting, especially when they are stocked at hot (> 30°C) and humid (RH > 80%) conditions resulting in greater risk for PSE pork production (Santos et al., 1997). When the temperature is above 30°C, lairage time should be shortened as much as possible to limit the risk of PSE production (Fraqueza et al., 1998).

Spraying pigs with cold water (approximately 10°C) in the lairage pen results in a 2°C drop in body temperature, leading to a reduced incidence of PSE pork (Long and Tarrant, 1990). However, at environmental temperatures below 5°C, showering is not recommended as it may lead to DFD pork due to shivering and muscle glycogen depletion in response to cold stress (Knowles et al., 1998).

When compared to cattle and sheep, pig lairages are the noisiest place in slaughterhouses (Weeks et al., 2009), with noise ranging from 76 to 108 dB, with the highest peaks (120 dB) being

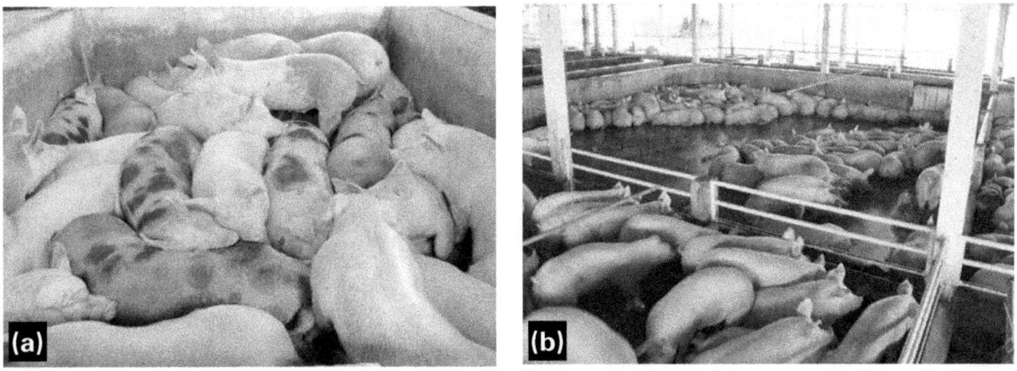


**Figure 14.2:** Pigs kept in lairage at high (a) and low (b) stocking densities

recorded in the pre-stun area (FAWC, 2003; Rabaste et al., 2007; Talling et al., 1996). Lairage noise is mostly represented by gates clanging and pig vocalization (Weeks, 2008; Weeks et al., 2009), which produce a fear response in pigs resulting in a more rapid drop in early *post-mortem* muscle acidification (lower initial pH) and increased production of exudative pork (van de Perre et al., 2010; Warriss et al., 1994). To reduce the risk for PSE meat the average sound level during lairage has to be lower than 85 dB (Vermeulen et al., 2015a,b).

### *14.3.2 Moving pigs to stunning*

The combination of higher slaughter speed, poorly designed handling systems and large groups during the short period between the exit of the lairage pen and stunning may result in greater proportion of slips and high-pitched vocalization, and increased use of electric prods, all associated with greater skin lesions score, and lower $pH^{24}$ and higher drip loss in pork meat (Chevillon, 2001; Dokmanović et al., 2014; Hambrecht et al., 2005; Rabaste et al., 2007; Rocha et al., 2016; Vermeulen et al., 2015a; van der Wal et al., 1999). To ease handling and keep a smooth and consistent flow of pigs to the stunner, it is recommended to move pigs in groups of 6–8 to 18–20 to maintain slaughter throughput rates from 150 to 900 pigs/hour, respectively (Chevillon, 2001). Moving large groups of pigs (45 pigs) through the lairage alleys may result in a greater incidence of PSE pork and blood splashing (Barton-Gade et al., 1992; Figure 14.3a,b).

Critical factors at this point are the entrance into the race and the "stop–start" forward motion of pigs towards the stunner, which are both observed in races feeding electrical and $CO_2$ stunners and both increase the need for the use of electric prods to encourage pig movement (Faucitano 2010). The entrance of pigs into a $CO_2$ stunner has been significantly improved by the group-wise stunning system (Christensen and Barton-Gade, 1997). Compared to the traditional double race system, the group-wise handling system resulted in lower proportion of PSE and blood splashed pork due to reduced prodding and muscle exercise (Christensen and Barton-Gade, 1997; Franck et al., 2003).

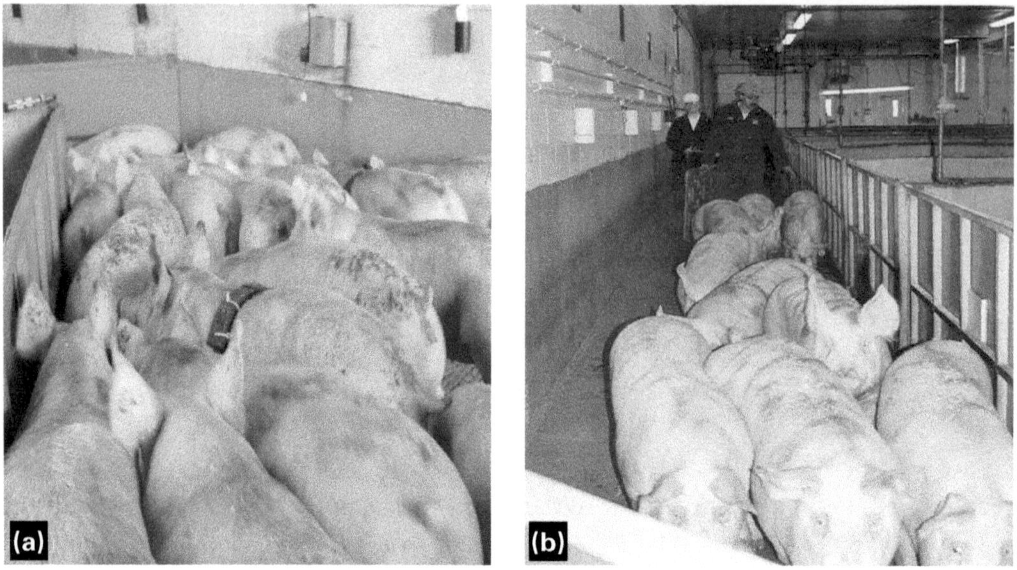


**Figure 14.3:** Moving pigs in large (a) versus small (b) groups through the lairage alleys

At medium-high throughput commercial slaughterhouses, pigs are moved into a restrainer, either V- or band-type, which lifts them up from the ground limiting their movements and conveys them to the point of electrical stunning, that is, head-only or head-to-chest type, respectively. Comparative studies between these two restrainer types reported that pigs restrained in the band conveyor produce less blood splashes (–30%) and PSE pork than those conveyed with the V-type restrainer (Griot et al., 2000; Lambooij et al., 1992).

### *14.3.3 Stunning of pigs*

Electrical and carbon dioxide ($CO_2$) gas are the most common stunning systems for pigs (see Chapters 7 and 8).

#### *14.3.3.1 Electrical and CO2 stunning*

When compared with carbon dioxide stunning, head-only electrical stunning increases the rate of the *post mortem* muscle glycolysis (PSE pork) in pigs due to increased post-stun convulsions and produces a 20-fold increase in the incidence of haemorrhages in the muscle due to increased blood pressure and muscle activity immediately after stunning due to the occurrence of tonic–clonic seizures or convulsions (Lambooij and Sybesma, 1988; Wotton et al., 1992).

The head-to-body system minimizes the negative effects of the head-only electrical stunning as the application of the body electrode inhibits the post-stun convulsions (Gilbert et al., 1984). However, when compared with $CO_2$ stunning, head-to-body stunning still increases the incidence of PSE and blood splashing on the loins and hams (Velarde et al., 2000a, 2001).

However, the effects of $CO_2$ stunning on pork quality depend on the gas concentration. Nowak et al. (2007) showed that meat quality was superior (higher $pH_{24}$ values) after exposure to 90% compared to 80% $CO_2$ for 100 seconds in a deep lift system which may be explained by the lower rate of convulsions, indicator of stunning efficiency, just prior to or post-slaughter.

## 14.4 POULTRY

In the period between leaving the farm and the moment of slaughter as other animals, poultry will undergo a range of different factors that affect the welfare as well as product quality. This will start with the withdrawal of food and a period of low water intake, to catching and crating, transport conditions, conditions during the waiting period at the slaughter plant and ending with shackling and stunning before slaughter. According to Savenije et al. (2002), feed withdrawal and transport quickly exhaust the main energy supplies of chickens. Energy exhaustion compromises the welfare of the birds and makes them progressively less capable of coping with further stressors. Catching, crating and transport are stressful stimuli but, it was found that neither feed deprivation nor transport under good conditions for short periods of time significantly affected meat quality.

### *14.4.1 Lairage of poultry*

Birds that arrive at the slaughterhouse in poor conditions, that is, suffering from injuries due to catching and crating or that are exhausted due to poor transport conditions, will suffer during lairage. For bird welfare reasons it is better to limit the waiting period at the slaughterhouse to a minimum. However, distinguishing between birds in good or poor condition is not easy while large numbers of them are held in transport containers. A good qualification system to check the state of birds based on weather conditions, duration of the fasting period, transport duration and visual observations could be helpful to minimize the risks of poor animal welfare and of quality losses due to the stress experienced during the pre-slaughter period.

Weight loss due to high temperature at lairage may range from 0.1 to 5% at the processing plant (Bilgilli et al., 1989). Good temperature management during lairage will benefit animal welfare as well as product quality and yields.

### *14.4.2 Handling poultry before stunning*

Handling before stunning can be divided into two methods: hanging birds into shackles at the slaughterline before electrical water bath stunning and tilting or tipping birds on to a conveyor belt for gas stunning.

To present birds to an electrical water bath stunning system, birds are taken out of the transport containers or crates manually and suspended by their legs at shackles. Hanging the birds into the shackles is performed at high speed resulting in a considerable number of bruised legs or even broken bones (Sparrey and Kettlewell, 1994; Veerkamp and de Vries, 1983). To minimize negative effects of suspending birds to the shackle line it is important that personnel handle birds with care and avoid forceful suspension. Furthermore, the size of the shackles should be appropriate to the size of the birds and their legs. Design of the slaughterline, that is, length, corners and breast support systems, can possibly help to limit the negative effects of shackling.

In case of gas stunning, birds are presented to the stunning system in their transport crates or container or on a conveyor belt. To present the birds on a conveyor belt, the transport containers are tilted and the birds slide or fall out of the containers on to the conveyor belt. After gas stunning, birds are suspended on the shackle line fully relaxed, eliminating the negative effects of live bird shackling (Uijttenboogaart, 1997).

## *14.4.3 Stunning of poultry*

### *14.4.3.1 Electrical stunning*

Electrical stunning in a multiple bird water bath stunner is the universally used method for stunning poultry at slaughter. The exact amount of current delivered to individual bird depends on the electrical impedance of that bird in the system. The effectiveness of electrical stunning depends on the frequency and waveforms of the current (Raj et al., 2006a,b). The higher the frequency, the more current is required to induce unconsciousness (Hindle et al., 2010). Birds with high impedance in the system will not receive enough current to become immediately unconscious and birds with a lower than average impedance will receive more current than necessary. Receiving more current is not an animal welfare issue, but the main problem with effective electrical stunning is that it very often goes with detrimental effects on meat and carcass quality. The study of Hindle et al. (2010) clearly indicated that effective electrical stunning in a commercial multibird water bath stunner always involves a number of carcasses having bloodspots and other quality defects (Figure 14.4). When the

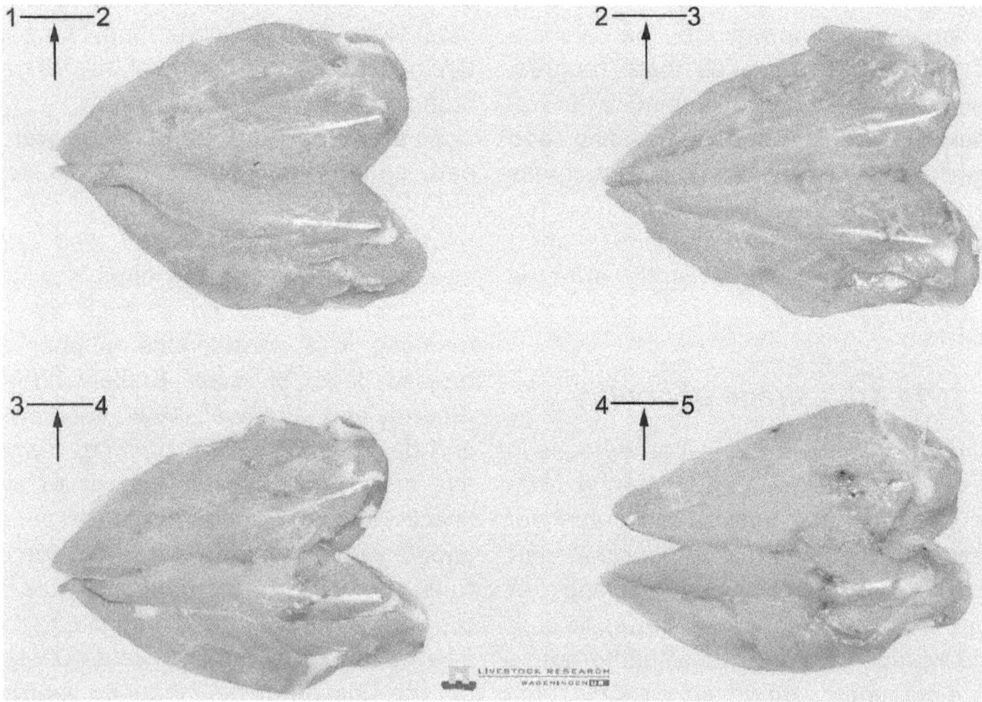


**Figure 14.4:** Different levels of bloodspots in breast meat

stunning frequency is increased and the applied current remains the same, the incidence of carcass damage such as bloodspots will decrease but, as a consequence, the percentage of effectively stunned birds will also decrease. Therefore, to induce effective stunning with higher frequencies demands higher currents at the same time and that will again result in increased level of bloodspots.

It is therefore obvious that effective stunning intervention in a multibird water bath stunner will conflict with product quality.

*14.4.3.2 Gas stunning*

The most common gas stunning method is a two-phase controlled atmosphere stunner (CAS) or a multi-step carbon dioxide stunning system (MCAS), but also systems containing 30% $CO_2$ with 70% $N_2$ are used in practice. The advantage of gas stunning is that birds do not need to be handled or shackled before stunning and that, at correct settings, they have minimal negative effects on product quality (Raj et al., 1990). Furthermore, exposure to increasing $CO_2$ levels will not affect meat quality as with electrical stunning. Bloodspots and broken wings will hardly occur due to stunning intervention. However, the occurence of convulsions or uncontrolled wing flapping can increase the risk of wing damage (Figure 14.5). Exposure to high $CO_2$ concentrations or to very low $O_2$ concentrations as with $N_2$ or argon stunning systems induce serious uncontrolled muscle contractions (convulsions). These convulsions can lead to wing damage like bruises and broken or dislocated wings which lead to economic losses.

In experiments using several different gas mixtures, McKeegan et al. (2007) concluded that a two-step system using 40% $CO_2$, 30% $O_2$ and 30% $N_2$ in the first phase and 80% $CO_2$, 5% $O_2$ and 15% $N_2$ in the second phase was to be preferred compared to a one-step system using either argon or argon mixed with 30% $CO_2$ from both a bird welfare and a meat quality perspective. In studies containing

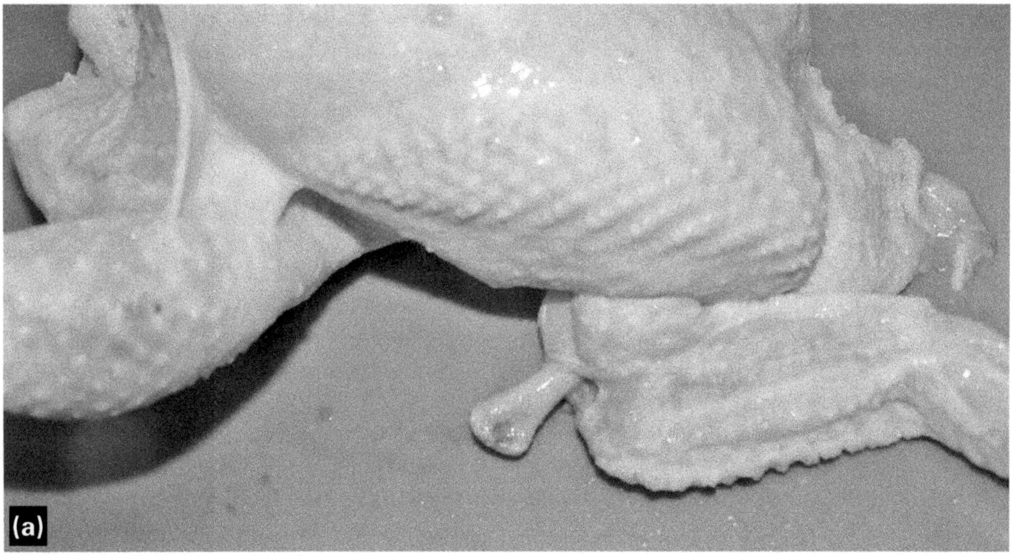


**Figure 14.5:** Two types of wing damage: (a) open fracture; (b) severe bleeding

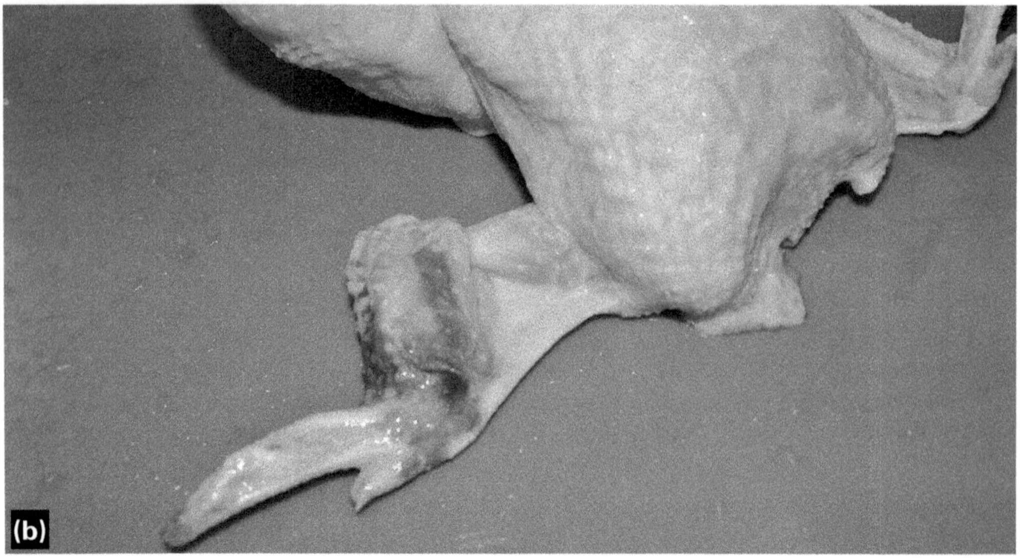


**Figure 14.5b**

multi-stepwise increase of $CO_2$, convulsions were not observed or only at a very moderate level (Gerritzen et al., 2013).

## 14.5 REFERENCES

Amtmann, V.A., Gallo, C., Van Schaik, G. and Tadich, N. (2006) Relaciones entre el manejo antemortem, variables sanguíneas indicadoras de estrés y pH de la canal en novillos. *Archivos de Medicina Veterinaria* 38 (3), 259–264.

Barton-Gade, P., Blaabjerg, L.O. and Christensen, L. (1992) New lairage system for slaughter pigs: effect on behaviour and quality characteristics. In: *Proceedings of the 38th International Congress of Meat Science and Technology*. Clermont-Ferrand, France, pp. 161–164.

Bilgilli, S.F., Egbert, W.R. and Hoffman, D.L. (1989) Research note: effect of postmortem ageing temperature on sarcomere length and tenderness of broiler Pectoralis major. *Poultry Science* 68(11), 1588–1591.

Carter, L. and Gallo, C. (2008) Effect of long distance transport by road and sea crossing on ferry on live weight losses and carcass characteristics in lambs. *Archivos de Medicina Veterinaria* 40, 259–266.

Chevillon, P. (2001) Pig welfare during pre-slaughter and stunning. In: *Proceedings of the 1st International Virtual Conference on Pork Quality*. Embrapa, Concordia, Brazil, pp. 145–158.

Christensen, L. and Barton-Gade, P. (1997) New Danish developments in pig handling at abattoirs. *Fleischwirtshaft* 77, 604–607.

Dalla Costa, F.A., Devillers, N., Paranhos da Costa, M.J.R. and Faucitano, L. (2016) Effects of applying preslaughter feed withdrawal at the abattoir on behaviour, blood parameters and meat quality in pigs. Meat Science 119, 89-94.

De la Fuente, J., Pérez, C., Vieira, C., Sánchez, M., González de Cháverri, E., García, M., Álvarez, I. and Díaz, M. (2006) The effect of transport on pH evolution of different muscles in suckling lambs. In: *Proceedings of the 52nd International Congress of Meat Science and Technology* (ICoMST), Dublin, Ireland, pp. 177–178.

De Jong, I.C., Lambooij, E., Mechiel Korte, S., Blockhuis, H.J. and Koolhaas, J.M. (1999) Mixing induces long-term hyperthermia in growing pigs. *Animal Science* 69, 601–605.

Del Campo, M., Brito, G., Montossi, F., Soares de Lima, J.M. and San Julián, R. (2014) Animal welfare and meat quality: The perspective of Uruguay, a "small" exporter country. *Meat Science* 98, 470–476.

Díaz, M.T., Vieira, C., Pérez, C., Lauzurica, S., González de Chávarri, E., Sánchez, M. and De la Fuente, J. (2014) Effect of lairage time (0 h, 3 h, 6 h or 12 h) on glycogen content and meat quality parameters in suckling lambs. *Meat Science* 96, 653–660.

Dokmanović, M., Velarde, A., Tomović,V., Glamočlija, N., Markoviić, R., Janjiić, J. and Baltiić, M.Z. (2014) The effects of lairage time and handling procedure prior to slaughter on stress and meat quality parameters in pigs. *Meat Science* 98, 220–226.

Ewbank, R., Parker, M.J. and Mason, C.W. (1992) Reactions of cattle to head restraint at stunning, a practical dilemma. *Animal Welfare* 1, 55–63.

Faucitano, L. (2010) Effects of lairage and slaughter conditions on animal welfare and pork quality. *Canadian Journal of Animal Science* 90, 461–469.

Faucitano, L. and Geverink, N.A. (2008) Effects of preslaughter handling on stress response and meat quality in pigs. In: Faucitano L. and Schaefer, A.L. (eds.) *The Welfare of Pigs – from Birth to Slaughter.* Wageningen Academic Publishers, Wageningen, The Netherlands, pp. 197–224.

FAWC (2003) Report on the welfare of farmed animals at slaughter or killing. Part 1: Red meat animals. Farm Animal Welfare Council, London, UK.

Ferguson, D.M. and Warner, R.D. (2008) Have we underestimated the impact of pre-slaughter stress on meat quality in ruminants? *Meat Science* 80, 12–19.

Franck, M., Figwer, P., Jossel, A., Poirel, M.T. and Pasteur, X. (2003) Incidence of preslaughter stress on meat quality of non sensible pigs. *Revue Médicine Vétérinaire* 154, 199–204.

Fraqueza, M.J., Roseiro, L.C., Almeida, J., Matias, E., Santos, C. and Randall, J.M. (1998) Effects of lairage temperature and holding time on pig behavior and on carcass and meat quality. *Applied Animal Behaviour Science* 60, 317–330.

Gallo, C. (2009) Bienestar animal y buenas prácticas de manejo animal relacionadas con la calidad de la carne. In: Bianchi G. and Feed O. (eds.) *Introducción a la ciencia de la carne.* 1st ed. Editorial Hemisferio Sur, Montevideo, Uruguay, pp. 455–494.

Gallo, C. and Gatica, C. (1995) Efectos del tiempo de ayuno sobre el peso vivo, de la canal y de algunos órganos en novillos. *Archivos de Medicina Veterinaria* 25, 69–77.

Gallo, C., Lizondo, G. and Knowles, T. (2003a) Effects of journey and lairage time on steers transported to slaughter in Chile. *Veterinary Record* 152, 361–364.

Gallo, C., Teuber, C., Cartes, M., Uribe, H. and Grandin, T. (2003b) Mejoras en la insensibilización de bovinos con pistola neumática de proyectil retenido tras cambios de equipamiento y capacitación del personal. *Archivos de Medicina Veterinaria* 35 (2), 159–170.

Gerritzen, M.A., Reimert, H.G.M., Hindle, V.A., Verhoeven, M.T.W. and Veerkamp, W.B. (2013) Multistage carbon dioxide gas stunning of broilers. *Poultry Science* 92, 41–50.

Geverink, N.A., Engel, B., Lambooij, E. And Wiegant, V.M. (1996) Observations on behaviour and skin damage of slaughter pigs and treatment during lairage. *Applied Animal Behaviour Science* 50, 1–13.

Gilbert, K.V., Devine, C.E., Hand, R. and Ellery, S. (1984) Electrical stunning and stillness of lambs. *Meat Science* 11, 45–58.

Gispert, M., Faucitano, L., Guárdia, M.D., Oliver, M.A., Coll, C., Siggens, K., Harvey, K. and Diestre, A. (2000) A survey on pre-slaughter conditions, halothane gene frequency, and carcass and meat quality in five Spanish pig commercial abattoirs. *Meat Science* 55, 97–106.

Grandin, T.A. (1990) Design of loading facilities and holding pens. *Applied Animal Behaviour Science* 28, 187–201.

Grandin, T. (2007) Handling and welfare of livestock in slaughter plants. In: Grandin T. (ed.) *Livestock handling and transport* (Chapter 20). CABI, Wallingford (3rd edition), pp. 329–353.

Grandin, T. and Shivley, C. (2015) How farm animals react and perceive stressful situations such as handling, restraint, and transport. *Animals (Basel)* 5(4), 1233–1251.

Gregory, N. (1998) *Animal welfare and meat science*. CABI Publishing, Oxford, UK, pp. 398.

Griot, B., Boulard, J., Chevillon, P. and Kérisit, R. (2000) Des restrainers à bande pour le bien être et la qualité de la viande. *Viandes et Produits Carnés* 3, 91–97.

Guàrdia, M.D., Estany, J., Balash, S., Oliver, M.A., Gispert, M. and Diestre, A. (2005) Risk assessment of DFD meat due to pre-slaughter conditions and RYR1 gene in pigs. *Meat Science* 70, 709–716.

Guàrdia, M.D., Estany, J., Balash, S., Oliver, M.A., Gispert, M. and Diestre, A. (2009) Risk assessment of skin damage due to pre-slaughter conditions and RYR1 gene in pigs. *Meat Science* 81, 745–751.

Hambrecht, E., Eissen, J.J., Newman, D.J., Smits, C.H., Verstegen, M.W. and den Hartog, L.A. (2005) Preslaughter handling effects on pork quality and glycolytic potential in two muscles differing in fiber type composition. *Journal of Animal Science* 83, 900–907.

Hindle, V.A., Lambooij, E., Reimert, H.G.M., Workel, L.D. and Gerritzen, M.A. (2010) Animal welfare concerns during the use of the water bath for stunning broilers, hens, and ducks. *Poultry Science* 89, 401–412.

Hoffman, K. (1988) El pH, una característica de calidad de la carne. *Fleischwirtschaft Español* 1, 13–18.

Honkavaara, M. (1989) Influence of lairage on blood composition of pig and on the development of PSE pork. *Journal of Agricultural Science in Finland* 61, 433–440.

Hood, D.E. and Tarrant, P.V. (1980) *The problem of dark-cutting in beef.* Martinus Nijhoff, The Hague. 501 pp.

HSA, Humane Slaughter Association. (2006a) Captive bolt stunning of livestock, Guidance notes N°2, 4th edition. Humane Slaughter Association, Wheathampstead, UK.

HSA, Humane Slaughter Association. (2006b) Insensibilización eléctrica de animales de carne roja. (Guía N°4). 4th edition. Humane Slaughter Association, Wheathampstead, UK.

Huertas, S.M., Gallo, C. and Galindo, F. (2014) Drivers of animal welfare policies in America. *Revue Scientifique et Technique–Office International des Epizooties* 33 (1), 55–66.

Jacob, R.H., Pethick, D.W. and Chapman, H.M. (2005) Muscle glycogen concentrations in commercial consignments of Australian lamb measured on farm and post-slaughter after three different lairage periods. *Australian Journal of Experimental Agriculture* 45, 543–552.

Jones, S.D.M., Cliplef, R.L., Fortin, A., McKay, R.M., Murray, A.C., Pommier, S.A., Sather, A.P. and Schaefer, A.L. (1994) Production and ante-mortem factors influencing pork quality. *Pig News and Information* 15, 15N–18N.

Kannan, C., Terrill, T.H., Kouakou, B., Gelaye, S. and Amoah, E.A. (2002) Simulated preslaughter holding and isolation effects on stress responses and live weight shrinkage in meat goats. *Journal of Animal Science* 80, 1771–1780.

Knowles, T.G., Brown, S.N., Edwards, J.E. and Warriss, P.D. (1998) Ambient temperature below which pigs should not be continuously showered in lairage. *Veterinary Record* 143, 575–578.

Lambooij, E. and Sybesma, W. (1988) The

effect of environmental factors such as preslaughter treatment and electrical stunning on the occurrence of haemorrhages in the shoulder of slaughter pigs. In: *Proceedings of the 34th International Congress of Meat Science and Technology*. Brisbane. Australia, pp. 1–3.

Lambooij, E., Merkus, G.S.M. and Hulsegge, B. (1992) A band restrainer for slaughter pigs. *Fleischwirtshaft* 7, 1271–1272.

Lammens, V., Peeters, E., De Maere, H., De Mey, E., Paelinck, H., Leyten, J. and Geers, R. (2007) A survey of pork quality in relation to pre-slaughter conditions, slaughterhouse facilities, and quality assurance. *Meat Science* 74, 381–387.

Long, V.P. and Tarrant, P.V. (1990) The effect of pre-slaughter showering and post-slaughter rapid chilling on meat quality in intact pork sides. *Meat Science* 27, 181–195.

McKeegan, D.E., Abeyesinghe, S.M., McLeman, M.A., Lowe, J.C., Demmers, T.G.M., White, R.P., Kranen, R.W., Van Bemmel, H., Lankhaar, J.A.C. and Wathes, C.M. (2007) Controlled atmosphere stunning of broiler chickens. II. Effects on behaviour, physiology and meat quality in a commercial processing plant. *British Poultry Science*, 48 (4), 430–442.

Moss, B.W. (1978) Some observations on the activity and aggressive behaviour of pigs when penned prior to slaughter. *Applied Animal Ethology* 4, 323–339.

Muñoz, D., Strappini, A.C. and Gallo, C. (2012) Indicadores de bienestar animal para detectar problemas en el cajón de insensibilización de bovinos. *Archivos de Medicina Veterinaria* 44, 297–302.

Nanni Costa, L., Lo Fiego, D.P., Dall'Olio, S., Davoli, R. and Russo, V. (2002) Combined effects of pre-slaughter treatments and lairage time on carcass and meat quality in pigs of different halothane genotype. *Meat Science* 61, 41–47.

Nowak, B., Mueffling, T.V. and Hartung, J. (2007) Effect of different carbon dioxide concentrations and exposure times in stunning of slaughter pigs: impact on animal welfare and meat quality. *Meat Science* 75, 290–298.

OIE. World Organization for Animal Health (2014) Terrestrial Animal Health Code. Chapter 7. Animal Welfare.

Paulick, C., Stolle, F.A. and von Mickwitz, G. (1989) The influence of different stunning methods on meat quality of sheep meat. *Fleischwirtschaft* 69 (2), 27–230.

Paranhos Da Costa, M., Huertas, S., Gallo, C. and Dalla Costa, O. (2012) Strategies to promote farm animal welfare in Latin America and their effects on carcass and meat quality traits. *Meat Science* 92 (3), 221–226.

Pérez, M.P., Palacio, J., Santolaria, M.P., Aceña, M.C., Chacón, G., Verde, M.T., Calvo, J.H., Zaragoza, P., Gascón, M. and Garcia-Belenguér, S. (2002) Influence of lairage time on some welfare and meat quality parameters in pigs. *Veterinary Research* 33, 239–250.

Rabaste, C., Faucitano, L., Saucier, L., Foury, D., Mormède, P., Correa, J.A., Giguère, A. and Bergeron, R. (2007) The effects of handling and group size on welfare of pigs in lairage and its influence on stomach weight, carcass microbial contamination and meat quality variation. *Canadian Journal of Animal Science* 87, 3–12.

Raj, A.B.M., Grey, T.C. and Gregory, N.G. (1990) Effect of electrical and gaseous stunning on the carcass and meat quality of broilers. *British Poultry Science* 31, 725–733.

Raj, A.B.M., O'Callaghan, M. and Knowles, T.G. (2006a). The effects of amount and frequency of alternating current used in water bath stunning and of slaughter methods on electroencephalograms in broilers. *Animal Welfare* 15 (1), 7–18.

Raj, A.B.M., O'Callaghan, M. and Hughes, S.I. (2006b) The effects of amount and frequency of pulsed direct current used in water bath stunning and of slaughter methods on spontaneous electroencephalograms in broilers. *Animal Welfare* 15(1), 19–24.

Rocha, L.M., Velarde, A., Dalmau, A., Saucier, L. and Faucitano, L. (2016) Can the monitoring of animal welfare parameters predict pork meat quality variation through the supply chain (from farm to slaughter)? *Journal of Animal Science* 94, 359–376.

Santos, C., Almeida, J.M., Matias, E.C., Fraqueza, M.J., Roseiro, C. and Sardinha, L. (1997) Influence of lairage environmental conditions and resting time on meat quality in pigs. *Meat Science* 45, 253–262.

Savenije, B., Lambooij, E., Gerritzen, M.A. and Korf, J. (2002) Effects of feed deprivation and electrical, gas, and captive needle stunning on early post-mortem muscle metabolism and subsequent meat quality. *Poultry Science* 81, 561–571.

SCAHAW (2002) The welfare of animals during transport (details for horses, pigs, sheep and cattle). Report of the Scientific Committee on Animal Healh and Animal Welfare, European Commission.

Shen, Q.W., Means, W.J., Thompson, S.A., Underwood, K.R., Zhu, M.J., McCormick, R.J., Ford, S.P. and Du, M. (2006) Preslaughter transport, AMP-activated protein kinase, glycolysis, and quality of pork loin. *Meat Science* 74, 388–395.

Sparrey, J. M. and Kettlewell, P. J. (1994) Shackling of poultry: is it a welfare problem? *World's Poultry Science Journal* 50,167–176.

Strappini, A.C., Metz, J.H.M., Gallo, C., Frankena, K., Vargas, R., de Freslon, I. and Kemp, B. (2013) Bruises in culled cows: when, where and how are they inflicted? *Animal* 7(3), 485–491.

Tadich, N., Gallo, C., Brito, M. and Broom, D. (2009) Effect of weaning and 48 h transport by road and ferry on some blood indicators of welfare in lambs. *Livestock Science* 121, 132–136.

Talling, J.C., Waran, N.K., Wathes, C.M. and Lines, J.A. (1996) Behavioural and physiological responses of pigs to sound. *Applied Animal Behaviour Science* 48, 187–202.

Tarumán, J. and Gallo, C. (2008) Bruising in lamb carcasses and its relationship with transport. *Archivos de Medicina Veterinaria* 40, 275–279.

Uijttenboogaart, T.G. (1997) Effect of gas and electrical stunning methods on meat quality. In: E. Lambooij (ed.) Proceedings of Satellite Symposium "Developments of new humane stunning and related processing methods for poultry to improve product quality and consumer acceptability". ID-DLO report 97.026, ID-DLO, Lelystad, The Netherlands, pp. 25–33.

Van de Perre, V., Permentier, L., de Bie, S., Verbecke, G. and Geers, R. (2010) Effect of unloading, lairage, pig handling, stunning and season on pH of pork. *Meat Science* 86, 931–937.

Van der Wal, P.G., Engel, B. and Reimert, H.G.M. (1999) The effect of stress, applied immediately before stunning, on pork quality. *Meat Science* 53, 101–106.

Veerkamp, C.H. and de Vries, A.W.(1983) Influence of electrical stunning on quality of broilers. In: Eikelenboom G. (ed.) *Stunning Animals for Slaughter.* M. Nijhoff, The Hague, The Netherlands, pp. 187–199.

Velarde, A., Gispert, M., Faucitano, L., Manteca, X. and Diestre, A. (2000a) The effect of stunning method on the incidence of PSE meat and haemorrhages in pork carcasses. *Meat Science* 55, 309–314.

Velarde, A., Ruiz-de-la-Torre, J.L., Stub, C., Diestre, A. and Manteca, X. (2000b) Factors affecting the effectiveness of head-only electrical stunning in sheep. *Veterinary Record* 147, 40–43.

Velarde, A., Gispert, M., Faucitano, L., Alonso, P., Manteca, X. and Diestre, A. (2001) Effects of the stunning procedure and the halothane genotype on meat quality and incidence of haemorrhages in pigs. *Meat Science* 58, 313–319.

Velarde, A., Gispert, M., Diestre, A. and

Manteca, X. (2003) Effect of electrical stunning on meat and carcass quality in lambs. *Meat Science* 63, 35–38.

Vergara, H. and Gallego, L. (2000) Effect of electrical stunning on meat quality of lamb. *Meat Science* 56, 345–349.

Vergara, H., Linares, M.B., Berruga, M.I. and Gallego, L. (2005) Meat quality in suckling lambs: effect of preslaughter handling. *Meat Science* 69, 473–478.

Vermeulen, L., Van de Perre, V., Permentier, L., De Bie, S., Verbeke, G. and Geers, R. (2015a). Pre-slaughter handling and pork quality. *Meat Science* 100, 118–123.

Vermeulen, L., Van de Perre, V., Permentier, L., De Bie, S., Verbeke, G. and Geers, R. (2015b). Sound levels above 85dB pre-slaughter influence pork quality. *Meat Science* 100, 269–274.

Warner, R.D., Truscot, T.G., Eldridge, G.A. and Franz, P.R. (1998) A survey of the incidence of high pH beef meat in Victorian abattoirs. In: 34th International Congress of Meat Science and Technology, Brisbane, Australia, pp.150–151.

Warriss, P.D. (1987) The effect of time and conditions of transport and lairage on pig meat quality. In: Tarrant, P.V., Eikelenboom, G. and Monin, G. (eds.) *Evaluation and Control of Meat Quality in Pigs.* Martinus Nijhoff Publisher, Boston, MA, pp. 245–264.

Warriss, P.D. (2000) The effects of live animal handling on carcass and meat quality. In: Warriss P.D. (ed.) *Meat Science: An Introductory Text,* Chapter 7. CABI, Wallingford, UK, pp. 131–155.

Warriss, P.D., Kestin, S.C., Brown, S.N. and Wilkins, L.J. (1984) The time required for recovery from mixing stress in young bulls and the prevention of dark cutting meat. *Meat Science* 10, 53–68.

Warriss, P.D. and Brown, S.N. (1985) The physiological responses to fighting in pigs and the consequences for meat quality. *Journal of the Science of Food and Agriculture* 36(2), 87–92.

Warriss, P.D., Brown, S.N., Bevis, E.A., Kestin, S.C. and Young, C.S. (1987) Influence of food withdrawal at various times preslaughter on carcass yield and meat quality in sheep. *Journal of Science of Food and Agriculture* 39, 325–334.

Warriss, P.D., Brown, S.N., Edwards, J.E., Anil, M.H. and Fordham, D.P. (1992) Time in lairage needed by pigs to recover from the stress of transport. *Veterinary Record* 131, 194–196.

Warriss, P.D., Brown, S.N., Adams, S.J.M. and Corlett, I.K. (1994) Relationships between subjective and objective assessments of stress at slaughter and meat quality in pigs. *Meat Science* 38, 329–340.

Warriss, P.D., Brown, S.N., Edwards, J.E. and Knowles, T.G. (1998) Effect of lairage time on levels of stress and meat quality in pigs. *Animal Science* 66, 255–261.

Weeks, C. A. (2008) A review of welfare in cattle, sheep, and pig lairages, with emphasis on stocking rates, ventilation and noise. *Animal Welfare* 17, 275–284.

Weeks, C.A., Brown, S.N., Lane, S., Heasman, L., Benson, T. and Warriss, P.D. (2009) Noise levels in lairages for cattle, sheep, and pigs in abattoirs in England and Wales. *Veterinary Record* 165, 308–314.

Wirth, F. (1987) Tecnología para la transformación de carne de calidad anormal. *Fleischwirtschaft Español* 1, 22–28.

Wotton, S.B., Anil, M.H., Whittington, P.E. and McKinstry, J.L. (1992) Pig slaughtering procedures: head-to-back stunning. *Meat Science* 32, 245–255.

# Concluding remarks

Farm animal welfare concern is based upon the belief that animals can suffer, and it is now clearly an important issue for ordinary people, especially across Europe, who demand that animals be bred, reared, transported and slaughtered in a humane way. Animal welfare is indeed seen as attribute of an "overall food quality concept", which is strictly correlated to other issues such as animal health, food safety and food quality. Conditions that harm animal welfare negatively affect animal health and productivity, and damage specific quality aspects thereby jeopardizing economic return, profitability and the ultimate product quality.

Slaughtering of animals is seen as the most critical stage of animal production. Council Regulation No 1099/2009 on the protection of animals at the time of killing lays down rules for the killing of animals bred or kept for the production of food, wool, skin, fur or other products as well as the killing of animals for the purpose of depopulation and for related operations. It aims to minimize avoidable pain and suffering caused to animals through the use of properly approved stunning methods, based on scientific knowledge and practical experience. The scope of this regulation is to protect welfare of all vertebrates including fish. Concern over the welfare of animals is by no means restricted to Europe. Animal welfare has been identified as a priority of the World Organisation for Animal Health (OIE), an international organization with 184 member countries. Currently, this organization has welfare standards for slaughter, transport and killing of animals for disease control agreed among its member countries.

The development of the existing legislation and standards is based on the best available scientific advice, taking into account public expectations, socio-economic consequences and trade concerns. The European Food Safety Authority (EFSA) provides scientific advice and technical support for the development and implementation of legislation and policies in the field, which have a direct or indirect impact on food and feed safety. Important work was recently carried out to support the implementation of the Regulation on the protection of animals at the time of killing, in particular regarding the scientific elements needed for the development of monitoring procedures at slaughterhouses and for the assessment of new or modified stunning methods.

The onus of maintaining good animal welfare is on the business operators, who are required to put in place standard operating procedures and monitoring procedures, including animal-based indicators, to evaluate the efficiency of stunning. Indeed, stunned animals will have to be regularly monitored to ensure that they do not regain consciousness till death, and no further processing can begin until death is confirmed in animals. In i, business operators are requested to appoint an animal welfare officer who is accountable for implementing animal welfare measures. People with responsibility for handling, movement, lairage, stunning and slaughter of animals should be trained and provided with certificates of competence for the tasks they are expected to carry out, by an independent body.

This book provides comprehensive practical information concerning key stages of operation, risk factors associated with each stage and animal-based indicators as monitoring tools. It is intended to be a reference book for food business operators, veterinarians and animal welfare officers for everyday activity ensuring welfare of animals.

# Index

www.ingramcontent.com/pod-product-compliance
Ingram Content Group UK Ltd.
Pitfield, Milton Keynes, MK11 3LW, UK
UKHW051450290926
15471UKWH00008B/1113

9 781789 184099